WHERE TO WATCH BIRDS IN
SURREY & SUSSEX

MATT PHELPS & ED STUBBS

H E L M
LONDON • OXFORD • NEW YORK • NEW DELHI • SYDNEY

HELM
Bloomsbury Publishing Plc
50 Bedford Square, London, WC1B 3DP, UK
29 Earlsfort Terrace, Dublin 2, Ireland

BLOOMSBURY, HELM and the Helm logo are
trademarks of Bloomsbury Publishing Plc

First published in the UK 2024

A catalogue record for this book is available from the British Library
Library of Congress Cataloguing-in-Publication data has been applied for

ISBN: PB: 978-1-3994-0423-5; ePub: 978-1-3994-0422-8;
ePDF: 978-1-3994-0421-1

2 4 6 8 10 9 7 5 3 1

Typeset in the UK by Mark Heslington
Maps by Brian Southern

Printed and bound in Great Britain by CPI Group (UK) Ltd, Croydon CR0 4YY

MIX
Paper | Supporting
responsible forestry
FSC® C013604

To find out more about our authors and books visit
www.bloomsbury.com and sign up for our newsletters

Cover photographs. Front: Turtle Dove (t), John Richardson;
Little Ringed Plover (b), Kit Day; Back: Goshawk (l), Helge Sørensen;
Dartford Warbler (c), Ed Stubbs; Pomarine Skua (r), Sam Viles

CONTENTS

Contents

ACKNOWLEDGEMENTS

Both of us grew up with various editions of *Where to Watch Birds In Kent, Surrey and Sussex*, so we were honoured when we were approached by Bloomsbury to write this updated version for Surrey and Sussex. We are both passionate when it comes to birding in these counties – bar three years at university, Ed has spent his whole life in Surrey, while Matt has lived in Sussex for six years after living in Surrey for six years (and previously in Hampshire, near to the Surrey border). We have each travelled widely in both counties, while also paying close attention to our local areas over the course of many years. Furthermore, we have been or are involved with either the Surrey Bird Club or Sussex Ornithological Society in some capacity.

At the time of writing (2023), it has been nearly 15 years since the last edition of a *Where to Watch* publication covering Surrey and Sussex. As a result, we were both keen for this book to provide a fresh take on the suite of classic, established birding locales, as well as to include various 'new' or less-visited ones that may pique the interest of the more explorative birder.

During the writing process we made a dedicated effort to visit every site mentioned in this book. However, we are still greatly indebted to the various different birders, site managers and others who helped us hugely with their advice on their local patches and reserves. We couldn't have done it without them. These individuals are as follows:

SURREY
Wes Attridge, Rich Bonser, Dave Brassington, David Campbell, John Clark, Arjun Dutta, Mark Elsoffer, Jeremy Gates, Thomas Gibson, Dave Harris, Gordon Hay, Sam Jones, Jake Klavins, Zach Pannifer, Janet Parr, Shaun Peters, Isaiah Rowe, Ed Sames, Josephine Snell, Chris Turner, Penny Williams.

We would also like to pay respect to Steve Chastell who sadly died just before this book went to print, and who was a good birding friend to both of us and a pillar of the Surrey birding community, not least in his work as County Recorder.

SUSSEX
Bola Akinola, Anna Allum, Chris Ball, Joe Bassett, Val Bentley, Tony Benton, Tom Burns, David Campbell, Elliot Chandler, Paul Cole, Nigel and Debbie Colegate, Richard Cowser, Liam Curson, Martin Daniel, Paul Davy, Andy Daw, Cliff Dean, Matt Eade, Sim Elliot, Jake Everitt, Bernie Forbes, Malcolm Freeman, Jake Gearty, John Gowers, Alastair Gray, Penny Green, Richard Grimmitt, Pete Hughes, Gareth James, Paul James, Simon Linington, Alan Loweth, Stuart Malcolm, Mark Mallalieu, Lesley Milward, Owen Mitchell, Chris and Juliet Moore, Alan Parker, Charlie Peverett, Laurence Pitcher, David Rogers, Ken Smith, Fran Southgate, Graeme Spinks, Steve Tillman, Martyn Waller, Barry Yates.

Thanks also to the photographers who generously provided excellent images: John Richardson, Helge Sørensen, Kit Day, Josh Jones and Sam Viles.

Ed would like to offer further thanks to his partner, Nadia, whose patience and enthusiasm have been crucial not just during this book-writing process, but far beyond. He'd also like to thank his mum and dad, Judith and Michael, for their unwavering support over the years. Finally, a special thank you to David Campbell and Owen Mitchell for their particularly extensive assistance with proof-reading.

Matt would like to offer further thanks to his wife Kate, for her unwavering support and tolerance throughout not only the writing of the book, but also during twelve years of listening to him talk endlessly about birds, and his daughter Bryony for always making him smile. He'd also like to dedicate this book to the memory of his dad, who sadly died during the writing period, and who first planted the seed of his fascination with birds when he was a child. Lastly, thank you to David Campbell, Paul James and Mark Mallalieu for always being on hand to answer myriad questions.

INTRODUCTION

South-East England offers much for both the resident or visiting birder – and a great deal of the habitat and species diversity of the region can be enjoyed across the counties of Surrey and East and West Sussex. Note that three counties are listed there, rather than the two in the title of this book. Strictly speaking, 'Sussex' no longer exists in any form other than a colloquial term for the now distinct counties of East and West Sussex. Nonetheless, for the purpose of easier classification, we will often refer to 'Sussex' as a single entity.

There are many differences in terms of the geography, population and birding opportunities across this region, but also common features throughout. The most notable difference between Surrey and Sussex is, of course, the cumulative 140km coastline of East and West Sussex. Surrey does, however, hold the greater extent of woodland and heathland, and some considerable reservoirs, former gravel pits and other wetlands.

Both Surrey and Sussex have become significantly more developed and heavily inhabited since the early 2010s, with the cumulative population of the region growing by around 200,000 people in that time. This undoubtedly has placed greater strain on the countryside and green spaces that we birders cherish and has presented more challenges to the ongoing protection of nature in what is already a built-up part of the country. Nonetheless, this boom in the population has also led to a growing number of people desiring a connection with nature and, in particularly populated parts of this region, an encouraging increase in the number of young people developing an interest in birding and other natural history-related pursuits. The arrival of the COVID-19 pandemic in 2020, while devastating in so many ways, also led to a huge increase in people's awareness of their natural surroundings, with many finding solace in the study and appreciation of birds while confined to their immediate local area. This highlighted the need for more wild places and greater access to such spaces, as the landscape of the South-East becomes ever more developed.

Thankfully, there are still swathes of undeveloped habitat in this region that should remain as havens given their vital wildlife value, not just nationally but on an international scale. We will look in more detail at the regional highlights of the respective counties shortly but, briefly, the Thames Basin and Wealden Heaths, the chalk grasslands of the North and South Downs and the various river valleys, in particular the Arun Valley SPA in West Sussex, are all rightly offered the highest levels of environmental protection, with initiatives such as SANGs (Suitable Alternative Natural Greenspace) put in place to try and protect the heaths, especially, from the inevitable pressure that comes from nearby development. Surrey and West Sussex are the first and second most wooded counties in England, respectively. East Sussex is somewhat less so, but the High Weald towards the north of the county offers some of the best woodland birding in the region.

SURREY

At 1,663 square kilometres, Surrey is the smallest of the three counties covered in this book, but it is also the more heavily populated and home to some 1.2 million people (albeit most residing in the boroughs closest to London). Despite its size and proximity to the capital, there are no cities in Surrey, with Woking and Guildford being the largest conurbations in terms of population.

Surrey has undergone a fair amount of boundary shuffling in the past few decades. The Borough of Spelthorne was added in 1965, dragging the likes of Staines Reservoir across the new border into Surrey from the now defunct Middlesex, while other areas once considered part of the county, such as Barnes, Battersea and Surrey Docks, have been lost to Greater London. For ornithological and other natural history recording purposes the old vice-county boundary, as it stood in the late nineteenth century, is maintained by the Surrey Bird Club with the new Greater London areas retained and Spelthorne exempted. In this book, however, sites from the entire 'Greater Surrey' region are included – in other words, those within both the vice-county and political county boundaries.

Anyone who has ever flown over Surrey or climbed to the top of Leith Hill tower won't be surprised to know that the county has 22 per cent woodland cover, which is almost double the national average. The greatest tracts of woodland are found along the North Downs – particularly north of the centre of this ridge, in the area around Effingham and Bookham – and along the Greensand Ridge and Surrey Hills through the south and south-west of the county, from Dorking down towards Godalming, Haslemere and the 840ha Chiddingfold Forest in the West Weald. These areas, perhaps unsurprisingly, offer the best chances to encounter some of the otherwise more elusive woodland species in the county, such as Marsh Tit, Lesser Spotted Woodpecker, Hawfinch and, increasingly, Goshawk and Honey Buzzard, though the latter remains hard to find in Surrey. Goshawk, however, has increased, not just in Surrey but right across the south-east. Gone are the days when birders would travel to the New Forest in Hampshire to see this mighty *Accipiter*, with the Wealden woodlands of Surrey and Sussex now offering just as much chance of an encounter. Indeed, such has been its dramatic increase in recent years that, for the first time in any such guide to birding in Surrey and Sussex, we provide specific viewpoints and directions to some of the best places in the region to see this species.

The highest point in Surrey and, indeed, South-East England, is Leith Hill, south of Dorking. At 294 metres, it is the highest of the 'three peaks of the Surrey Hills' – in addition to Pitch Hill and Holmbury Hill. Box Hill, situated just east of the 'Mole Gap' along the North Downs, is arguably the most famous of all the Surrey Hills and as such is also the most popular with tourists. The chalk ridge of the North Downs cuts right across the middle of Surrey from Reigate to Dorking and offers productive downland birding, including Skylark, Linnet, Red-legged Partridge, Raven and some of the remaining strongholds of Yellowhammer in the county. Both the North Downs and the South Downs through Sussex are not just of note for their birdlife but also their botanical and entomological diversity. Chalk grassland such as that found in these areas is of international significance and supports a host of rare plants, including many species of orchid, as well as butterflies found only in such locations, like Small Blue and Silver-spotted Skipper.

Surrey is perhaps best known by naturalists for its expansive swathes of heathland. The 3,500ha found in the county represent some 13 per cent of the remaining lowland heath in the UK and hold nationally significant numbers of species such as Dartford Warbler, Woodlark and Nightjar. Surrey is also home to the only remaining heathland site in the South-East for breeding Curlew (Thursley Common, itself supporting the largest mire in the South-East). With the increasingly hot and dry summers we are now experiencing, wildfires are sadly becoming an ever more frequent event and one can't help but worry for the future of these Curlews and other species that rely so heavily on these precious landscapes. Likewise, development in the region risks placing ever greater strain on these already fragmented habitats.

Southern and eastern areas of Surrey are largely dry and often wooded, so birders in search of wetland species generally head to the western, north-western and northern regions of the county, to the old gravel pits and other waterbodies in the Blackwater Valley (Tice's Meadow, Frimley Gravel Pits, Mytchett Lakes), the Wey Valley (Frensham Ponds and the water meadows between Godalming and Woking) and the reservoirs dotted along the Thames in south-west London plus, of course, Beddington Farmlands near Croydon and the London Wetland Centre in Barnes.

SUSSEX

Although officially split in two in 1974, the historic county of Sussex has roots dating back to the South Saxons in the fifth century. Cumulatively, it covers a total area of 3,783km^2 and is home to around 1.7 million people. West Sussex is the larger of the two counties at 1,991 square kilometres, compared to East Sussex's 1,792km^2; West Sussex is just ahead in terms of population too, with 882,000 people compared to just over 850,000 in East Sussex. Though it holds considerably fewer people for the size of the area, there are two cities in Sussex: Chichester and Brighton & Hove, which together are home to over 300,000 people or around 20 per cent of the total population of both East and West Sussex. One only need look at a population distribution map to see that there is considerably more space per person in Sussex than Surrey, hence there is more open space for wildlife, for the time being anyway, as the inevitable spread of London ever further into the South-East continues.

Sussex has a rich association with the sea and with migration, and not just the avian kind, as the Sussex coastline has always been an arrival point for travellers from the near Continent. It is somewhat fitting, then, that the coast remains the focus of much of the birding here, particularly for birders from inland regions of East and West Sussex and coastless Surrey, as well as those travelling from further afield. Many of the main sites featured in this book are names steeped in the history of southern coastal birding, such as Pagham Harbour and Beachy Head – both often attracting scarce and rare species to their landscapes and the hordes of birders that follow.

Seawatching is a particularly popular feature of Sussex birding throughout the year, but especially from late winter into spring, when departing flocks of divers, auks and wildfowl give way to arriving terns and skuas. One of the most desired prizes is Pomarine Skua, with many birders making the annual pilgrimage to the coast in the hope of connecting with this iconic species passing close by in May. Though Selsey Bill near Pagham is the most southerly point in Sussex and a popular seawatching spot in its own right, its position somewhat tucked in by the Isle of Wight means it is not quite as productive as it otherwise would be. Instead, sites such as Splash Point near Seaford are more favoured, especially for 'Poms'.

There are essentially three distinct geological bands in Sussex: the flat and fertile coastal plain, the rolling chalk hills of the South Downs and the sandy ridge of the High Weald (with a slice of the clay of the Low Weald separating the latter two). The highest point in Sussex and, indeed, the whole of the South Downs National Park, is Black Down on the Surrey/West Sussex border. At just short of 280m in elevation, it is considerably higher than other Sussex high points such as Ditchling and Firle Beacons. The South Downs run east to west across West Sussex from west of Midhurst towards the border with East Sussex, where they edge ever closer to the sea, eventually reaching the coast between Brighton and Eastbourne. Here the chalk ridge is cut off abruptly by the English Channel, forming the famous white cliffs, with Beachy Head being the tallest feature of its kind in England. Along this iconic stretch of coast can be found some of the only nesting Kittiwakes and Fulmars in the South-East.

The South Downs were given National Park status in 2010 – the most recent area in England to have been awarded this accolade. Though National Parks are arguably more about recreation than conservation these days, there is no doubt that the South Downs offers some of the most spectacular scenery in the South-East – and the downland birding can be excellent, too. Species now largely or totally absent from Surrey, such as Corn Bunting and Grey Partridge, remain here in reasonably good numbers, along with healthy populations of Yellowhammer, Skylark and Red-legged Partridge. This is largely thanks to the land use, with relatively small-scale arable farming still practised widely, and some landowners giving wildlife a helping hand where they are able to do so. A superb example of this is the Norfolk Estate near Arundel, which has won awards for its commitment to conservation through sensitive management of its farmland. The Burgh, one of the most celebrated areas of downland farmland in the South-East, is among several such prime South Downs birding sites featured in this book.

A number of large rivers run through East and West Sussex, including stretches of the Thames and the Medway. Those that flow through the Downs to the south coast – namely the Ouse, the Cuckmere, the Arun, the Eastern Rother and the Adur – offer some excellent wetland and estuary birding sites along their lengths, with famous sites such as Pulborough Brooks, Amberley Wildbrooks, Lewes and Iford Brooks, the Adur Estuary, Shoreham Harbour and Rye Harbour all featured in this book.

As already mentioned, woodland forms a substantial portion of the habitat in Sussex (around 20 per cent in total). The most wooded areas are towards the north of both East and West Sussex, following a band roughly from Midhurst east to Ashdown Forest, as well as another band across West Sussex from South Harting, near the Hampshire border, east to Houghton Forest on the edge of the Arun Valley. The High Weald holds 7 per cent of all ancient woodland cover in the UK. Unique to this region are areas of ghyll woodland, where small streams cut through sandstone ridges, flanked by steep-sided wooded valleys. Some 6 per cent of the woodland in the High Weald is classed as ghyll woodland, with much of it found in East Sussex, particularly the areas around Ashdown Forest, Broadwater Warren and Hastings Country Park. In addition to upwards of 50 breeding pairs of Goshawk at the time of writing, Honey Buzzard is also increasing in East and West Sussex, with Sussex cumulatively holding the largest breeding population of this elusive and enigmatic migrant raptor anywhere in the UK. As with Goshawk, we have for the first time here provided details of some key sites to encounter this species, although as these birds are still far from numerous in the UK, we have been rather more discreet and selective in choosing which sites to include.

Though not quite as famous for its heathlands as Surrey, Sussex holds a considerable amount of this valuable habitat, although over half of it is at Ashdown Forest. Away from this core area there are pockets of heathland dotted across the north of East Sussex, and north and west West Sussex, with the Heathlands Reunited project – spearheaded by the South Downs National Park Authority – aiming to reconnect them in the coming decades. In addition to all the previously mentioned specialist heathland bird species of the region, here can also be found a few scattered populations of the rare Field Cricket.

BIRD CLUBS

The Surrey Bird Club (SBC) was formed in 1957, with the key aims of educating people about the birds of Surrey, promoting and participating in ornithological studies, publishing an annual report and, from time to time, publications on the birds of the county, as well as supporting and fostering the preservation and conservation of birds and places of ornithological interest.

The SBC also encourages people to record the birds that they see and pass on those records so that they end up in the county's database. In order to encourage knowledge about the birds in the county, the Club runs field trips, publishes information showing where certain species can be found, runs and attends events where birds can be seen and encourages participation in bird surveys.

The Sussex Ornithological Society (SOS) was formed in 1962 and now enjoys a membership of around 2,000 people. The aims of the society are to record and study wild birds in the county of Sussex, to assist in the preservation of wild birds in the UK and the education of its members and the general public in ornithological science, and to raise awareness of the need for the protection of wild birds and their habitats. Since its inception, the SOS has donated hundreds of thousands of pounds to many conservation projects, including new hides at RSPB reserves such as Pulborough Brooks and Pagham Harbour.

Both the SBC and the SOS have, as part of their respective committees, a person in the role of County Recorder. This is a crucial role and one that is replicated in every ornithological club or society around the UK. This person has the unenviable task of gathering and collating reams of data, often from many tens of thousands of bird records submitted each year through the likes of BTO BirdTrack, Cornell's eBird system, BirdGuides or through direct correspondence with the recorder. In the event of records of rare and scarce species, the recorder will then disseminate these to a records committee for approval or rejection on the basis of a description and/or photos or other documentation. If accepted, these unusual records will be added to the summaries which are compiled together each year for the respective county's annual bird report – an eagerly awaited publication sent out to members documenting all the notable bird records for the year, early and late migrant species arrivals and departures, and any particularly large gatherings. Links to the Surrey Bird Club and Sussex Ornithological Society websites can be found in the list of organisations at the end of the book.

CHANGES IN SPECIES DISTRIBUTION AND ABUNDANCE

Time moves fast in terms of birding and bird populations. Since a book of this nature was last published, in 2009, there have been winners and losers in the populations of wild birds in Surrey and East and West Sussex. Back then, Buzzards were only just re-establishing themselves in South-East England; now they are abundant, along with similarly increasing numbers of Red Kite. In Sussex, there are now also breeding Marsh Harriers. Both Goshawk and Honey Buzzard have gone from strength to strength too, as already mentioned.

It's not just raptors that are doing well, as herons and egrets are also rapidly increasing in both Surrey and Sussex. Cattle Egret and Great White Egret have both rapidly gone from rarity level to fairly run-of-the-mill in most of the south of England now, with successful breeding of the former having already occurred in Sussex in recent years. It's surely only a matter of time before Glossy Ibis does the same, as records of this species continue to increase, with counts of multiple birds together now not uncommon on the coast. Firecrest is another species that has continued to expand its range across Surrey and Sussex, with some sites featured in this book now boasting tens of breeding pairs of this iconic little passerine. Common warblers such as Chiffchaff and Whitethroat are thriving now more than ever, as well. As the climate continues to warm, overshooting species from the Continent are becoming more regular, with Bee-eater now annual in Sussex and often several records of Hoopoe a year too. It's surely only a matter of time until the former attempts to breed somewhere in the south-east again after two pairs successfully bred on the Isle of Wight in 2014.

Various reintroduction projects have taken place in the region, or nearby, in the past decade, most notably the White Stork Project, based at the Knepp Estate near Horsham in West Sussex, and the White-tailed Eagle programme on the Isle of Wight – young adult birds from which have now taken up permanent residence in West Sussex and have become an increasingly frequent sight in both East Sussex and Surrey too. In the 20 years since the start of the Knepp project, rewilding has gone from being a rather niche, radical concept to a much more mainstream idea, with many traditional conservation organisations now adopting some of the principles in their own land management, such as the impressive landscape-scale tidal recovery project at RSPB Medmerry near Chichester. As these landscapes evolve and the reintroduced species rediscover them, birders in this region may, in years to come, bear witness to the sight of a White-tailed Eagle predating a White Stork – something that would have seemed unthinkable just a decade or two ago. There is talk of other possible future reintroductions too, such as Black Grouse and Red-backed Shrike, although only time will tell if these actually go ahead or are successful.

Sadly, there have been a number of losers in the past decade. Turtle Dove and Lesser Spotted Woodpecker continue to decline rapidly, especially the former, though an increase and stabilisation in the number of singing males recorded each year on the Knepp Estate offers some hope – while Willow Tit and Wood Warbler (as a breeding species) have been lost entirely from Sussex and Surrey now and Tree Sparrow is very close to going the same way, with former colonies at Beddington Farmlands lost altogether and other former sites in the far south-east of East Sussex (Scotney Gravel Pits and Rye Harbour) now clinging by a thread. Many northern species still considered 'common' in some parts of Surrey and Sussex, such as Tree Pipit and Willow Warbler, are also on a seemingly unretrievable decline, as our warming planet continues to reshape our avifauna.

HOW TO USE THIS BOOK

THE REGION

The region under discussion consists of the counties of Surrey, East Sussex and West Sussex. These counties have been divided into sub-regions, which generally correspond to fairly distinctive geographical and/or geological areas. The map on page 19 shows how the region has been divided.

CRITERIA FOR SITE INCLUSION

Since the region supports such a rich diversity of habitats for birds, the choice of sites for this book has not been easy. An attempt has been made to include all those places which either hold a wide variety of species and/or support certain specialised and uncommon birds. Further sites have been chosen either because they are particularly good examples of their habitat type or to ensure that the less well-known and more remote areas of the region are adequately represented in the book. The final criterion for inclusion in the book is access. Although most of the places in the book are open to the public, entry to some of the reserves may require a permit obtained in advance of a visit. Private sites of ornithological interest have been included only if they can be viewed sensibly and safely from a public place, e.g. a footpath or road.

For a variety of reasons, including the disappearance of notable species and changes in access, some sites included in earlier books have been omitted. Others have been included for the first time, such as Tice's Meadow, Shackleford Farmland, Medmerry RSPB and West Rise Marsh.

MEASUREMENTS

Throughout, the text measurements are given in those units most readily understood by British readers. Directions for accessing sites are given in miles.

HABITAT

This section gives a brief description of the site or sub-region, concentrating on features that are particularly important for the birds. Details are also given of any historical features of interest, other aspects of the flora and fauna and ownership or reserve status.

SPECIES

This section describes the more significant and interesting aspects of a site's birdlife. The text is arranged roughly in chronological order starting with the season that is generally best for birdwatching. It has sometimes been very difficult to decide which species to include for each site. Common birds are generally excluded unless they are of particular significance, e.g. a large roost of Black-headed Gulls. The presence of certain rare and vulnerable breeding species has either been ignored or their whereabouts treated with suitable discretion. Rare migrants and vagrants have been included for interest and to demonstrate the potential of the site for attracting such birds. It has been impossible to avoid giving long lists of species, but these have been reduced as much as possible, sometimes referring to groups of birds collectively, e.g. rarer winter grebes, seaduck, migrant landbirds, etc. These terms are fully defined in the Glossary. Some attempt has been made to give the scale of numbers to be expected, whether single individuals, small parties or flocks of hundreds of birds, the frequency with which the various species occur and any circumstances which might be related to such occurrences, e.g. hard weather, gales, etc.

TIMING

This section gives information regarding the best months or season, time of day, weather and, if coastal, state of the tide to visit the site. Recommendations are also made with regard to the popularity of the site and potential problems of disturbance.

ACCESS

This section aims to provide practical and useful information for those visiting a site for the first time. Directions are largely based on recent visits to the sites. The facilities available at each site, including car parks, toilets, visitor centres, cafés, well-marked footpaths, hides and viewing screens, are summarised at the end of each section. A list of the most accessible sites for disabled visitors and those using public transport can be found on page 296.

Directions to the site generally start from the nearest town, village or major road. These are followed by further details regarding access into and around the site. Where appropriate, road names and useful landmarks, such as public houses, are mentioned to further help guide visitors. Wherever possible, public rights of way have been used and details given of restrictions on access at the time of writing. Visitors are warned, however, that on private land the owner's permission should always be sought before deviating from public paths. Outline maps have been carefully prepared to complement the text. In addition, the 1:25000 scale Ordnance Survey Explorer and Outdoor Leisure series of maps are strongly recommended for detailed exploration of the sites.

Most reserves and country parks are well signposted and have visitor/information centres. In addition, detailed maps showing footpaths and hides are usually available either in leaflets and/or on notice boards. Visitors are urged to take heed of all such relevant information available on-site. For reserves, details are also given of any special arrangements for access such as obtaining the necessary permits. It should be noted that at some reserves where permits are not required for individuals, organised visits by groups of birdwatchers must be booked in advance. Even where this is not a necessity, those arranging group visits to an 'open' reserve are advised to contact those responsible for the site. The websites of national, regional and county organisations responsible for the management of sites included in the book are given in the List of Organisations section. This section also includes the websites of local and county birdwatching groups and clubs.

CALENDAR

This is a quick reference section giving the most interesting species and groups of birds that can be expected throughout and during different periods of the year. To simplify the calendars, species and groups of birds that may only occasionally be seen have been excluded.

KEY TO THE MAPS

▦	Sea/inland water
▦	Area of interest, eg. reserve
Ⓟ	Car parking
Ⓗ	Birdwatching hide
★	Viewpoint/screen
⊕	Church
■	Building
⚇	Sewage farm
Ⓥ	Vistor centre
Ⓦⓒ	Toilet
⊶⊷	Railway
╱	Main roads
╱	Minor roads
- - -	Footpath
♧♧	Woodland
▰	Towns

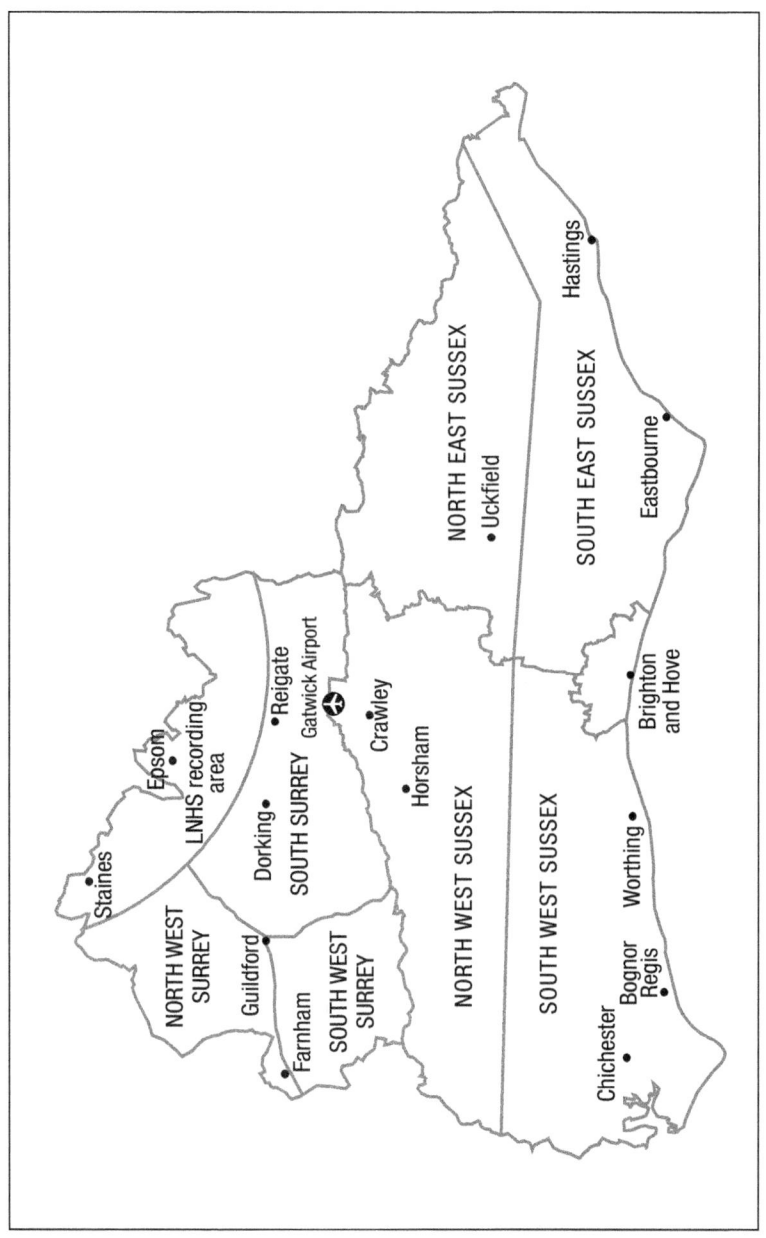

NORTH EAST SUSSEX
• Uckfield

SOUTH EAST SUSSEX

Hastings

Eastbourne

Reigate
Gatwick Airport

Crawley

Horsham

Dorking
SOUTH SURREY

Epsom
LNHS recording area

Staines

NORTH WEST SURREY

Guildford

Farnham
SOUTH WEST SURREY

NORTH WEST SUSSEX

SOUTH WEST SUSSEX

Brighton and Hove

Worthing

Bognor Regis

Chichester

SPELTHORNE

MAIN SITES
1 Staines Reservoir
2 Staines Moor

OTHER SITES
A1 Queen Mary Reservoir
A2 Wraysbury Reservoir
A3 King George VI Reservoir
A4 Shepperton Gravel Pits/Laleham
A5 Bedfont Lakes Country Park

Sitting north of the River Thames and in the far north-west corner of modern-day Surrey, Spelthorne Borough used to be part of the historic county of Middlesex. It was incorporated into Surrey in 1965 – and its own boundaries have been tweaked several times since. As a result, Spelthorne does not form part of the Surrey vice-county, which means virtually all natural history recording groups exclude the region from their records (including the Surrey Bird Club). This would normally be a minor point, but Spelthorne includes some of the best-known wetland and reservoir sites in the modern county of Surrey. Unsurprisingly, this makes it popular with birders, even if it doesn't fall within the Surrey vice-county recording area. To the uninitiated such intricacies can cause confusion, especially when it comes to that popular component of birding: listing! It means there are a

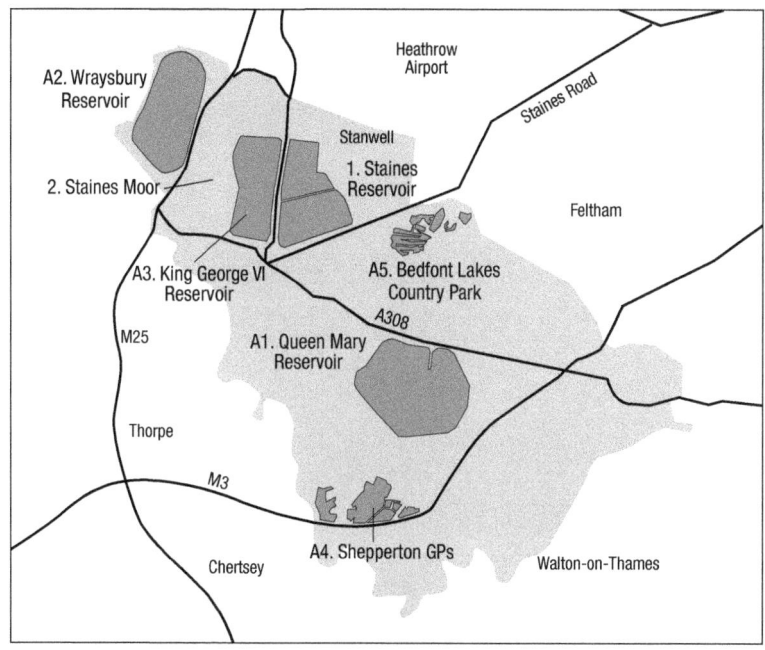

handful of species that have occurred in Spelthorne but aren't included on the Surrey list. Spelthorne does sit within the London Natural History Society recording area, but features separately in this book due to its distinct regional status.

Spelthorne contains some of the biggest reservoirs in the London region – five in total, which take up more than 15 per cent of the area. Apart from their main use of ensuring a stable and energy-efficient drinking water supply to London, they are important sites for birds and, in the case of the Queen Mary Reservoir, a sailing training centre. Unfortunately, there is limited access to these reservoirs, and Staines Reservoir is the only one that has public access. As a result, the big, largely inaccessible reservoirs have been given limited coverage in this book compared to Staines. In fact, the latter is the only publicly accessible large reservoir in both modern-day and vice-county Surrey, rendering it an appealing locale for many birders. A further 15 per cent (approximately) of Spelthorne is covered by other lakes, mostly former gravel pits.

The vast majority of the remaining space in Spelthorne is concrete. Heathrow Airport provides a noisy backdrop to the district, as does the M25. An exception to this is Staines and Stanwell Moors, which are situated along the River Colne and form the southernmost part of the Colne Valley Regional Park. The expanse of green space in an otherwise urban landscape makes it a relative haven for various forms of wildlife.

Naturally, the reservoirs and pits are the main focus for birders in Spelthorne. Winter and passage seasons are particularly productive at Staines Reservoir, which hosts nationally important numbers of certain waterbird species (it is one of the most reliable sites in Britain for Black-necked Grebe, for example). Gulls are present in large numbers year-round. Ring-necked Parakeets are a noisy, blatant presence across the district. Raptors have increased too – it is no longer surprising to see a Red Kite circling over the area, or a Peregrine dashing through. The accessibility of Staines Reservoir – and indeed Staines Moor – means both sites have a good history of birding coverage and, as a result, a fine list of rarities and scarcities. Compared with other Surrey regions in this book, there is a distinct lack of mature woodland, heathland or farmland. Species like Treecreeper and Bullfinch are not numerous; others such as Woodlark and Yellowhammer, for instance, are very rare.

1 STAINES RESERVOIR

OS Explorer 160
OS grid ref: TQ 051731
Postcode: TW19 7JP

HABITAT

Officially called Staines Reservoir, this site is actually two large, pumped storage reservoirs (known as North Basin and South Basin), separated by a narrow causeway with a public footpath. Together the basins comprise some 172ha of water. They lie just north of Staines and Ashford and south-west of Heathrow Airport. Being so close to Heathrow and the M25 creates a noisy backdrop, and the site is unlikely to ever win any prizes for scenery. It is, however, a popular birding destination, no doubt due to the fact that it's easily the most accessible of the south-west London/north Surrey reservoirs.

Due to its important numbers of wintering wildfowl, the area has been designated as a Site of Special Scientific Interest (SSSI). The water depth of 3m is attractive to diving duck in particular. The edges of the basins are sloping concrete, which attract passage waders. Very occasionally, one of the basins is drained for maintenance, which transforms the site from an open expanse of water to a rich habitat of shallow pools, shingle bars and mud, almost reminiscent of an estuary. Fence lines and areas of cropped grass along the causeway are often used by insectivores. Staines also benefits from having a lot of open sky.

SPECIES

Winter can be an unforgiving time out on the causeway, but it is normally a productive time to visit. Nationally important numbers of Black-necked Grebe winter here, with peaks counts often reaching low double-figures. Sometimes good views can be obtained close to the causeway – birds often linger through the spring, when display and mating can be witnessed. Slavonian Grebe is a rare visitor. In most winters a Great Northern Diver will take up temporary residence. Gadwall, Shoveler and Tufted Duck are usually the most numerous wildfowl species. Several hundred of the latter species can be counted and it's worth scanning through the rafts for Scaup. Goldeneye winter in decent numbers as well and, on milder days, the drakes can be seen performing their comical display. Staines used to be one of the best sites in Britain for Ruddy Duck but, after a DEFRA cull, records are few and far between. Smew has become trickier here, as it has in much of the South-East, but is still a midwinter possibility, along with other species such as Goosander. Passerine numbers are unremarkable in the winter, though Water Pipits from Staines Moor will often be seen on the shore. Peregrine is a fairly frequent visitor too.

Spring passage commences in March and Staines has good form for early Wheatears and hirundines. Black Redstart is also a possibility at this time – check the area around the towers at the west end of the causeway. Wildfowl passage includes Common Scoter and Garganey, with a chance of more unusual species such as Brent Goose or Red-breasted Merganser. In the right conditions, April and May can produce exciting birding. Wader passage peaks from mid-April through early May with a variety of species possible, including inland goodies such as Bar-tailed Godwit, Grey Plover, Turnstone and Sanderling. Staines is particularly reliable for the overland passage of terns: Arctic and Black are expected annually and sometimes large flocks pass through, with Sandwich also fairly regular; Little Tern is rarer but a distinct possibility. Little Gull is also a likely visitor, with Kittiwake an outside possibility. On particularly lively days of passage (usually involving inclement weather) a real surprise can occur, such as Arctic Skua or Fulmar. Gatherings of Swifts and hirundines can be impressive and usually attract the attention of a migrant Hobby. Yellow Wagtail is regular along the causeway in the second half of April; White Wagtail is possible too.

Midsummer is typically quiet, but return wader passage can get going as early as July, which is often when the first Black-necked Grebes reappear. August is the best month for wader passage with the typical, 'common' inland species the most likely: think Oystercatcher, Ringed Plover, Dunlin and Common Sandpiper. Black-tailed Godwit is regular, too, and sometimes fairly large flocks (for London!) can drop in. Scarcer possibilities include Spotted Redshank, Knot and Little Stint.

September through to November offer an outside chance of a seabird, especially if there's been an easterly gale into the Thames. Skuas are naturally very rare but can occur, as can Gannet, Leach's Petrel and Manx Shearwater. Rock Pipit is recorded annually on the concrete shores, with October the most productive month. Both regular species of phalarope have been recorded, too, with Grey the more likely. From November, cold weather or prolonged easterlies offer a chance of seaduck, with Red-breasted Merganser and Common Scoter always more likely than Velvet Scoter, Eider and Long-tailed Duck, all of which have occurred.

Staines is one of the best sites in London for rarities and has built up an impressive list of unusual species down the years. This includes a phenomenal selection of waders, such as Baird's, Buff-breasted and Sharp-tailed Sandpipers, Long-billed Dowitcher, Wilson's Phalarope, Lesser Yellowlegs and Collared Pratincole. Other notable rarities here have included Whiskered and Caspian Terns, Horned Lark and Tawny Pipit.

It's worth noting that a telescope is frequently essential at Staines, with birds often distant.

TIMING

Staines Reservoir is renowned for being windy at any time of year – wrap up warm in the winter, especially if you plan on staking out the site for a while. Both basins are very large and, on bright days, it can be tough to scan the south basin, which is better viewed in dull conditions. Wader numbers are much better when one of the basins is drained, but this happens sporadically at best. In the summer, clouds of insects can be encountered along the causeway. Early mornings and late afternoons are often best during the passage seasons, with waders most likely to stop off during these times. Inclement weather, such as precipitation or a headwind, is also useful for dropping in passage terns, gulls and waders.

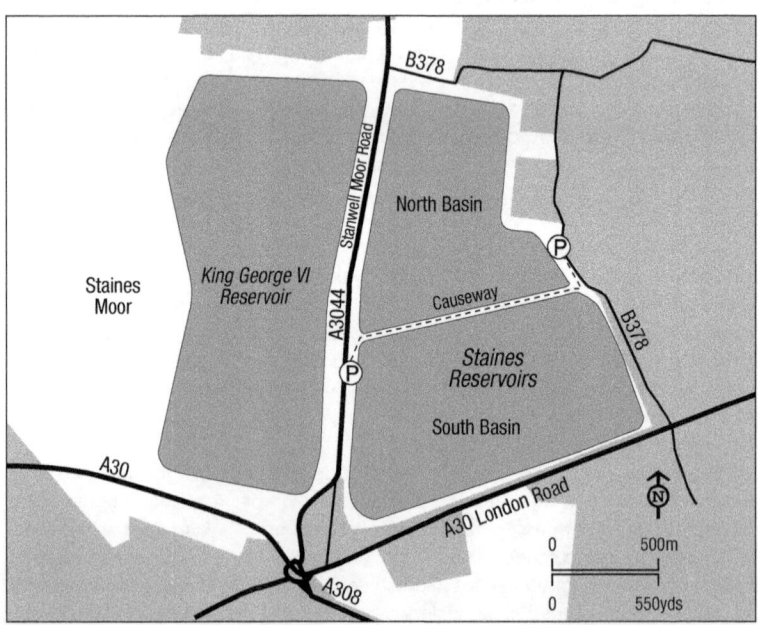

ACCESS

Open access to the reservoir is via the causeway (a public footpath) that bisects the two basins. There is strictly no access to any other part of the reservoirs.

If coming by train, Ashford is the nearest station and is a 20-minute walk away (turn left on Church Road and keep going straight). You can also walk from Staines, but this means following the very busy A3044 (Stanwell Moor Road) and is not recommended. If coming from London, take the Piccadilly Line to Hatton Cross and catch a 203 bus from Stop D in the adjacent bus station. It generally runs every 20 minutes Monday to Saturday and every 30 minutes on Sundays. Get off after about 5km at St Anne's Avenue and you will see the reservoir in front of you. Walk forward and turn right to find the path up to the causeway.

Parking is available on the west side of the reservoir on the A3044, at TQ 045728. A small car park on the east side of the reservoir, along Town Lane at TQ 055734, has recently been closed due to antisocial behaviour. It is still possible, however, to park in front of the locked gates or on the nearby grass verges.

FACILITIES: The nearest facilities are in Staines and Ashford.

CALENDAR

All year: Common wildfowl, Red Kite, Black-necked Grebe.

November–February: Goldeneye, Scaup, outside chance of scarcer duck like Smew, Black-necked Grebe numbers highest, Great Northern Diver, Water Pipit.

March–June: Garganey, Common Scoter, wader passage from April to May including Bar-tailed Godwit, Grey Plover, Turnstone and Sanderling, Arctic, Black and Sandwich Terns, chance of Little Tern, Little Gull, hirundines, Black Redstart, Wheatear.

July–October: Black-necked Grebe from July, wader passage between July and September including Black-tailed Godwit, Spotted Redshank, Knot and Little Stint, chance of storm-driven seabirds, Rock Pipit in October.

2 STAINES MOOR

OS Explorer 160
OS grid ref: TQ 032730
Postcode: TW19 6EQ

HABITAT

Situated in the flat valley of the River Colne between Wraysbury and King George VI Reservoirs, the 117ha Staines Moor is a large area of alluvial meadow and is a designated SSSI. Here, chalky residue brought down by the Colne mixes with the silty Thames Valley soil during times of flood, and as a result the moor has a rich and diverse flora, which is helped by the grazing of cattle and horses. Staines Moor is one of England's largest areas of neutral grassland that has never been extracted for gravel or agriculturally improved – the moor has remained unploughed, and no use of fertilizer or pesticides for 1,000 years makes for an ancient and unique habitat. Flanked by large reservoirs, the moor is crossed by the Colne and Wraysbury rivers and also features ponds, ditches, marsh, scrub and woodland.

SPECIES

Staines Moor is perhaps the best site in Surrey and Sussex for Water Pipit. Present from November until late March/early April, numbers vary – one or two can be expected, but 10 or more have been recorded. Generally numbers peak in March. The southern end of the moor is best: check the iris beds in the south-west corner, the floods in the south-east corner and along the Colne. Staines Moor is a good site for Short-eared Owl, too, with multiple birds present in some winters. Barn Owl is also recorded on occasion. A Jack Snipe or two may be flushed from damper patches. Golden Plover is occasionally present at this time of year but has become rare on the moor. Goosander can sometimes be found on the Colne.

Lapwing and, impressively, Redshank attempt to breed in some years. The latter was lost as a breeding species in virtually all of Surrey many years ago and, sadly, both these waders may be lost from the moor in time. More sporadic in their breeding attempts are Grasshopper Warbler and typically there are none on the moor, but every few years a male or two will hold territory and breeding has been suspected on several occasions.

Passerine passage can be good, with Ring Ouzel, Wheatear and Yellow Wagtail possible in the spring. Whinchat can occur but is better sought in autumn, when several birds can be dotted around the site. There is a remote chance of Wryneck, too. Famous scarcities and rarities at Staines Moor include Barred Warbler and Brown Shrike.

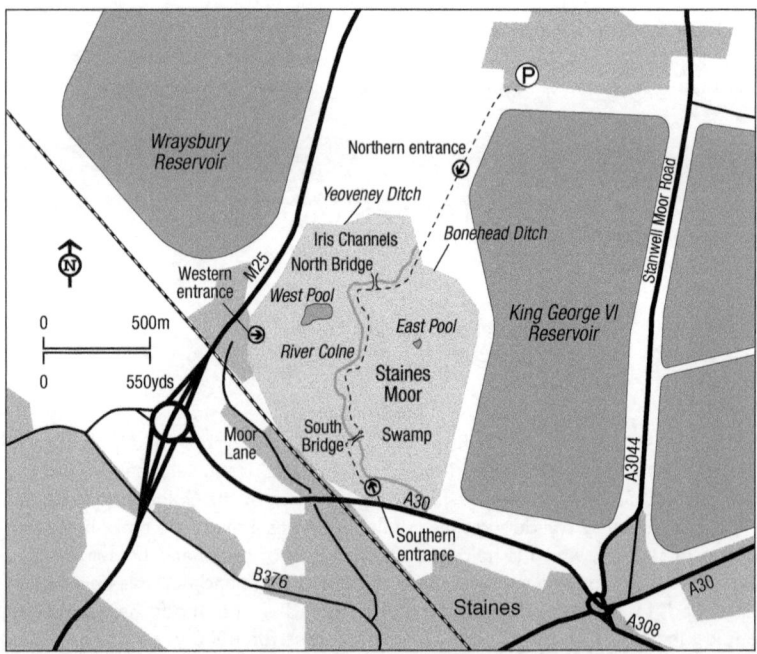

TIMING

Early morning is recommended – Staines Moor is popular with dog walkers, especially in the drier areas, as well as joggers and a few anglers. In winter, parts of the site can flood or at least become very wet, so waterproof footwear is recommended.

ACCESS

If arriving by public transport to Staines town, walk north-west until you get to Moor Lane. At TQ 030722, you can take a public footpath over the railway line then under the A30 that comes out on the southern end of the moor.

If arriving by car, the best place to head for is the northern end of the site. Leave the M25 at Junction 14 and take the small local road to Stanwell Moor (the next exit after A3113). At a bend by The Anchor pub, turn right into Hithermoor Road and park when the banks of the King George VI Reservoir come into view (roughly TQ 040743). Follow the concrete footpath south by the reservoir fence, fork right through a kissing gate after 200m or so, past fields with ponds, to the open moor. Some parking is also available in a layby near the A30 bridge at the south end of the moor.

The moor is unfortunately not practical for wheelchair or pushchair users.

FACILITIES: The nearest facilities are in Staines.

CALENDAR

All year: Little Egret, Red Kite, Barn Owl, Cetti's Warbler.

October–March: Water Rail, Golden Plover, Common and Jack Snipe, Short-eared Owl, Water Pipit.

April–June: Redshank, occasional breeding Grasshopper Warbler, migrant passerines including Wheatear and Ring Ouzel.

July–September: Passage waders, migrant passerines including Whinchat and Yellow Wagtail.

OTHER SITES IN SPELTHORNE

A1 QUEEN MARY RESERVOIR

More than a mile wide, this massive reservoir is split in two, the western half of which is designated as a Site of Importance for Nature Conservation. The other half is used for sailing. There is a major gull roost in the winter, which is also a good time for scarcer divers and grebes. The wildfowl line-up is similar to that of Staines Reservoir, though Goosander is more regular here. There is a ringing group based at the site too. Access is permit only, with a limited number available, and those with permits can only visit within certain hours. Contact Thames Water for further details. Access is in the north-east corner, off Ashford Road (TQ 079703).

A2 WRAYSBURY RESERVOIR

Due to its nationally important numbers of wintering waterbirds this 205ha SSSI is listed here (TQ 025745), but there is strictly no access from Thames Water, aside from for WeBS counters. On a clear day and with high-magnification optics, one can pick up larger species on parts of this reservoir from the high point of the Queen Mary Reservoir car park.

A3 KING GEORGE VI RESERVOIR

As with Wraysbury Reservoir, there is no access to this site, which sits between Staines Reservoir and Staines Moor (TQ 041733). Occasionally, larger birds can be seen with a telescope from the causeway at Staines.

A4 SHEPPERTON GRAVEL PITS/LALEHAM

A series of pits south of Queen Mary Reservoir which are fairly complicated to navigate and access. Winter is the best time to visit, with Smew a possibility. The large fields in the area can sometimes attract grey geese in cold weather. The best way to access the pits is from Fairview Drive (TQ 068677) – a public footpath leads out from there between the Littleton Lane East and Sheepwalk pits.

A5 BEDFONT LAKES COUNTRY PARK

Bedfont Lakes Country Park is a 72.5ha Local Nature Reserve north of Ashford. At the north end is a private nature reserve, including hides. Friends of Bedfont Lakes Country Park get access included with their membership, which costs £15 per year. However, much of the site is visible from public areas. Smew is an occasional winter visitor and there are records of Bittern. Common Terns often visit in the summer. There are car parks in Bedfont Road (TQ 076724) and Clockhouse Lane (TQ 084729).

NORTH-EAST SURREY/ SURREY-IN-LONDON

MAIN SITES
3 Thorpe Park
4 Walton-on-Thames area reservoirs
5 Richmond Park
6 WWT London Wetland Centre
7 River Thames: Kew to Rotherhithe
8 Beddington Farmlands
9 Canons Farm
10 Holmethorpe area
11 Bookham and Leatherhead
12 Ashtead and Epsom Commons

OTHER SITES
B1 Wey Manor Meadows
B2 Esher Common and Claremont
B3 Bushy Park
B4 Hogsmill Sewage Farm
B5 Wimbledon and Putney Commons
B6 Wandsworth Common
B7 Tooting Common
B8 Morden Hall Park
B9 Poulter Park
B10 South Norwood Country Park
B11 Chelsham and Woldingham area
B12 Farthing Downs
B13 Little Woodcote
B14 Nonsuch Park
B15 Hogsmill Local Nature Reserve
B16 Epsom and Walton Downs
B17 Priest Hill

The London Natural History Society's (LNHS) recording area is everywhere within a 20-mile radius of St Paul's Cathedral. As a result, part of Surrey – both the vice-county and modern-day – falls within it, as can be seen in the map on page 30.

Central London is not what most people picture when they think of Surrey, but the county used to extend all the way to the River Thames in what is now Inner London. This can still be identified in place names such as Surrey Docks, near Rotherhithe. This heavily built-up part of the city is a far cry from the rural land-scapes in the south of the vice-county. This is one of the quirks of the vice-county boundary, though, and several key sites – including Beddington Farmlands and WWT London Wetland Centre – fall within the area and are very much established 'Surrey' sites. Every site in this section is within the vice-county with the sole exception of Bushy Park.

The River Thames is a significant feature in this region, running from the east of

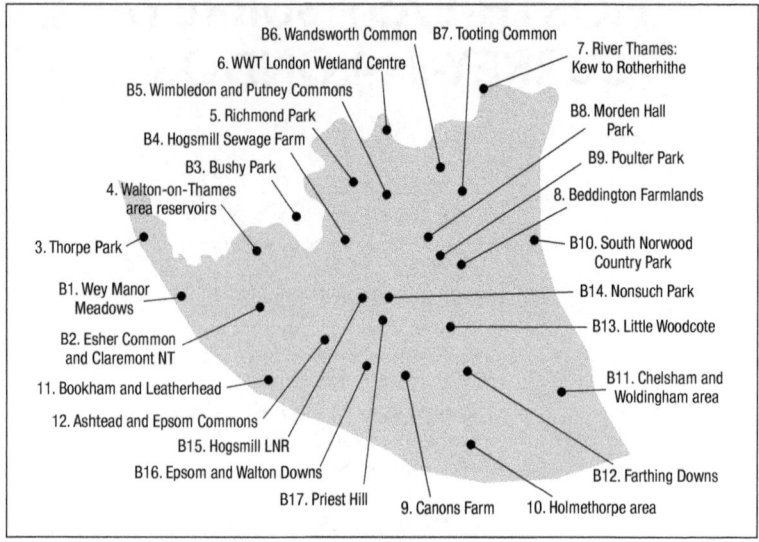

the area (Rotherhithe) west to the M25 at Staines-upon-Thames, and the border with Spelthorne. Much of this stretch of the Thames is tidal, with old industrial buildings and increasing river traffic. WWT London Wetland Centre sits on its banks. As the river enters modern-day Surrey, it runs north of the Walton-on-Thames cluster of reservoirs and gravel pits – one of the most important collections of waterbodies in the county. In the far north-west of this region, where the Thames exits the vice-county, is Thorpe Park, which is another notable area of gravel pits.

The area south of the Thames within modern-day London is built-up, with sprawling suburbia taking up most of the land. The western part of this area holds several well-known parks and commons – Richmond Park, for example, is a veritable oasis of green amid the ever-increasing urban landscape. Beddington Farmlands is a similarly striking area of hospitable habitat amid a sea of concrete to the east near Croydon.

As you move further from the centre of the capital into the south London suburbs and beyond, the number of scattered parks, woods and commons increases. They serve as a reminder of the rural past of the area. The size and quality of these sites grow as you reach the southern boundary of the LNHS recording area. Indeed, the commons around Bookham and Epsom wouldn't look out of place in the far south of the county, in terms of habitat. There are even a few heathland species, such as Dartford Warbler and Nightjar, on the lowland heath at Esher Common. Patches of arable farmland can be found as well, with notable areas at Canons Farm near Burgh Heath and around the North Downs at Chelsham and Woldingham.

The North Downs serves as a natural border along part of this southern boundary, from Box Hill (site 18, page 67) east to the Kent border. Indeed, one can stand on top of the downs and feel like you're positioned right between two very different areas, with the south London suburbs and beyond to your north, and the scenic, greener Surrey countryside to your south.

This range of habitats presents itself nicely to the north Surrey/south London birder. Unsurprisingly, though, many focus their efforts on the wetland sites – places like Beddington Farmlands, Holmethorpe and the Walton-on-Thames area reservoirs are some of the best-covered Surrey hotspots, whether by groups of birders or dedicated individuals. These three in particular are among the best birding sites in Surrey. The density of the human population in this region means, by default, there are more birders. And this means there are often 'new' birding locations being discovered by the more explorative birder, whether it be a previously unwatched park, stretch of river or pocket of farmland.

The composition of common species varies in this region. Close to central London, countryside species such as Rook and Skylark, for example, are very rare. Indeed, gulls, Feral Pigeons and Ring-necked Parakeets often seem like the only species on show in the capital. This line-up transitions as you move south, in line with the habitat shift outlined above. The 'oasis' effect of some of the parks and pockets of farmland – often isolated among suburbia – make it easier to find many species, including migrants, especially compared with the more needle-in-a-haystack expanses of habitat in rural Surrey.

Some of the more specialist species that are fairly widespread in outer Surrey tend to be localised in this region; for example Esher Common is the only place you're likely to find Nightjar, Marsh Tit is restricted to a handful of sites in the south of the area, and Goshawk is virtually unheard of. On the other hand, North-East Surrey/Surry-in-London is very productive for waterbirds. Gulls are one example – Caspian Gull is very rare in any other part of Surrey, Great Black-backed Gull is found far more readily here than elsewhere in the county and the first breeding attempt of Mediterranean Gull in Surrey took place in the region. The Thames flyway, along with the aforementioned waterbodies, mean that this area is by far the most productive in the vice-county for wayward seabirds, scarcer wildfowl species and waders. Cetti's Warbler is another species far more numerous in this region than elsewhere in Surrey and it also holds the only vice-county sites for Water Pipit (Beddington and WWT London Wetland Centre). Breeding specialists found in far lower numbers (or not at all) elsewhere in the county include Peregrine and Black Redstart.

3 THORPE PARK

OS Explorer 160
OS grid ref: TQ 028684
Postcode: KT16 8PH

HABITAT

The Thorpe Park area comprises a relatively complex network of old gravel-extraction pits. These have been landscaped and are now used for various activities, including watersports, pleasure craft and the well-known theme park. The core area is between the M25 and A320 – this includes the productive Manor Lake (the only lake where watersports don't take place), as well as Abbey, Fleet and St Ann's Lakes. Longside Lake is tucked along the west side of the M25, while a further two pits are to be found east of the A320 at Abbey Mead. There are some small islands and areas of reeds across the series of pits, but there are few areas of shoreline for waders. Surprisingly, the area has received somewhat limited attention from birders down the years.

SPECIES

Winter is by far the best time to visit. Between November and March, Manor Lake in particular can be packed with wildfowl. Between 50 and 100 Pochard is a typical count, while Tufted Duck numbers can reach more than 500 across the whole site. It is perhaps not surprising that recent finds here have included both Ferruginous Duck and Scaup. A few Goldeneye winter across the complex, with Goosander a relatively regular visitor. Red-crested Pochard is rarer. Numbers of dabbling duck are lower, but Wigeon, Shoveler and Gadwall are all present in numbers. Wildfowl numbers are less impressive on Abbey and St Ann's Lake but both are still worth checking; St Ann's sometimes has a Goldeneye or two and occasionally a Great Black-backed Gull will drop in. Thorpe Park used to be one of the best sites in Surrey for Smew, but this beautiful species has become decidedly rare these days. Their favoured pits are the two at Abbey Mead (which are shallower than the main complex) or Manor Lake. All three scarcer grebes have occurred. A few Snipe winter (occasionally a Jack Snipe is among them), visible on the west side of Manor Lake, along with a small Lapwing flock. Cetti's Warbler is resident.

Wader passage is fairly minimal, though both Oystercatcher and Little Ringed and Ringed Plovers have attempted to nest before, with the latter a very rare breeding bird in Surrey. Common Terns breed on rafts on Manor and St Ann's Lakes. Reed Warbler also breeds. Little Gull and Black Tern have passed through on passage before and there is doubtless potential for some good finds here.

TIMING

Early morning is best, especially at weekends when disturbance on the pits from watersports must be considered, though note that none takes place at Manor Lake. Generally these pits are quiet, though.

ACCESS

Chertsey train station is about a 20-minute walk from the Abbey Mead pits (over the M3 bridge). However, a car is best for accessing this area. For the main pits

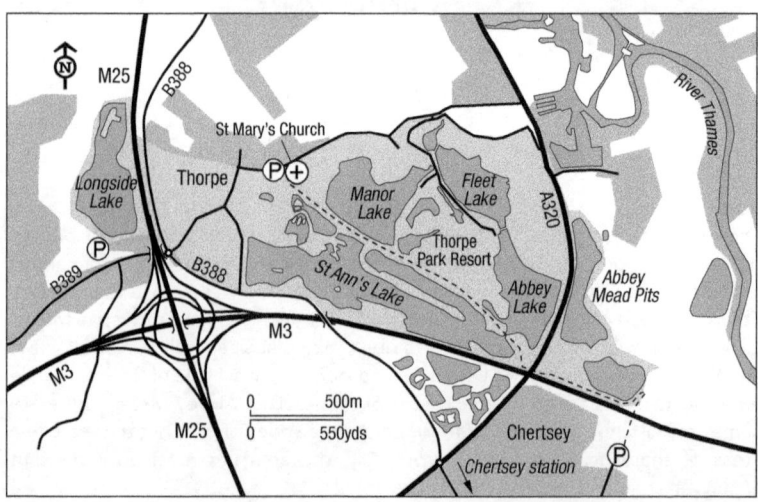

(including Manor Lake), park sensibly in Thorpe village and take the footpath near St Mary's Church running south-east (TQ 023687). This path runs between Manor Lake and St Ann's Lake, all the way past Abbey Lake up to the A320. Fleet Lake can only be viewed from inside Thorpe Park Resort.

For the main Abbey Mead pit, it's possible to park off the A320 at TQ 039678 and scan. The Ferry Lane footpath that runs north of Chertsey provides access to the smaller pit from the east side. Alternatively, both of these pits can be reached by continuing along the footpath from St Mary's Church.

For Longside Lake, there is limited parking at the end of Green Road (TQ 014682).

None of these sites is particularly good for wheelchair or pushchair users.

FACILITIES: The nearest facilities are in Thorpe village.

CALENDAR

October–February: Good numbers of *Aythya* duck sometimes attracting scarcer species like Scaup, Goldeneye, Goosander, outside chance of Smew, possible Jack Snipe and Green Sandpiper, Lapwing, Red Kite, Kingfisher, Cetti's Warbler.

March–September: Little Ringed Plover, Common Sandpiper, occasional Oyster-catcher and Ringed Plover, Common Tern.

4 WALTON-ON-THAMES AREA RESERVOIRS

OS Explorer 161
OS grid ref: TQ 128672
Postcode: KT8 2LF

HABITAT

These reservoirs and pits in the Walton-on-Thames area are some of the most important waterbird sites in Surrey, although access to most is limited by owners Thames Water. The largest is Queen Elizabeth II Reservoir (QE2), which is a mighty 129ha and reaches a maximum depth of 18m. Its design has an effect on the fish population – the concrete shores mean that only European Perch and Ruffe breed. A solar panel farm covers a tenth of the surface.

Immediately to the north of QE2 are Knight and Bessborough Reservoirs, which are separated by a narrow causeway. These are deep and among the last to freeze over in hard weather. North of Bessborough Reservoir are Chelsea and Lambeth Reservoirs (sometimes referred to as Molesey Reservoirs). These were taken out of use in 1999; the land was then used for the extraction of aggregates and it is now a relatively dynamic site with reedbed, shingle, patches of willow and scrub, and some areas of mud.

South-east of all these sites, and south of Molesey, is Island Barn Reservoir. Some 50ha in size, the River Ember runs along the south and east side. A sailing club hosts regular training and races. An area of wet ground to the south is sometimes good for wildfowl. Between Island Barn and QE2 is Molesey Heath – an area of rough ground and scrub – as well as the remnant Field Common and Molesey Gravel Pits. The former Hersham Gravel Pit is to the south.

SPECIES

Winter wildfowl include a small flock of Goldeneye that can move between the reservoirs. Goosander is fairly rare these days. A Scaup is usually present and, like the Goldeneye, can range between sites. Cold weather may increase wildfowl numbers and with it the chance of a rarer species. A wintering flock of feral Barnacle Geese, sometimes totalling more than 200 birds, often roosts on Chelsea and Lambeth and feeds in the day on the wet meadows south of Island Barn Reservoir. A Great Northern Diver will often winter, tending to favour Island Barn Reservoir; of the scarce grebes, Black-necked is most likely. Winter gull roosts at QE2 and Island Barn can often hold scarcities such as Caspian and Yellow-legged Gulls or, better still, Glaucous or Iceland Gulls. Chelsea and Lambeth will sometimes host wintering Bittern.

Spring passage usually provides a few surprises, with Little Gull, Kittiwake and Arctic, Black and Sandwich Terns annual; Little Tern is less frequent. March and April are good months for Garganey or Common Scoter. Wader passage is typically best at QE2 (and less so at Chelsea and Lambeth). Species otherwise very rare in Surrey are annual in the Walton area and include Grey Plover, Turnstone, Sanderling and Bar-tailed Godwit, as well as the suite of typical inland passage waders. Passerine migration is usually limited to Wheatears and Yellow Wagtails on the causeway (plus the odd Black Redstart). Impressive hirundine counts can be made in rainy or windy weather.

Breeding species include Pochard, Common Tern, Cetti's Warbler and, in recent years, Mediterranean Gull – the first breeding attempt for Surrey was among the Black-headed Gull colony at Chelsea and Lambeth. Less regular breeders include Shelduck and Oystercatcher.

A similar assortment of passage species move through in the autumn, including flocks of Black-tailed Godwit, Wood Sandpiper and Ruff. From September, stormy weather – or north-easterlies – can sometimes blow a seabird in, with Arctic and Great Skuas recorded in recent years. Petrels and shearwaters used to be more regular at the reservoirs than they are now – it's thought the increase in local Peregrines has seen the amount of time tubenoses hang around much reduced! Seaduck are possible, too, with Common Scoter the most likely. Brent Goose is annual. Late September and October are the best months for Rock Pipit, which occurs in most years.

The Walton area has a fine list of rarities to its name, including many firsts for Surrey, such as Bonaparte's Gull, Penduline Tit, Squacco Heron and White-rumped Sandpiper. Indeed, something ultra-rare is found in most years.

Breeding birds at Molesey Heath include Lesser Whitethroat and Cetti's Warbler. Rarities here have included Yellow-browed Warbler. The remaining pool at Molesey Gravel Pit has attracted Great White Egret before.

TIMING

Sailing at Island Barn can cause disturbance, so try to avoid weekends and visit early in the morning. The exposed nature of the reservoirs means it's unwise to bird them in especially wet and windy weather. That said, inclement weather can be best for dropping in passage waders, gulls and terns.

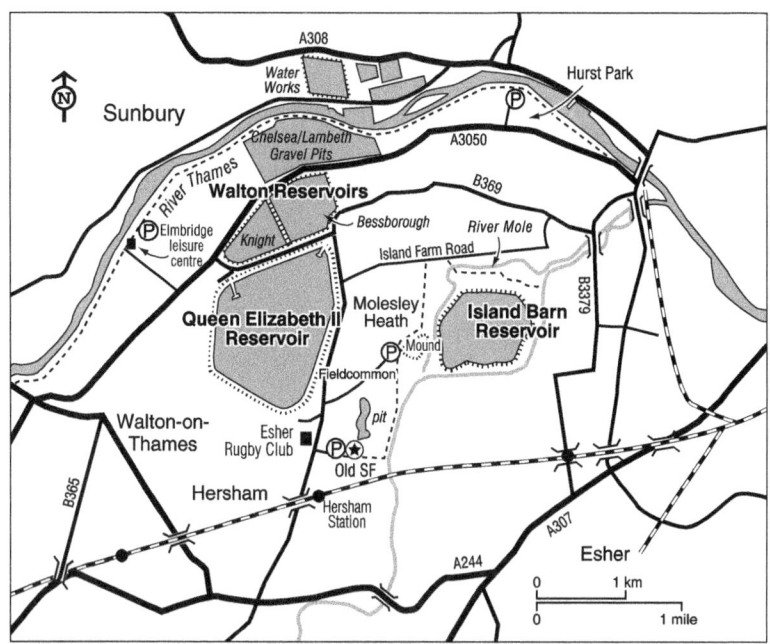

ACCESS

Unfortunately, access to QE2, Knight and Bessborough, and Chelsea and Lambeth Reservoirs is limited to a small group of permit holders who have permission to visit from Thames Water. For several years there has been talk of Chelsea and Lambeth being converted into a public nature reserve but, at the time of writing, there are no signs of this happening.

Island Barn Reservoir can be accessed by members of the Surbiton & District Bird Watching Society, which costs £20 a year to join (£5 for under-16s), at the time of writing. There is parking along Ray Road (TQ 135676); walk south along the reservoir entrance track then up the steps.

Alternatively, Island Barn can be viewed from high ground at Molesey Heath (a telescope is recommended). Park on Approach Road at the entrance to Molesey Heath (TQ 133676) and walk to the obvious mound. Hersham Gravel Pit can be accessed via public footpaths running south from Molesey Heath.

FACILITIES: The nearest facilities are in Walton-on-Thames and Molesey.

CALENDAR

All year: Shelduck, Mediterranean Gull, Peregrine, Cetti's Warbler.

November–February: Barnacle Goose, Goldeneye, Scaup, chance of Black-necked Grebe and Great Northern Diver, gulls including Caspian, Yellow-legged and white-wingers.

March–June: Garganey, Common Scoter, wader passage from April to May including Bar-tailed Godwit, Grey Plover, Turnstone and Sanderling, Arctic, Black and Sandwich Terns, chance of Little Tern, Little Gull, hirundines, Black Redstart, Wheatear, Yellow Wagtail.

July–October: Wader passage between July and September including Black-tailed Godwit and scarcer species, chance of storm-driven seabirds from September, Rock Pipit in October.

5 RICHMOND PARK

OS Explorer 161
OS grid ref: TQ 200729
Postcode: TW10 5HX

HABITAT

At an impressive 950ha, Richmond Park is the largest of London's Royal Parks and an NNR, SSSI and SAC. It can offer a peaceful retreat from the bustle of south London and is famous for its herds of Red and Fallow Deer (approach with caution during the rutting season in the autumn). The park is dominated by ancient oak woodlands, acid grassland and some extensive areas of bracken. Pen Ponds are two connected waters located in the centre of the site; Upper Pond has a small reedbed. There are several smaller ponds elsewhere in the park and Beverley Brook runs through the north-eastern section.

SPECIES

Typical woodland birds can be found in the park. Among the more notable species is Lesser Spotted Woodpecker, which just about hangs on as a resident. It can be elusive, with early spring the best time to find one and Jubilee and Saw Pit Plantations often favoured. Little Owl is another fairly shy resident but is generally more cooperative, often sitting out on favoured oaks at dawn and dusk. Holly Lodge, Richmond Gate and Spankers Hill Wood are good places to look. Skylarks still manage to breed despite extraordinary levels of disturbance from dogs. Reed Buntings breed and are sometimes tame around Pen Ponds, where Water Rail is a year-round – but very elusive – presence. Hobby is a regular visitor in the summer and early autumn, with Kestrel, Sparrowhawk and Buzzard resident. Good numbers of Mandarin and Ring-necked Parakeet add colour to proceedings year-round – more than 30 of the former can be found on Peg's Pond in Isabella Plantation in the winter.

Stonechat also breeds but is far more evident on passage. Indeed, Richmond Park can be good for chat passage, with Whinchat and Wheatear present in small numbers at peak times. Lawn Field and Sawyer's Hill are good places to look; both can hold wintering Dartford Warbler, too. Other passerines to keep an eye out for during migration periods include both flycatcher species (Spotted is annual in early autumn), Firecrest, Redstart and, occasionally, Ring Ouzel and Wood Warbler. Scarcities recorded before have included Wryneck, Golden Oriole and Ortolan

Bunting. Hawthorn Valley, The Bog and Holly Lodge Paddocks and Pond Slade are good places to go searching for migrant passerines.

Pen Ponds have good form for attracting unusual species, too, with scarce grebes and dabbling duck visiting on occasion. Mandarin can be numerous but Red-crested Pochard is a more irregular presence. In the winter, common wildfowl such as Shoveler and Pochard can show superbly; Gadwall and Wigeon are present in lesser numbers. Great Crested Grebe breeds – usually Common Tern does as well, but it can often be outcompeted by Black-headed Gull. That said, recent work to extend the banks on the island on Upper Pen Pond has helped. Reed Warbler can be found in the summer; Cetti's is rarer. Passage waders shouldn't be expected, but Common Sandpiper is regular in late summer and early autumn and other occasional visitors include Greenshank, Dunlin and Black-tailed Godwit. If water levels are low, double-figure counts of Little Egret may grace the ponds.

Many common species are quite tame in Richmond Park and, coupled with the presence of Little Owls and showy waterbirds, it is a good site for photography.

TIMING
The park is open to vehicles from 7:00 am in the summer and 7:30 am in the winter. Vehicle gates close at dusk all year-round. Pedestrian gates are open 24 hours except during the six-week deer culls from November to early December and February to early March. During these months, pedestrian gates open at 7:30 am and close at 8:00 pm.

It can become very crowded at weekends, especially in the summer, so early or late visits are advised.

ACCESS

From Richmond Station (National Rail or London Underground), catch the 371 or 65 bus to the pedestrian gate at Petersham, which is on the west side of the park.

There are a number of car parks within Richmond Park: Pembroke Lodge, Isabella Plantation, Kingston Gate, Pen Ponds and Roehampton Gate all have accessible parking (Isabella Plantation is Blue Badge only). There are other car parks at Sheen Gate, Robin Hood Gate and Broomfield Hill. Spaces are often limited at weekends and on public holidays. Don't park on the roads or grassland.

Many of the footpaths are suitable for wheelchairs and pushchairs. There is also a free minibus service to help visitors with limited mobility get better access to the park and, in particular, Isabella Plantation. The free service operates every Wednesday between late April and late October, from 9:40 am until 4:10 pm, and runs between all the Richmond Park car parks, Ladderstile Gate, Richmond Gate, Pembroke Lodge and Isabella Plantation.

FACILITIES: A visitor centre (including a public toilet) is open daily from 10:00 am until 4:00 pm between April and October, and from 11:00 am until 3:00 pm otherwise. It is shut on Christmas Day. There are various cafés, food/drink kiosks and toilets.

CALENDAR

All year: Mandarin, occasional Red-crested Pochard, Great Crested Grebe, Water Rail, Little and Tawny Owls, Lesser Spotted Woodpecker, Stonechat, Reed Bunting.

October–March: Wintering wildfowl (with a chance of a scarce waterbird), Woodcock, Dartford Warbler, Firecrest.

April–September: Common Tern, Hobby, warblers, Wheatear and Whinchat on passage, migrant passerines including flycatchers.

6 WWT LONDON WETLAND CENTRE

OS Explorer 161
OS grid ref: TQ 229768
Postcode: SW13 9WT

HABITAT

Opened in 2000, WWT London Wetland Centre is a fantastic wetland reserve right in the heart of the capital and on the northern extremity of the Surrey vice-county. The site occupies 44ha of land which was formerly Barn Elms Reservoirs – a group of four rectangular basins tucked into a loop in the River Thames. These reservoirs were converted into a wide range of wetland habitats before the centre opened and, in 2002, some 29.9ha of the site was designated as an SSSI.

A large collection of captive birds is situated on the west side of the reserve; a much larger area is maintained across the rest of the site for wild birds. This includes extensive reedbeds, wet meadows, dykes, pools of various sizes and depths, scrapes, islands and scrub. There are various hides, viewing screens and an observation tower from which to watch the birds.

SPECIES

A winter visit can produce excellent birding, often including some high-quality species. Perhaps the star of the show is Bittern – a few birds (as many as seven) will winter in the reedbeds, usually around Main Lake, and a bit of patient scanning from Dulverton, Headley or WWF Hides can produce nice views. Plenty of wildfowl winter on site with Gadwall, Wigeon, Shoveler, Teal, Pochard and Tufted Duck the most numerous. There is usually a handful of Shelduck, Pintail and Goldeneye as well – the latter two species have become scarce in Surrey. The Grazing Marsh and Wader Scrape are worth checking for Water Pipit, with one or two present each winter. Jack Snipe is another winter speciality but can be harder to detect among the Snipe – sometimes one will show on Grazing Marsh, Wader Scrape or one of the Main Lake islands. Water Rail (as many as 15) and Lapwing (up to 100) are regular at this time of year too. It's worth checking through the loafing gulls on Main Lake in the winter, too, as Caspian and Yellow-legged Gull irregularly drop in, with even Iceland Gull possible. High tide on the Thames is the best time to check. Wildside can be good for finches, usually Siskin, but sometimes Lesser Redpoll and Brambling – check the feeders.

Little Ringed Plover and Sand Martin are often the first summer visitors to return and both species breed, with the latter using a purpose-built nest bank on the southern side of Wader Scrape. The reedbeds support good numbers of Cetti's, Reed and Sedge Warblers. Scarce breeders in a Surrey context include Shelduck, Pochard, Oystercatcher, Lapwing (as many as 14 pairs) and Redshank (most years) and, occasionally, Teal and Shoveler. Avocet has bred before, too, but is typically a rare passage visitor.

Both passage seasons invariably turn up scarcities, such is the extent of the habitat and reserve's location along the Thames. Brent Goose and Garganey are virtually annual; wild swans and grey geese are very rare and more confined to late autumn, as are diving duck species such as Scaup and Smew. Wader passage is notable here and usually includes Ringed Plover, Common and Green Sandpipers, Dunlin, Greenshank, Whimbrel and Black-tailed Godwit. Sometimes relatively large groups of the latter species can drop in during July and August. Wood Sandpiper passes through most autumns. Grey Plover, Spotted Redshank, Ruff and Bar-tailed Godwit are rarer. Pectoral Sandpiper and Temminck's Stint have occurred more than once. Despite its position next to the Thames, terns, Little Gull and Kittiwake are not regular on passage. Passerine migration is fairly limited, although Yellow Wagtails, Wheatears and Whinchats may be found on Grazing Marsh, and perhaps a flycatcher can be discovered around Wildside in the autumn.

Since it opened, London Wetland Centre has already garnered a fine list of scarcities and rarities, including Ring-necked Duck, Night and Purple Herons, Spotted Crake, Red-rumped Swallow, Bluethroat and Surrey's first Savi's Warbler.

TIMING

The reserve is open every day bar Christmas. Opening and closing times vary depending on the time of year – check the centre's page on the WWT website – with last admissions an hour before. Note that, often, the reserve isn't open until 9:30 am.

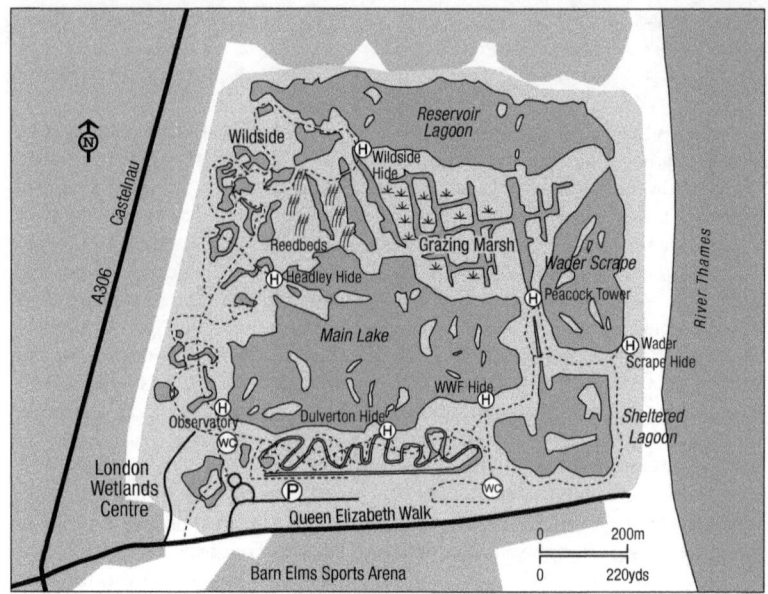

ACCESS

Entry via the visitor centre (TQ 226768) is free for WWT members. For non-members (at the time of writing), adults cost £14.09, concessions £11.81 and children between the ages of four and sixteen are £8.63. Children under four and carers assisting disabled visitors enter for free.

The reserve is easily accessible by private and public transport and is a 15-minute walk from both Barnes and Barnes Bridge train stations. Buses (33, 209, 378, 419 and 533) from Barnes station, Putney Bridge and Hammersmith underground stations stop at the Red Lion Pub on Castlenau, which is a five-minute walk away. The 485 bus from Hammersmith Bridge comes directly to the reserve, though it is not a frequent service (twice an hour with no service on Sundays or bank holidays). The reserve is situated on Sustrans Cycle Route 4.

If travelling by car, once in Barnes follow the brown tourist signs to the reserve, which is situated outside the London Congestion Charging Zone (but inside the Ultra Low Emission Zone charge area). There is free parking for visitors in the large car park adjacent to the visitor centre (TQ 228767), where four public electric-vehicle charging points can be found.

FACILITIES: The visitor centre restaurant is open 9:30 am–5:00 pm. There are numerous toilets on site, including accessible toilets and baby changing facilities. In total there are five hides, as well as the multistorey Peacock Tower.

CALENDAR

All year: Shelduck, Pochard, Lapwing, Cetti's Warbler.

November–February: Wildfowl including Pintail and Goldeneye, Water Rail Bittern, Jack Snipe, outside chance of a scarce gull.

March–June: Chance of Garganey, Little Ringed Plover, Oystercatcher, Redshank, wader passage sometimes including Wood Sandpiper, Mediterranean Gull, breeding Sand Martin, Reed and Sedge Warblers

July–October: Wader passage including Black-tailed Godwit and Greenshank (and a chance of Wood Sandpiper).

7 RIVER THAMES: KEW TO ROTHERHITHE

OS Explorer 161
OS grid refs: TQ 186775 to TQ 365798
Postcodes: TW9 3AB to SE16 7UF

HABITAT

Central London may not feel like modern-day Surrey, but it is an important part of the vice-county, the border of which straddles the southern side of the River Thames. The Thames is tidal here and stretches of mud are exposed at low tide. Plentiful bridges cross the water, which is increasingly busy with river craft. The Thames can funnel birds flying upriver or downriver and, as a result, species otherwise very rare in Surrey are sometimes encountered.

Situated along the southern side of this stretch of the river are various parks, lakes and river frontages – many of these are covered collectively in this Inner London chapter, including Kew Gardens, Lonsdale Road Reservoir, Battersea, Burgess and Southwark Parks, South Bank and Russia Dock Woodland.

SPECIES

From the perspective of a Surrey birder, the Thames holds most interest here. The vast majority of vice-county records of Guillemot have occurred on the section of the river covered here. Winter or autumn gales, or simply periods of strong north-easterlies, are best, ideally coupled with a big tide. There are also a handful of records of Puffin and Razorbill – both species with mega status in Surrey. More regularly occurring – but still rare – are Shag, Kittiwake and Sandwich Tern. At low tides, waders can occasionally be found on the shoreline, typically Common Sandpiper and Oystercatcher.

Working from west to east, the Royal Botanic Gardens at Kew have some wildfowl, including Mandarin and feral Red-crested Pochard (birds have been released here before). The Thames at Kew often holds a wintering Common Sandpiper.

Further east, towards Barnes, the former Lonsdale Road Reservoir (also known as Leg o' Mutton Reservoir) has been converted into a nature reserve. Decent numbers of wildfowl winter and Pochard occasionally breeds. Reed Warbler and Common Tern can be found too. Surrey's only record of Eastern Subalpine Warbler occurred here in April 2003. The Thames foreshore at Hammersmith Bridge can be productive for large gulls, including Yellow-legged.

Battersea Park can be a good spot for Peregrine and the boating lake often proves attractive to wildfowl, which occasionally includes scarcer visitors. Passage migrants pass through. In recent years Tawny Owl has bred successfully. Battersea Power Station sometimes hosts Black Redstart. Burgess and Southwark Park have waterbodies that can attract wildfowl – the former has recorded Ferruginous Duck, Common Scoter, White-fronted Goose and regular wintering Mediterranean Gull.

The stretch of river from Waterloo Bridge to Rotherhithe offers the best chance for a storm-driven seabird, as well as large gulls. At South Bank, Tate Modern can attract passerine migrants that have included Ring Ouzel, Redstart and Firecrest. Black Redstart is possible here too. The Tate Modern chimney is a regular perch for Peregrine and one or two birds can often be observed.

There are a couple of places of note at Rotherhithe. Russia Dock Woodland, which includes Stave Hill Ecological Park, is an urban mosaic of grassland and woodland interspersed with small ponds. Firecrest winters annually, and there are a couple of recent records of Yellow-browed Warbler too. Passerine migrants are to be expected at the right time of year and there are a few occurrences of Pied Flycatcher and Redstart, with Wood Warbler also recorded. The best places to search for passerines are along Watermans Walk and the areas of woodland immediately to the east of Stave Hill.

The Thames in Rotherhithe (formerly Surrey Docks), especially around Greenland Pier, and further north by the Hilton Doubletree hotel is good for large gulls with Yellow-legged Gull recorded regularly, and Caspian Gull occurring with increasing frequency. Late summer can be productive for juvenile Yellow-legged in particular – being armed with a loaf or two of bread can be useful! Mediterranean Gulls have also become an increasingly regular sight, particularly in late summer, while Kittiwake and Sandwich Tern have been recorded on more than one occasion in recent years. Wader records have included Turnstone, Curlew, Whimbrel and Ruff. Greenland Dock attracts a Tufted Duck flock that has in years gone by included Scaup and Ring-necked Duck. The nearby waterbodies of Canada Water and Surrey Water have attracted Long-tailed Duck, and Reed Warbler and Common Tern often breed.

TIMING

This whole area is generally busy, especially on weekends. Early weekday mornings are best. High tide (ideally following easterlies of some description) is best if searching the Thames for seabirds, though low tide is better if looking for gulls and waders.

ACCESS

It is best to access this part of London by public transport, with the London Underground the most useful form. Kew Gardens underground station is on the District Line and is an 800m walk. Bus route 110 stops near Kew Gardens station.

For Lonsdale Road Reservoir, the nearest stations are Barnes Bridge (National Rail) and Hammersmith (London Underground). From either station, pedestrians can approach the site mainly along the Thames Path rather on roads. Bus routes include 33, 72, 209, 283, 419 and 485.

Battersea Park is a short distance from the Battersea Park and Queenstown Road mainline stations. From the former, exit to the right along Battersea Park Road, walk 50m to the traffic lights and turn right into Queenstown Road; the park is on the left after 150m. From the latter, exit to the right along Queenstown Road and the park is on the left after 300m. The nearest London Underground station is Sloane Square, from which there is a 1km walk south via Sloane Street and Chelsea Bridge Road or alternatively, take a 137 or 452 bus.

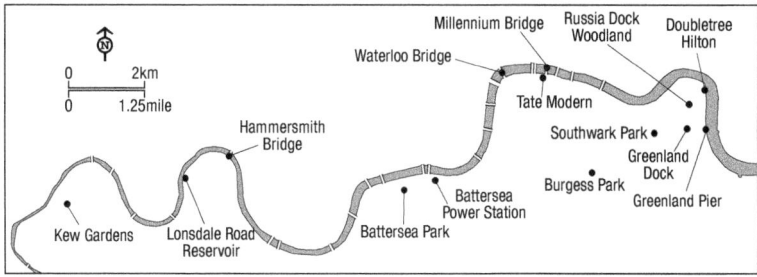

For Burgess Park, the nearest railway station is Elephant & Castle – 1.5km to the north – which can be reached by Underground or National Rail. More than a dozen bus routes from Elephant & Castle serve Burgess Park: route 343 passes along Albany Road, routes 12, 35, 40, 45, 68 and 468 pass the Camberwell Road entrance, and routes 53, 63, 78,168, 172, 363 and 453 pass the Old Kent Road entrance. Southwark Park is a five-minute walk from Canada Water underground (Jubilee Line) or a two-minute walk from Surrey Quays overground.

The Thames at South Bank is a five-minute walk from London Waterloo, and for the most part is along the river.

Canada Water underground is on the Jubilee Line (there is also an overground service to here) and is the best place to access the Rotherhithe area, including Stave Hill and Russia Dock Woodland.

FACILITIES: There are numerous facilities in this area, including toilets and cafés at Kew Gardens and Battersea, Burgess and Southwark Parks.

CALENDAR

All year: Wildfowl including Mandarin at Kew, Peregrine, Ring-necked Parakeet, Black Redstart.

September–March: Chance of scarcer wildfowl on the waterbodies, Caspian and Yellow-legged Gulls along the Thames especially at Rotherhithe, chance of seabirds on the Thames, Firecrest at Stave Hill.

April–August: Yellow-legged Gull from early July, passage migrants in April, May and August.

8 BEDDINGTON FARMLANDS

OS Explorer 161
OS grid ref: TQ 289662
Postcode: SM6 7NN

HABITAT

One of the most famous birding sites in London and Surrey, Beddington Farmlands has been an ornithological hotspot for more than 100 years. Situated along the Wandle Valley, several miles west of Croydon, the habitat has changed with the increasing urbanisation of the surrounding area. In the late nineteenth century it was wet farmland (Beddington Farm), with some sewage disposal. From the 1940s, there was more open field sewage treatment and in 1969 a sewage treatment works was opened. In the 1970s, the wet meadows were replaced by sludge beds. Gravel extraction began in 1998. From the late 1990s, the landowners and Sutton Council began plans to develop the brownfield site into a public nature reserve.

Today, the 161ha Beddington is a large expanse of sludge lagoons, pools, wet grassland and two flood-relief lakes, as well as an ex-landfill site. Some habitat development is still underway. At the time of writing Beddington is only partially open to the public, via a public footpath and series of hides along the west side of the site, with full access only granted to the Beddington Farm Bird Group. Beddington boats a bird list of more than 260 species, with an annual total of over 150 species.

SPECIES

For decades Beddington has been Surrey's premier gull hotspot, with thousands of birds collecting on the pools and around the landfill in the winter. However, since the closure of the landfill numbers have dropped considerably. Despite this, it's still the best place in the county for Caspian Gull, with at least one or two usually present. Yellow-legged Gull is relatively regular too; Mediterranean is more likely in spring or late summer. Small numbers of Great Black-backed Gull are usually present. White-wingers were once virtually annual but have become rarer. Along with the London Wetland Centre, Beddington is the only regular site in Surrey for Water Pipit – these occasionally reach the wet grassland visible from the public section, but are typically in the permit-only area, where several Jack Snipe also winter. A few Green Sandpipers normally spend the winter months here too. Resident species often detectable in winter include Cetti's Warbler, Peregrine and Water Rail. Typical winter wildfowl include Shoveler, Gadwall and Teal; Shelduck is usually around – and sometimes breeds – but Pintail is uncommon. A few Pochard can be found with the Tufted Duck, but rarer diving duck occur very infrequently.

Beddington has a long history of rare and scarce species and, as a result, spring and autumn passages are the most likely to turn up surprises. Winds from the east are best for waterbirds – especially waders, the passage of which can be dynamic at times. Common and Green Sandpipers are the default species but Wood Sandpiper is virtually annual, with autumn more likely than early May. Species that tend to peak in spring include Ringed Plover, Whimbrel and Bar-tailed Godwit. Late spring has produced a few Temminck's Stints at Beddington over the years as well, though this is very much a Surrey rarity. Greenshank and Black-tailed Godwit occur in both seasons, with Dunlin and Grey Plover slightly more typical of the autumn (the latter is rare). Species like Knot, Little Stint and Curlew Sandpiper are possible, but rare.

Garganey and Brent Goose occur in most years, the latter usually just flying through. Great White Egret is an increasing presence at Beddington with several records annually. Spoonbill and Glossy Ibis have occurred more than once. Raptors can sometimes be detected moving through the airspace and this is probably the most likely place in Surrey for a Marsh Harrier to fly over. The open nature of much of the site makes it appealing to migrant chats – Wheatear, Whinchat and Yellow Wagtail are regular on passage. Some of the more memorable national rarities that have occurred down the years include Britain's first Glaucous-winged Gull, Killdeer, Lesser Yellowlegs, Pacific Golden Plover, Rustic Bunting and Citrine Wagtail.

Summer is generally quiet but breeding birds include Shelduck, Pochard, Little Ringed Plover and Sedge and Reed Warblers. Lapwing attempt to nest most summers but are rarely successful. Teal is a sporadic breeder. Sadly, the site's famous Tree Sparrow colony is no more – and with its loss the species is now extinct in Surrey.

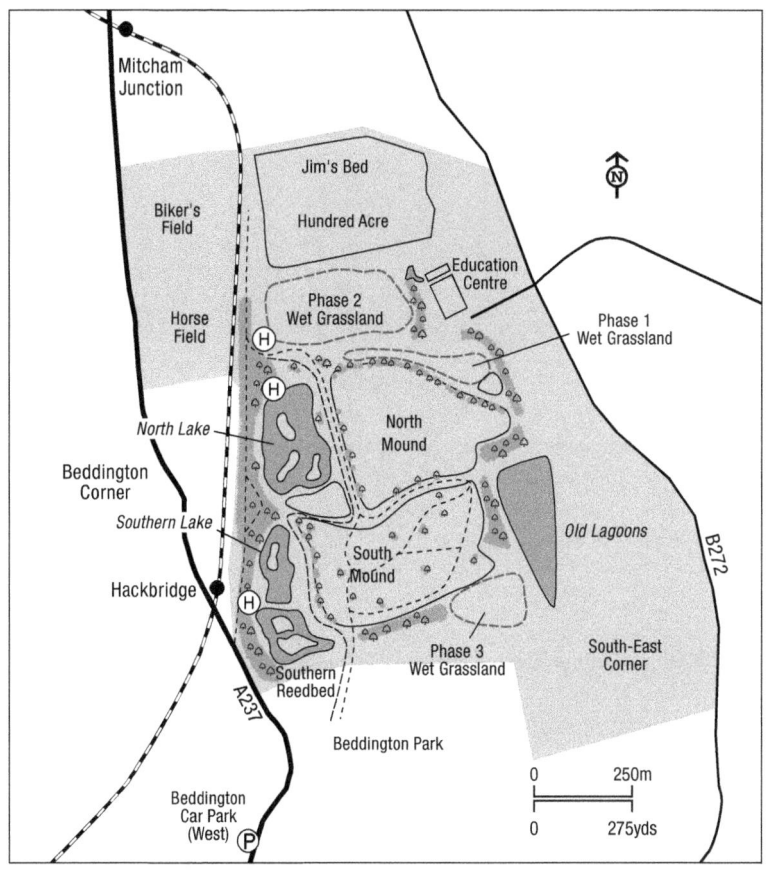

TIMING

Any time of day can be worthwhile, but early mornings are best. If you're hoping to find a passage goodie, winds from the east are the most productive (especially in drizzle or cold spells).

ACCESS

Public access is restricted to a footpath on the west side of the reserve, accessible by bike and on foot. There are three entrances: one via a gate in the north-west corner of Beddington Park (TQ 287656); another via a path from Hackbridge that takes you east over a footbridge (TQ 284661); the third via the field south of Mitcham Junction (TQ 286672). Three public hides overlook Phase 1 Wet Grassland, North Lake and South Lake.

The rest of the site is currently restricted to keyholding members of the Beddington Farm Bird Group. As part of Viridor's restoration planning, future public access to more of the reserve may be in place in the next few years.

The number 151 (Wallington <> Worcester Park) and 127 (Purley <> Tooting Broadway) buses stop close to all three public entrances to the Farmlands. If using these buses, get off at either Park Road (151) or at Hackbridge Corner (127/151).

Hackbridge train station is a five-minute walk to the Beddington Park and Hackbridge entrances; Mitcham Junction is a 10-minute walk from the northern entrance. Trams from Wimbledon to Beckenham/New Addington are also found at Mitcham Junction station.

If travelling by car, it's recommended to use the free parking available at Beddington Car Park West in Beddington Park, which is a short walk from the southern entrance to the farmlands. Parking is also available on a number of side roads to the west of Beddington Park and in Hackbridge, though not all of these are free.

FACILITIES: The nearest public toilets are found in Beddington Park, where a café is open 9:00 am to 4:30 pm, and in Hackbridge, where there is a Lidl.

CALENDAR

All year: Shelduck, Lapwing, Peregrine, Cetti's Warbler.

November–February: Occasional Pintail, Green Sandpiper, Jack Snipe, Caspian and Yellow-legged Gulls, outside chance of Glaucous or Iceland Gull, Water Pipit.

March–June: Little Ringed Plover, wader passage including Whimbrel, Greenshank and a chance of Bar-tailed Godwit and Wood Sandpiper, Hobby, breeding warblers including Lesser Whitethroat and Reed and Sedge Warblers, Wheatear, Yellow Wagtail, Ring Ouzel.

July–October: Wader passage between July and September including Black-tailed Godwit and scarcer species, Caspian and Yellow-legged Gulls from July, Whinchat.

9 CANONS FARM

OS Explorer 146
OS grid ref: TQ 249575
Postcode: KT20 6DG

HABITAT
Situated between the settlements of Banstead, Burgh Heath, Chipstead and Kingswood, Canons Farm is a small area of open, arable farmland, interspersed with small wooded areas. Bordering the farm on the east is Banstead Woods – an ancient woodland LNR that holds a population of Purple Emperor butterflies.

SPECIES
Canons Farm holds most interest during the passage seasons, especially early autumn, when migrant chats (including Black Redstart), pipits and wagtails pass through. Ring Ouzel is possible. The open nature of the site means it can be good for vis-mig, including scarce raptors. Resident breeders include Little Owl, Skylark and Yellowhammer; in the winter, good numbers of the latter two species can be found in the fields. Finch flocks sometimes contain Brambling. Barn Owl is a scarce winter visitor. Canons Farm has form for scarcities, too, with Dotterel perhaps the most impressive species on the site list.

Banstead Woods holds a tiny breeding population of Marsh Tits. Hobby attempts to nest most years.

TIMING
Canons Farm is relatively popular with dog walkers, so early mornings are advisable.

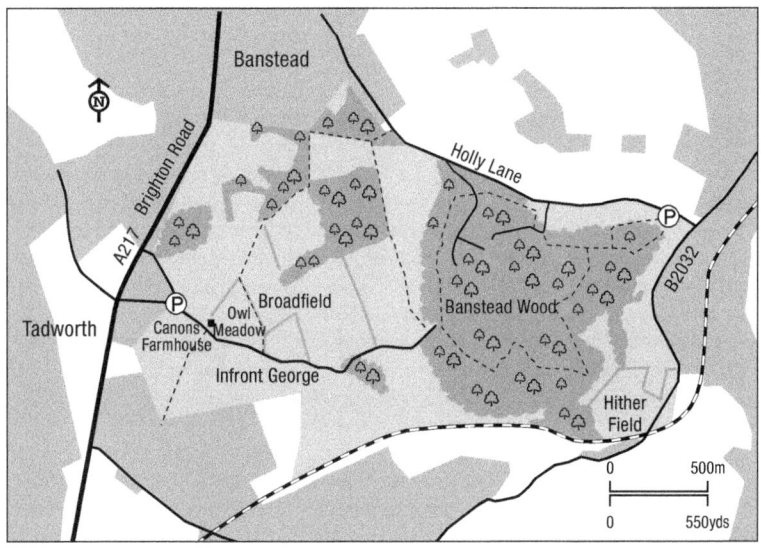

ACCESS

Kingswood railway station (Tattenham Corner Line; Southern operated) is a 15-minute walk from the farm. The 420 bus (Metrobus), which operates between Redhill and Sutton, stops at Lyme Regis Road, a five-minute walk from the north-west of Banstead Woods.

If travelling by car from the A217, park at the western end of Canons Lane (near Burgh Heath) at approximately TQ 245579. It is preferable to park along Ballards Green, which is a cul-de-sac off Canons Lane. If visiting the woods, or accessing the farm from the east end via the woods, there is a purpose-built car park at Banstead Wood at approximately TQ 273583.

There are many public footpaths, although it is probably a bit tricky for wheel-chair and pushchair users.

FACILITIES: The nearest facilities are in Burgh Heath.

CALENDAR

All year: Little Owl, Skylark, Marsh Tit, Yellowhammer.

October–March: Possible Barn Owl, chance of Brambling in finch flocks.

April–September: Migrant chats including Whinchat and Black Redstart, chance of Ring Ouzel, Hobby from mid-April.

10 HOLMETHORPE AREA

OS Explorer 146
OS grid ref: TQ 294517
Postcode: RH1 4EU

HABITAT

Situated between Merstham, Nutfield and Redhill, this area is one of the more noteworthy birding sites in Surrey. Sand and Fuller's Earth has been quarried from the area for centuries and the pits were first flooded in 1960. Today, Holmethorpe Sand Pits is a series of deep pits and floods of varying sizes, with the most significant areas Watercolour Lagoons, Spynes Mere, The Moors, Glebe Lake and Mercers West. These mostly comprise wetland, grassland and lakes.

Much of the area is managed by Surrey Wildlife Trust, including Spynes Mere and The Moors. The former has a few scrapes and islands and, since 2021, an arti-ficial sandbank to encourage nesting Sand Martins; the private Mercers Quarry has an established Sand Martin colony. The Moors has a fair-sized bulrush bed. Some 33 species of butterfly have been recorded and include annual Purple Emperor and regular Brown Hairstreak; 23 species of dragonfly are on the site list, with Willow Emerald Damselfly and Lesser Emperor among them.

Mercers Country Park, which contains Mercers Lake (the biggest waterbody in the Holmethorpe area), is a landscaped pit used for sailing. Other areas (such as Mercers Farm and Nutfield Ridge) are arable and grazing areas respectively, but there are plenty of footpaths.

SPECIES

Mercers Lake is particularly good for wildfowl and other waterbirds during the winter months and has form for attracting divers and scarcer grebes, as well as an impressive diving-duck list that includes Eider and Velvet Scoter, so it is worth scanning through the Tufted Duck. Dabbling duck tend to prefer the Watercolour Lagoons, The Moors and Spynes Mere. The close proximity of a large landfill site towards Redhill means that impressive numbers of gulls visit – sometimes thousands. Caspian Gull has increased markedly here in the winter, with a white-winger not impossible, along with Mediterranean and Yellow-legged Gulls. Occasionally, gulls loaf in the Watercolour Lagoons, when they're much easier to observe. However, the landfill is due to be closed in the coming years and this may impact gull numbers. The Moors has Water Rail in the winter and also a few Jack Snipe – sometimes they show on Railway Pools. At this time of year flocks of Skylark and Yellowhammer gather at Mercers Farm, where Little Owl can be found.

Spring usually starts with a trickle of hirundines from late March. Little Ringed Plover typically drops in but rarely attempts to breed. Migrant chats and wagtails move through Mercers Farm, and the mound between the two Watercolour Lagoons can yield Wheatear. Wader passage can be rather hit-and-miss at Holmethorpe but goodies may well be found, especially at Spynes Mere, with Wood Sandpiper, Spotted Redshank and Little Stint among the more unusual species recorded here before. Common and Green Sandpiper are most likely, though. Spynes Mere can be a good spot for Shelduck, which has bred, but appearances are more sporadic these days. With more than 215 species recorded, an unusual discovery is perfectly possible during the migration periods.

Good numbers of warbler breed, with Garden Warbler and Lesser Whitethroat in the scrubbier areas, and Cetti's, Reed and Sedge Warblers by the waterbodies. Common Terns nest at Mercers Lake and Hobby often overflies the site.

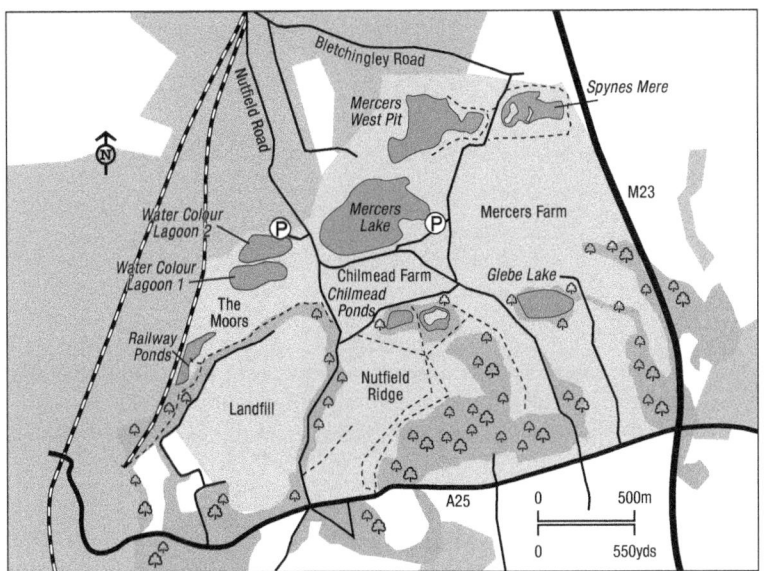

TIMING

Early mornings are best. The site is fairly popular with dog walkers and joggers and can get busy at weekends. The opening times of the car park at Mercers Country Park vary, but are typically 9:30 am until 4:30 pm.

ACCESS

Parking is best at Mercers Country Park lake, where a public car park is situated (but only open at selected times – see above). From here it is possible to walk the whole site via public footpaths (the Tandridge Border Path runs north from Mercers West and Spynes Mere).

Otherwise, park in residential roads north of the pits – Holmesdale Avenue and Nutfield Road are recommended. For access just to The Moors and Watercolour Lagoons, parking in the industrial estate on the west of the site is advised.

The complex can be accessed from Redhill train and bus stations via Cavendish Road.

Holmethorpe is not suitable for wheelchair users or those with limited mobility, although good telescope views of Mercers Lake can be had from the public car park.

FACILITIES: Hot drinks can be purchased at the sailing club at Mercers Country Park. A Tesco Express can be found in the Watercolours housing estate along Canalside – House Martins nest above the shop!

CALENDAR

All year: Water Rail, Peregrine, Little Owl, Raven, Cetti's Warbler, Yellowhammer.

October–March: Wildfowl including occasional scarcities, chance of Caspian, Mediterranean or Yellow-legged Gulls, Jack Snipe.

April–September: Little Ringed Plover, wader passage, Common Tern, *Acrocephalus* warblers, Wheatear and Whinchat on passage.

11 BOOKHAM AND LEATHERHEAD

OS Explorer 146
OS grid refs: TQ 126563 / TQ 159564
Postcodes: KT23 3LT to KT22 7UL

HABITAT

This section encompasses a few sites between the village of Bookham and town of Leatherhead. The Bookham Commons (Great Bookham Common and Little Bookham Common) are a National Trust-owned SSSI. Habitat types include woodland (covering two-thirds of the site), scrub, grassland and open water. Little Bookham Common is a mosaic of rough grassland and scrub. Both commons have a well-recorded array of invertebrate fauna, including more than 300 species of butterflies and moths.

East of the commons is an area of arable farmland (Barracks and Slyfield Farms).

At the north end of the farmland is the River Mole, which flows down towards Leatherhead. Fetcham Mill Pond is situated on the south side of the Mole between Fetcham and Leatherhead. Norbury Park is a Surrey Wildlife Trust-managed area of woodland, grassland and farmland along the Mole valley south of Leatherhead.

SPECIES

Bookham Commons was a stronghold for Nightingale until a population crash; thankfully habitat restoration has enticed birds back and in most summers now there is a male or two holding fort. Marsh Tit is among the resident woodland species. Garden Warbler and Lesser Whitethroat breed in the summer, when an appearance from a Hobby is not unusual. Hawfinches will occasionally visit in the winter.

Yellowhammer breeds on the farmland at Barracks and Slyfield Farms. Lesser Spotted Woodpecker hangs on along the Mole between Slyfield Farm and Leatherhead but is elusive and best looked for in areas of alder and birch. Towards Leatherhead, the river usually holds a few Goosander in the winter – they can often be seen from Common Meadow or Leatherhead Bridge (which is a good spot for Little Egret too). Grey Wagtail is a familiar sight too, along with wintering Teal and Chiffchaffs.

Fetcham Mill Pond has wintering Pochard and resident Water Rail and Kingfisher, with the latter species often giving good views. Norbury Park is another site where Lesser Spotted Woodpecker is a possibility, while fields and paddocks along the Mole can be good for migrant chats and wagtails. The stretch of river between here (Pressforward Bridge) and Leatherhead is very reliable for Kingfisher.

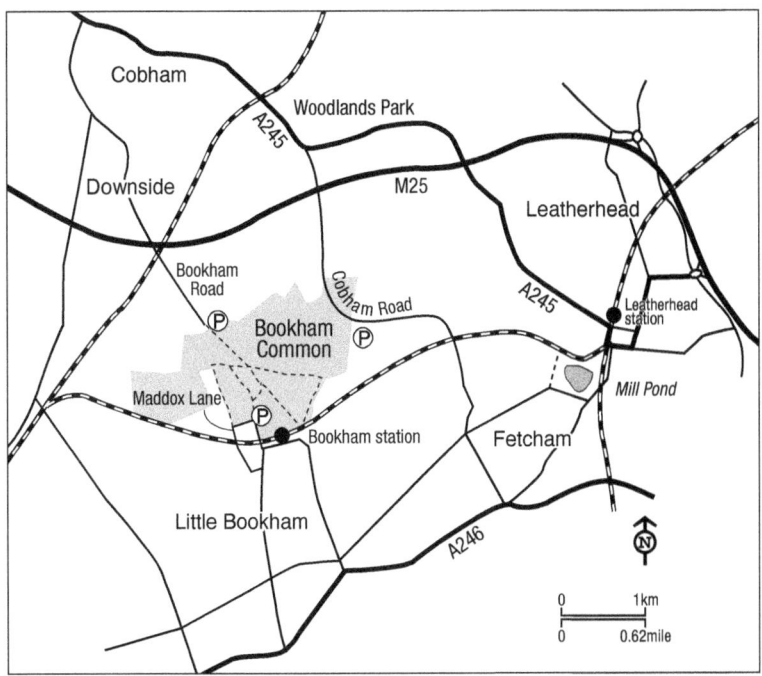

TIMING
Bookham Commons can sometimes be busy with dog walkers at the weekend.

ACCESS
Bookham Commons is readily accessible by train – Bookham station is right next to the south end of the site (historically Nightingales could be audible from the platform!). The Countryliner 479 bus (Epsom–Leatherhead–Guildford) alights at Bookham Station. There are car parks at Maddox Lane (TQ 12525582) and Cobham Road (TQ 133568).

Public footpaths north and east from Cobham Road provide access to Barracks and Slyfield Farms. Footpaths from Leatherhead run along the Mole and past Fetcham Mill Pond. Leatherhead train station is a 10-minute walk from Fetcham Mill Pond.

For Norbury Park, use Fetcham or Young Street car parks, located off the A246 each side of Bocketts Farm Park. It is not suitable for wheelchair users or those with limited mobility.

FACILITIES: There is a small hide at Bookham Commons overlooking one of the ponds (roughly at TQ 131562). It has access for mobility vehicles and wheelchairs. The nearest toilets and shops are in Fetcham and Leatherhead.

CALENDAR
All year: Water Rail at Fetcham Mill Pond, Kingfisher, a very outside chance of Lesser Spotted Woodpecker along the Mole and at Norbury Park, Marsh Tit, Grey Wagtail, Yellowhammer.

October–March: Pochard at Fetcham Mill Pond, Goosander on the Mole, a chance of Hawfinch at Bookham Commons.

April–September: Hobby, Nightingale, warblers.

12 ASHTEAD AND EPSOM COMMONS

OS Explorer 146/161
OS grid ref: TQ 183601
Postcode: KT21 1NW

HABITAT
Epsom and Ashtead Commons is a 360.4ha SSSI that boasts a range of habitats on the London Clay. It is one of the most important sites in Surrey for invertebrates – more than 1,000 species of beetle have been recorded and Purple Emperor is among the butterfly species present. Ashtead is mainly rough grass, conifer plantation and ancient oak woodland, while the adjacent Epsom Common has areas of heather, deciduous woodland and the Stew Ponds. Prince's Coverts is an area of plantation woodland just to the north-west. Rushett Farm is an area of arable farmland on the north side of Ashtead Common.

SPECIES

Breeding birds on Ashtead and Epsom Commons include Cuckoo, Woodcock, Garden Warbler and Lesser Whitethroat. Sadly, Nightingale is no longer present here, although Marsh Tit hangs on (one of the closest sites to London for this species) and Lesser Spotted Woodpecker is an outside possibility (also at Prince's Coverts). Hobby is a fairly regular fly-over visitor in the summer. Typical winter visitors include Siskin and Lesser Redpoll; in good years for Crossbill they can sometimes be found at Prince's Coverts. The Stew Ponds has a fairly typical selection of wildfowl in the winter and Mandarin is resident. It does have form for rarities, however, including Little Bittern and Garganey.

Rushett Farm is a small yet productive farmland site immediately north of Ashtead Common. Lapwing attempts to breed most years, Yellowhammer is resident and, very occasionally, Grey Partridge is present. Jack Snipe can be found in wetter areas in winter, when decent flocks of thrushes and finches may also be present.

TIMING

Visits early in the day are recommended – the commons are popular with dog walkers.

ACCESS

Ashtead train station is conveniently situated at the south end of Ashtead Common – various footpaths across the site (and to Epsom Common) can be reached from the level crossing. Epsom Common car park is situated on Christ Church Road, Epsom. Wheelchair and pushchair access is possible from here down to the Stew Ponds.

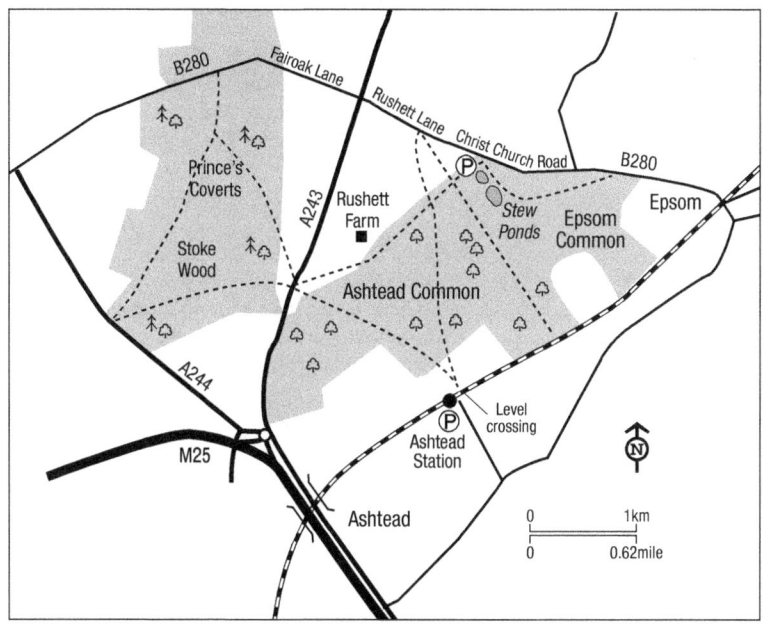

Prince's Coverts can be accessed by parking on Fairoak Lane or at the Star Pub. There are various tracks into the woodland.

For Rushett Farm, park sensibly on Rushett Lane or Leatherhead Road and take one of the public footpaths onto the farmland. Alternatively, you can walk west from Epsom Common car park.

FACILITIES: The nearest toilets are in Ashtead village.

CALENDAR

All year: Mandarin, outside chance of Lesser Spotted Woodpecker, Marsh Tit, Yellowhammer at Rushett Farm.

October–March: Wintering wildfowl, Jack Snipe at Rushett Farm, finches including a chance of Crossbill at Prince's Coverts.

April–September: Cuckoo, roding Woodcock, Hobby, warblers.

OTHER SITES IN NORTH-EAST SURREY/ SURREY-IN-LONDON

B1 WEY MANOR MEADOWS

Situated along the River Wey between Addlestone and Weybridge (TQ 061637), this site includes flood meadows, quarries and ponds. Green Sandpiper and Little Ringed Plover sometimes pass through, while Wigeon and Lapwing winter. Park on Byfleet Road and take the footpath east onto the meadows.

B2 ESHER COMMON AND CLAREMONT

In recent years, Esher Common (TQ 132622) has held Dartford Warbler and Nightjar. Hobby and Little Owl are sometimes recorded. Claremont (TQ 130631) is a National Trust site at which Lesser Spotted Woodpecker is occasionally seen, as it is at Esher Common. Both sites have car parks and the nearest railway stations are Esher and Oxshott.

B3 BUSHY PARK

Although not in the vice-county, this site (TQ 161693) is listed here. It holds a very similar range of species to Richmond Park, including wintering Dartford Warbler. Feral Red-crested Pochard are regular on the ponds. The nearest railway stations are Hampton Court, Hampton Wick and Teddington, all close to the edge of the park and reached via Waterloo.

B4 HOGSMILL SEWAGE FARM

This sewage works (TQ 194682) has an open lagoon and wintering waterfowl sometimes include Shelduck. Jack Snipe is possible at this time of year as well. Wader passage is fairly limited but usually involves Green Sandpiper and, on occasion, more unusual species. Gulls sometimes bathe in the lagoon and Caspian has been recorded. Access to the reserve (which includes a hide) is limited to members of Surbiton & District Bird Watching Society or Friends of

Hogsmill Sewage Treatment Works. The lagoon can be viewed from the north platform at Berrylands station.

B5 WIMBLEDON AND PUTNEY COMMONS

Wimbledon and Putney Commons (TQ 227718) support a variety of woodland and open-country birds. Chats pass through on passage, and Firecrest and Dartford Warbler may be found in the winter. Wimbledon station is a 15-minute walk away; the 93 bus passes the East side of the common. It is a 20-minute walk from Southfields underground station (District Line).

B6 WANDSWORTH COMMON

A typical south London park in terms of habitat and birds, Wandsworth Common (TQ 274735) has surprising form for rarities, including Night Heron and Pallas's Warbler. Balham station is adjacent to the park.

B7 TOOTING COMMON

Another park (TQ 291721) that can be good for passage migrants and water-birds on the lake. Mediterranean Gulls can sometimes be found among the Black-headed flocks. The nearest underground stations, both on the Northern Line, are Tooting Bec and Balham. National Rail stations near the north end of the site are Balham and Streatham Hill. Bus 315 crosses Tooting Bec Common along Bedford Hill.

B8 MORDEN HALL PARK

This National Trust property (TQ 263686) has wintering Water Rail and Firecrest and regular Kingfisher. Reed Warbler breeds. Scarcer passerines have included Yellow-browed Warbler and Siberian Chiffchaff. London Underground access is either a 500m walk along Aberconway Road from Morden (Northern Line) or a tram to Phipps Bridge from Wimbledon (District Line).

B9 POULTER PARK

This stretch of the River Wandle (TQ 274674) is sometimes productive for winter-ing insectivores including Siberian Chiffchaff and Firecrest. Mitcham Junction is a five-minute walk; alternatively, park on Peterborough Road.

B10 SOUTH NORWOOD COUNTRY PARK

Croydon's premier birding site (TQ 354683) has a range of habitats, including a large area of wet grassland and a sizeable lake. Reed Warbler breeds, Common Sandpiper is regular on passage and Water Rail and Jack Snipe usually winter (the latter species is less reliable). Whinchat and Stonechat are annual passage migrants. South Norwood has a fine list of rarities to its name, including Pied-billed Grebe, Twite and Red-backed Shrike. The nearest train stations are Elmers End, on the east edge of the site, and Birkbeck, at the northernmost corner of the site. A Tramlink route crosses the site, stopping at Harrington Road on the western edge of the park.

B11 CHELSHAM AND WOLDINGHAM AREA

The farmland and open countryside in this area is poorly birded but has produced relatively recent records of Corn Bunting, which is now a real rarity in Surrey but perhaps persists in the wider area. Check the various cover crops, which will often hold Yellowhammer and Brambling in the winter. Grey and Red-legged Partridges are possible too, along with good numbers of raptors and Raven. Access is best by car, with limited parking at Warren Barn Farm (TQ 376571). The fields around Upland Road, Beddlestead Lane and Chelsham Court Road are best for exploring.

B12 FARTHING DOWNS

The most extensive area of semi-natural downland left in Greater London, Farthing Downs (TQ 300579) supports a decent population of Skylark. Chats pass through on passage, while Hoopoe and Short-eared Owl have occurred. Coulsdon South train station is a five-minute walk away. There is also a car park.

B13 LITTLE WOODCOTE

This area of open country (TQ 284614) is good for migrant chats and, occasionally, Ring Ouzel. It also supports breeding Barn and Little Owls. Access from Oaks Park (good for wintering Firecrest) or Woodmansterne Road, where you can take Oaks Track east up to Telegraph Track.

B14 NONSUCH PARK

Nonsuch Park (TQ 231636) is a reliable spot for Hobby (in summer) and Little Owl, though the latter can be elusive. Good views of both species are possible, however. Passage migrants sometimes include Spotted Flycatcher. There are a number of access points to the park which include two car parks off London Road, Ewell and a car park off Ewell Road, Cheam.

B15 HOGSMILL LOCAL NATURE RESERVE

The stretch of the River Hogsmill here (TQ 212634) is a reliable spot for Kingfisher and, in the winter, Water Rail. Jack Snipe may be present in hard weather. There are various access points and there is normally room to park on Always Avenue.

B16 EPSOM AND WALTON DOWNS

Situated on high ground, Epsom and Walton Downs (TQ 218582) can be good for vis-mig and the diurnal passage of chats, including Whinchat. Barn Owl is sometimes seen at Epsom Downs racecourse. There is ample parking at Epsom racecourse (but avoid race days). Tattenham Corner train station is next to the site as well.

B17 PRIEST HILL

This small Surrey Wildlife Trust reserve comprising rough grassland, scrub and hedgerows (TQ 230615) is a good site for migrant chats, with Whinchat and Wheatear passing through annually. Passage Ring Ouzels also find it to their liking. Owls have included Long-eared.

SOUTH SURREY

MAIN SITES
13 Winterfold
14 Effingham Forest area
15 Ranmore Common
16 Dorking area waters
17 Leith Hill
18 Box Hill and Headley Heath
19 Hedgecourt Lake and Wire Mill

OTHER SITES
C1 Blackheath
C2 River Tillingbourne: Chilworth to Abinger Hammer
C3 Newlands Corner and Pewley Down
C4 Clandon Wood Natural Burial Ground
C5 Great Ridings Wood and Pennymead Lake
C6 Holmbury Hill
C7 Cranleigh area
C8 Vann Lane and Candy's Copse
C9 Capel and Newdigate
C10 Earlswood Lakes
C11 Burstow Park Farm
C12 Blindley Heath
C13 British Wildlife Centre
C14 Littlelake Farm
C15 Hurst Green Sewage Farm

The South Surrey region, as defined by the Surrey Bird Club, is the area of the vice-county south of the A3 (Cobham to Guildford), east of the A281 (Guildford to the Sussex border) and outside the LNHS boundary. In many ways it is 'South-East Surrey' or even 'South-Central Surrey', as the map shows. This is the biggest of the Surrey regions covered in this book, but it is also the most rural and

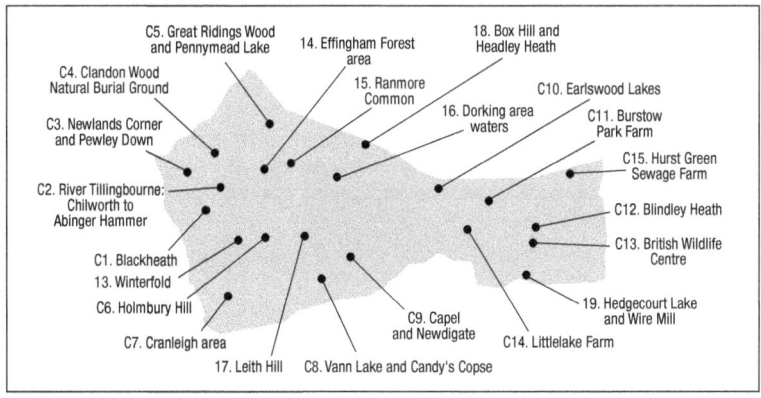

least birded. Dorking, Horley and Cranleigh are the largest settlements in the described area.

Hills and woodlands are a key habitat in South Surrey. The North Downs run across the north of the region from Pewley Down to Box Hill, with the Greensand Ridge a separate but equally imposing topographic feature to the south. Woodland on the North Downs is mainly deciduous, with areas of chalk downland and, at Headley Heath, heathland. The chalk is home to a variety of localised butterfly species and orchids, such as Silver-spotted Skipper, Adonis Blue, and Man and Musk Orchids. The Greensand hills are sandier with much more coniferous woodland. Leith Hill, the highest point in Surrey, has an area of open heath at Duke's Warren, as does Winterfold. There are no other areas of heathland in South Surrey away from the Greensand Hills and Headley Heath.

One thing South Surrey lacks is a major wetland site. The River Mole and the smaller River Eden and River Tillingbourne run through the area, but neither boast the riverside meadows and marshes found along stretches of the Wey in the west of the county. The Mole cuts through the North Downs from the north and moves east to cross the Weald at Dorking. The riverbed is stony and, in places, fast flowing. Of greater significance from a birder's perspective are a series of waterbodies around Dorking, comprised of disused chalk pits and fisheries. The Tillingbourne, which runs from Shalford to Wotton, has clear water from under the North Downs and has been dammed or diverted in places. The Eden is in the far south-east corner of Surrey and is a tributary of the River Medway. Old millponds along the waterways include Hedgecourt, which is the most significant waterbody in South Surrey.

There are large areas of farmland, with the most extensive tracts in the far south and east of the region. The arable land east of the M23 is unlike the rest of this wooded, hilly region. This far corner of Surrey, up to the Kent border, is the least-birded part of the county.

It is the hills and high points that offer the most attraction to birders in this region. All these vantages and topographic features, including the River Mole, mean that there are plenty of opportunities for vis-mig in South Surrey. A group of birders have begun regular watches from Leith Hill Tower – the highest point in Surrey – with productive results. Without doubt there are other, as yet undiscovered, watchpoints with similar potential in the area. South Surrey is the most heavily wooded region in Surrey and with that it's the best for species such as Goshawk, Marsh Tit, Firecrest and Hawfinch. Large tracts of woodland in this area are rarely birded, including some of the sites described, so the possibility for discovery of these forest dwellers – and others such as Honey Buzzard and Lesser Spotted Woodpecker – is high.

In fact, the region as a whole has probably the best potential for 'new' discovery in Surrey. Vast swathes of countryside, especially in the far south-east, receive very little coverage. It seems certain there are some areas of farmland or woodland, or perhaps a small waterbody or sewage works, that could offer productive results for the more open-minded birder.

13 WINTERFOLD

OS Explorer 145
OS grid ref: TQ 071431
Postcode: GU5 9EN

HABITAT

Winterfold is a wooded area atop the broadest plateau of the Greensand Ridge in Surrey, situated above the villages of Cranleigh and Ewhurst. Pines, gorse, bracken and ferns dominate the sandiest parts and scrub clearance at Winterfold Heath has regenerated areas of heather. The highest point is Pitch Hill at some 257m. Excellent views to the south can be seen from various points – indeed you can see the sea from the top of Pitch Hill on a clear day.

SPECIES

Winterfold Heath is perhaps the best site on the Greensand Ridge for heathland species, with Dartford Warbler, Nightjar, Stonechat, Tree Pipit and Woodlark all breeding, as well as Cuckoo, Woodcock and Willow Warbler. Pitch Hill is wooded and lacks the heathland species. Firecrest and Marsh Tit are resident throughout the area. Raven is a regular fly-over and there is a decent chance of Goshawk at the right time of the year, along with Hobby in the summer.

This can be a good site for Crossbill, especially outside the breeding season when large flocks roam around. Brambling is present in variable numbers. Unlike other Greensand Ridge sites in the county, there are few records of Hawfinch.

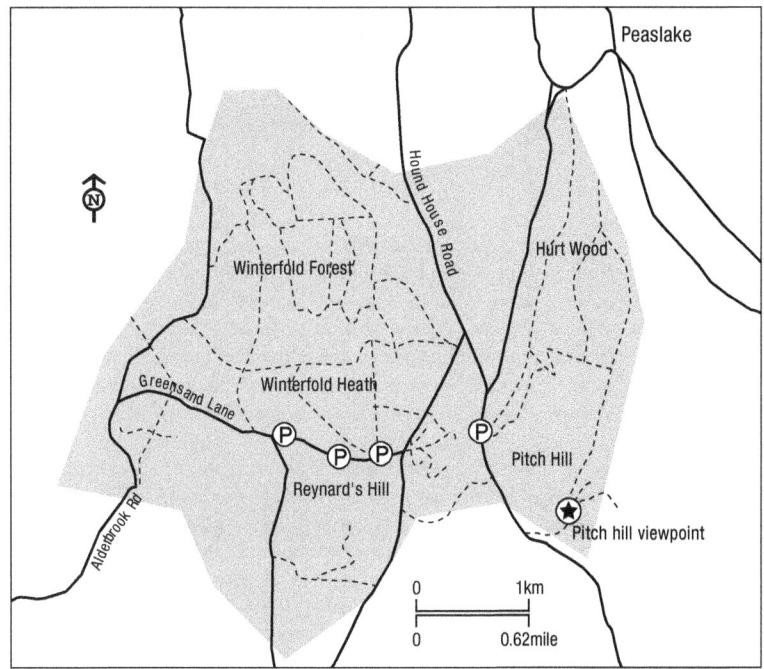

Ring Ouzel passes through in most autumns – the rowans and holly on Winterfold Heath are the best place to look.

TIMING

Spring and summer are best. The area is popular with dog walkers and mountain bikers at weekends.

ACCESS

Winterfold is far from any train station and travelling by car is the most practical way of getting to the site. Three nearby car parks along Greensand Lane at TQ 071424 (Reynards Hill), TQ 073425 (Winterfold Northside) and TQ 074425 (Horseblock Hollow) are best for accessing Winterfold Heath. For Pitch Hill, there's a car park at TQ 079426.

FACILITIES: The nearest facilities are in Cranleigh and Ewhurst.

CALENDAR

All year: Chance of Goshawk, Raven, Marsh Tit, Dartford Warbler, Firecrest.

April–September: Nightjar, Cuckoo, Woodcock, Hobby, Woodlark, Willow Warbler, Tree Pipit.

October–March: Ring Ouzel in October, Crossbill, Brambling.

---

14 EFFINGHAM FOREST AREA

OS Explorer 146
OS grid ref: TQ 097505
Postcode: KT24 5TD

HABITAT

Much of Effingham Forest is owned by the Forestry Commission, with extensive plantations of deciduous and coniferous trees. There are areas of ancient woodland too and a small pool at Dick Focks Pond. The habitat at Netley Heath to the south is similar, though the southern edge of the woodland there sits conspicuously along the North Downs. Sheepleas is managed by the Surrey Wildlife Trust and is largely ancient woodland, but there is also some chalk downland.

SPECIES

Effingham Forest is currently the best place in Surrey to connect with two desirable species: Goshawk and Hawfinch. Both can be seen from the viewpoint at Dick Focks Common (see access section). There's a chance of both species year-round, but Goshawk is best searched for between January and April from mid-morning to mid-afternoon. It is possible to witness multiple birds and sometimes displaying pairs. Hawfinch is typically most reliable from October to April. Small numbers gather at the tops of trees. Any time of day can yield results, but a couple of hours

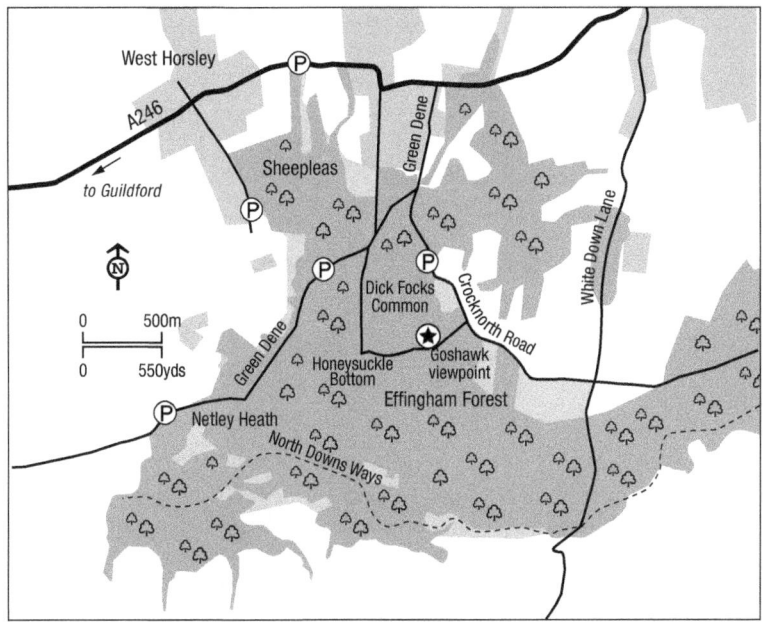

before dusk is advisable, as birds congregate ahead of roost. Hawfinch probably breeds in most years but is a lot more elusive at this time.

Both these species are possible at Netley Heath but are a lot less predictable (and harder to observe) there. Effingham Forest, Netley Heath and Sheepleas all have a healthy population of Marsh Tit and Firecrest, and both should be encountered (spring is best for Firecrest). Crossbill numbers vary, but this is one of the best sites in Surrey for the species and large gatherings can occur. Brambling are present in most winters – the beech woodland at Sheepleas can be especially reliable and, in good years for the species, birds will sometimes feed with Chaffinches at the Shere Road car park. Spotted Flycatcher can be found in Effingham Forest in the summer.

TIMING
The best times for Goshawk and Hawfinch are outlined in the species text. Otherwise mornings are best.

ACCESS
For Dick Focks Common, park sensibly at the Forestry Commission gate on the west side of Crocknorth Lane at TQ 100509, just north of the entrance track to Crocknorth Farm (on the other side of the road). From there, take the trail west then south before turning east to the viewpoint at TQ 099503.

For Netley Heath, there's a car park on Combe Lane at (TQ 088526). For Sheepleas, there are car parks at Green Dene (TQ 091509), St Mary's Church (TQ 088526) and Shere Road (TQ 084514).

FACILITIES: The nearest facilities are in West and East Horsley.

CALENDAR

All year: Woodcock, Goshawk (best in spring), Marsh Tit, Firecrest, Crossbill (variable), Hawfinch (best outside the breeding season).

October–March: Brambling.

April–June: Hobby, Spotted Flycatcher.

15 RANMORE COMMON

OS Explorer 146
OS grid ref: TQ 139504
Postcode: RH5 6SR

HABITAT

Ranmore Common SSSI is a 224ha SSSI north-west of Dorking that also incorporates Denbies Hillside, Hackhurst Downs and White Down. It's mainly woodland, some of it ancient, but there are areas of heath and rough pasture. The dominant tree species are Pedunculate and Sessile Oaks, with a shrub layer of Holly, Silver Birch and Yew. The chalk escarpment of Denbies Hillside offers panoramic views to the south. It is one of the better Surrey sites for orchids and other chalkland flora. Particularly impressive is the display of Early-purple Orchid in the woodland at the top of the slope. Both Adonis and Chalkhill Blue butterflies can also be seen in the open grassland areas.

SPECIES

Marsh Tit and Firecrest are among the resident species, the former in good numbers. As with the other wooded hills in the area, Hawfinch is possible year-round and sizeable flocks have been recorded. Cuckoo, Woodcock and Garden Warbler breed. The beech woodland at the north end of Ranmore Common occasionally holds singing Redstart and, in the winter, Brambling.

Skylark, Lesser Whitethroat and Yellowhammer breed at Denbies, where the expansive view can be productive for skywatching. Hobby is present in the summer, along with an outside chance of Honey Buzzard, and there's a chance of Goshawk year-round. Raven regularly flies over. Barn Owls are sometimes seen over the areas of rough grass. Denbies has form for scarcities, too, with Wryneck and Red-footed Falcon recorded before.

TIMING

Spring and summer are best. Ranmore doesn't tend to get as busy as Box Hill, but early mornings are still advised.

ACCESS

Dorking West station is 3km away and the Carlone 533 bus that stops at Ranmore Common (via Dorking railway station) runs on Tuesdays only (one bus per day). As a result, it's easiest to travel by car. There are four car parks. The National Trust car park (Ranmore East) is at TQ 141503 off Ranmore Common Road. Further west is Ranmore West Car Park (TQ 127502), Stony Rock car park (TQ 124504)

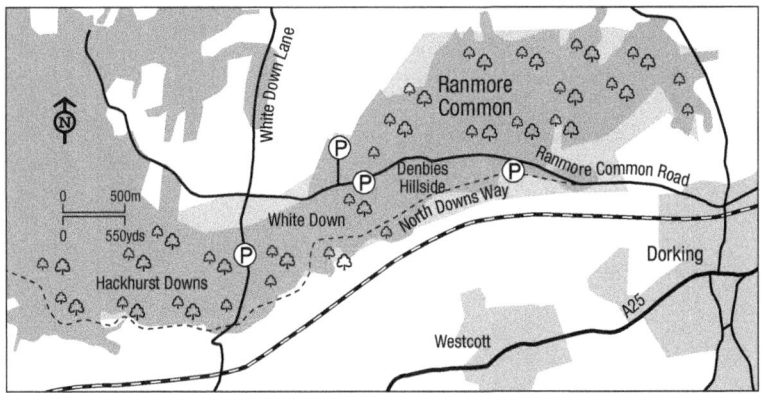

along Hogden Road and White Down car park (TQ 114494) along White Down Lane. This is not a suitable site for visitors with limited mobility.

FACILITIES: The nearest facilities are in Dorking.

CALENDAR

All year: Barn Owl, Raven, Marsh Tit, Firecrest, Hawfinch.

April–September: Woodcock, Cuckoo, Hobby, Skylark, Garden Warbler, Lesser Whitethroat, Yellowhammer.

---

16 DORKING AREA WATERS

OS Explorer 146
OS grid refs: TQ 150491 / TQ 151480 / TQ 224505
Postcodes: RH4 3JU, RH4 3LZ and RH3 7BQ

HABITAT

There are three waterbodies of note to the birder around Dorking: Buckland Park Lake, Bury Hill Fisheries and Milton Court Lake. Buckland Park Lake, part of the Buckland Estate, is a 20ha former sand pit that was only fully opened to the public in 2021. There are two hides. The water is deep, and there are small areas of reed and sandy shores. Bury Hill Fisheries is a popular angling site comprised of four lakes. It is private, but part of the site can be viewed. There are plans to rewild some of the land as well. The waters have a few islands, one of which contains a heronry, and there are large areas of reedbed. Milton Court Lake comprises two small, relatively wooded lakes just east of Milton Court Farm, Westcott.

SPECIES

The deep water at Buckland renders it attractive to diving duck in the winter, though never in large numbers. Notable records have included Eider and Smew. Great Crested Grebe and Reed Warbler breed and there is a colony of Sand Martins which can be viewed from the meadow near the car park. Doubtless the martins are partially behind the regular summer visits from Hobby. Wader passage is generally restricted to Common Sandpiper, but several other species have been recorded, typically flying through.

Bury Hill Fisheries too has breeding Reed Warbler and Great Crested Grebe, as well as a wintering *Aythya* flock that has drawn in Ring-necked Duck before. Ospreys occasionally visit on passage, with individuals sometimes lingering. Hawfinch is recorded sporadically outside the breeding season.

Milton Court Lake is the best of the three sites for Goosander, with wintering birds from along the Mole visiting with some regularity. Water Rail is also found here and there is even a record of Little Bittern.

TIMING

Buckland's opening time is usually 9:00 am but it can vary; check the Buckland Park Lake website. Bury Hill can be busy with anglers so early mornings are best.

ACCESS

Betchworth train station is 1km from Buckland Park Lake – walk south down Station Road and then east along Reigate Road until the park entrance opposite Lawrence Lane. The Redhill-Guildford bus (32) stops by the entrance. There is a spacious car park (TQ 226507), where you must purchase entrance tickets from the hut.

A public footpath at the end of Milton Street (TQ 150482) runs south through Bury Hill. The best views of the main lake can be had from the field at the south end.

For Milton Court Lake, take the footpath north from the A25 along Milton Court Lane at TQ 151488.

FACILITIES: Buckland Park Lake has toilets (including accessible toilets), a restaurant and a snack van.

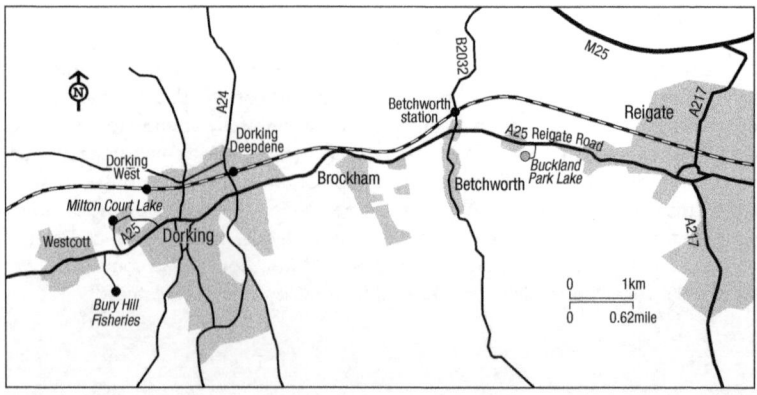

17 LEITH HILL

OS Explorer 146
OS grid ref: TQ 139432
Postcode: RH5 6LX

HABITAT

The highest point in Surrey and the second highest in South-East England, Leith Hill reaches some 294m above sea level and is one of the jewels in the crown of the Surrey Hills AONB. The addition of a folly in the 1700s increased the height of the summit by a further 20m, and wonderful views can be enjoyed from the top across the rest of the Greensand Ridge all the way to London and the English Channel. Indeed, on a clear day, it is said that 13 counties are visible from the top of the tower.

Most of Leith Hill is wooded – four areas of woodland surrounding the hill comprise a 337.9ha SSSI. The top is dominated by oak coppice, while the scarp woodland includes naturally regenerating holly, yew, rowan and beech. The dip slopes have been planted with conifers and a few Monkey Puzzle trees. The area at Duke's Warren has created some open heath. There is a cricket pitch and an old hillfort (Anstiebury Camp) on the eastern side near Coldharbour village.

SPECIES

Its significant height and position along the Greensand Ridge (of which it is the highest summit) means that visible migration can be very much in evidence during spring and autumn. A local group have access to the tower and, at peak times, thrush and finch passage can be impressive – on 13 October 2021, no fewer than 34,727 Redwings were logged flying west, a new Surrey record count. Several extraordinary fly-over records for a wooded hill in Surrey have been logged, such as Great Skua, Gannet and Common Scoter, so it's worth keeping an eye on the sky during the passage seasons. Leith Hill is probably the best site in Surrey for Ring Ouzel. Invariably, these are birds moving through on days of thrush passage, but often individuals or small groups will drop in, sometimes favouring the yews and holly immediately south of the tower or the rowans on Duke's Warren. October and early November are best.

The heathland at Duke's Warren has the full suite of 'classic' species, albeit in far smaller numbers than many of the western Surrey heaths. Still, Nightjar, Woodlark, Dartford Warbler, Tree Pipit and Redstart are present during the breeding season, along with Woodcock, Cuckoo, Willow Warbler and Spotted Flycatcher. Duke's Warren occasionally records migrant Pied Flycatcher in early autumn.

The location and elevation of Leith Hill in the wider wooded landscape make it

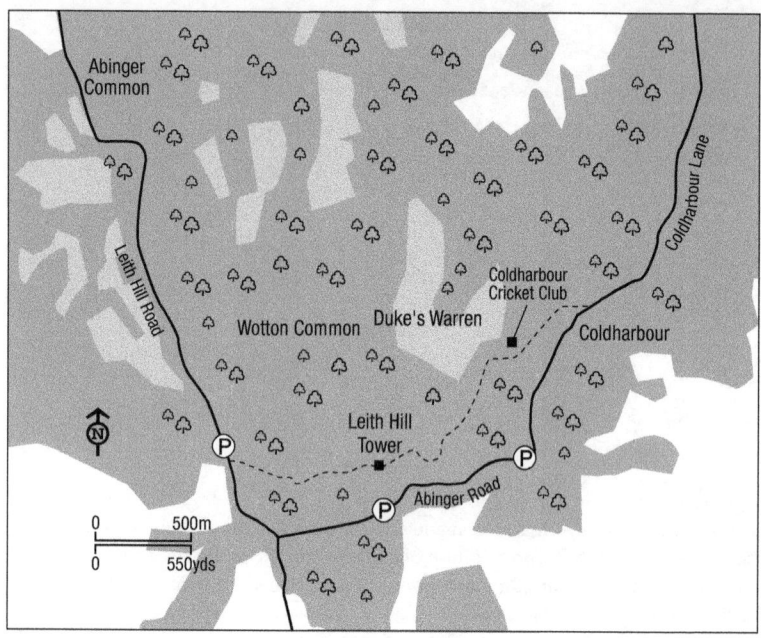

an excellent place for raptor watching. Goshawk is a real possibility here and scanning north from either the tower or the benches around it may produce results year-round, though early spring is best. Honey Buzzard is rare; Hobby is more likely to be seen in the summer. Firecrest and Marsh Tit are among the more common residents in the woodlands. Crossbill numbers fluctuate – it can be common in some years, with only a few around in others, but Leith Hill is still more reliable than most Surrey sites for this species. Puddles along the footpaths near Coldharbour cricket pitch sometimes lure birds down to drink, affording good views. Hawfinch is another presumed resident but it is very elusive outside the autumn and, even then, birds are more likely to be seen flying over. This is also one of the last places in Surrey that Lesser Redpoll can be semi-reliably found in the summer.

Winter can be very quiet, although Brambling favour the areas of beech and, occasionally, a Jack Snipe may be present on the wetter areas of Duke's Warren.

TIMING

Early mornings are strongly recommended for vis-mig. Raptor watching is best from mid-morning on dry, relatively still days. The tower is open to the public from 11:00 am until 4:30 pm on Fridays, Saturdays and Sundays. The whole site can become very busy with dog walkers, joggers and mountain bikers at weekends.

ACCESS

The nearest train station (Holmwood) and location served by a bus (Holmbury St Mary) are 4km away, so access by public transport is difficult.

There are three car parks. Starveall Corner car park (TQ 130432), where normally a Firecrest or two are present, is just under a mile from the summit and offers the easiest ascent. Windy Gap car park (TQ 139428) is a quarter of a mile walk via some very steep steps; Landslip car park (TQ 147432) is even steeper and three-quarters of a mile walk away.

FACILITIES: There are no toilets. A café at the base of the tower is open from 9:00 am until 5:00 pm at weekends and 10:00 am until 3:00 pm weekdays.

CALENDAR

All year: Woodcock, Goshawk, Woodlark, Marsh Tit, Dartford Warbler, Firecrest, Crossbill, Hawfinch.

October–March: Brambling.

April–September: Cuckoo, Nightjar, Hobby, Ring Ouzel on passage, Redstart, Spotted Flycatcher, Tree Pipit.

---

18 BOX HILL AND HEADLEY HEATH

OS Explorer 146
OS grid refs: TQ 182516 and TQ 201534
Postcodes: KT20 7LB and KT18 6NN

HABITAT

Box Hill is a summit of the North Downs mostly comprised of chalk downland and woodland. It is designated as a country park. The hill gets its name from the ancient box woodland found on the steepest west-facing slopes overlooking the River Mole. The western part of the hill is owned and managed by the National Trust. A viewpoint on the south side offers spectacular views across the wider Surrey and Sussex countryside. The north- and south-facing slopes support the area of chalk downland, which is well known for its large number of orchid species (and other rare plants) as well as its butterfly population (including Silver-spotted Skipper and Adonis and Chalkhill Blues).

The village of Box Hill separates the country park from Headley Heath, situated on high ground to the north-east, which is an SSSI made up of open acid heath, chalk downland slopes and mixed woodland. This mix of sandy soils and chalk grassland leads to a great diversity of plants and insects. Much of the site is owned by the National Trust.

SPECIES

Both sites are good for Hawfinch, especially Box Hill, which has recorded triple-figure flocks before. The north-east part is best (Ashurst Rough, Bramblehall Wood and Juniper Top); the west side of Headley Heath is the most productive area of that site. Brambling can be found in the beech woodland in the area; Siskin and Lesser Redpoll are more numerous in winter.

Breeding species at Box Hill include Firecrest, Marsh Tit and, on the chalk downland, Lesser Whitethroat. Raven and Peregrine will often use the thermals from the slopes and can be conspicuous. Woodcock, Woodlark and Dartford and Willow

Warblers breed at Headley Heath, along with Cuckoo, Stonechat and a good population of Garden Warbler. Redstart is an occasional breeder. Goshawk is possible at both sites.

TIMING

Both sites are best visited in the spring and summer, although Hawfinch is more readily found outside the breeding season. Box Hill can become incredibly busy at weekends in the summer.

ACCESS

Box Hill and Westhumble station is a 2.5km walk to Box Hill and a 5km walk to Headley Heath. Metrobus 21 goes from Epsom, Leatherhead, Dorking and Crawley to Box Hill east car park (not Sundays). The Surrey Connect 516, Dorking–Leatherhead–Epsom, stops at Headley.

The main car park at Box Hill (TQ 178513) is free for National Trust members, with an online payment system for non-members. Another car park on the east of the site is located at TQ 176519 along Zig Zag Road. For Headley Heath, the main car park is at TQ 204538 and Brimmer car park is at TQ 206532 – both are free for National Trust members and £4 a day for non-members at the time of writing.

At Box Hill, there is access for visitors with limited mobility from the main car park through the woods to the viewpoint.

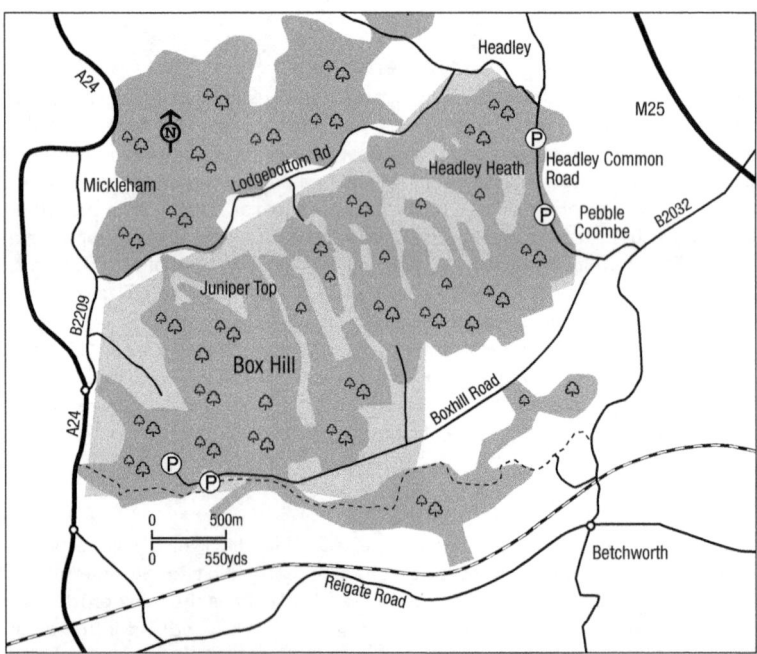

FACILITIES: The main Box Hill car park has toilets (including accessible toilets) and a café that's open from 9:00 am until 5:00 pm daily, as well as a hot drinks kiosk. At the main Headley Heath car park a kiosk serves hot drinks daily from 9:00 am until 4:30 pm.

CALENDAR

All year: Chance of Goshawk, Raven, Marsh Tit, Firecrest, Hawfinch.

April–September: Cuckoo, Woodcock, Woodlark at Headley Heath, Lesser White-throat, Garden Warbler.

19 HEDGECOURT LAKE AND WIRE MILL

OS Explorer 160
OS grid ref: TQ 351402
Postcodes: RH19 2PQ and RH7 6HJ

HABITAT

Hedgecourt Lake is an ancient mill pond and, at 17ha, is the largest body of semi-natural open water in east Surrey. It's the most important wildfowl refuge in the area. Habitats include reedbeds along the south and west side, areas with fen-type vegetation, willow and alder carr, and woodland. Twelve species of dragonfly have been recorded here and it's the only site in Surrey for the rare Touch-me-not Balsam plant. A Surrey Wildlife Trust reserve is situated on the west side of the lake. A sailing club operates on the lake. To the north-east, Wire Mill is smaller with stands of reed and sedge along its edges. It is used for water-sports and angling.

SPECIES

Wintering duck at Hedgecourt include a flock of Pochard and a few Mandarin. Dabbling species are usually in small numbers; Goosander visits occasionally. Bittern is probably a regular winter visitor but stays elusive in the reedbeds – occasionally one will sit by the water's edge or fly to another spot. Lesser Redpoll and Siskin can be found in areas of alder.

Passage birds include Common and Green Sandpipers and Osprey. Gulls drop in and loaf at the west end – normally the common species but Little and Mediterranean Gulls and Kittiwake have visited before, along with Black Tern. There are a few records of Great White Egret and the site has form for unusual species, including Slavonian Grebe and Ferruginous Duck.

Breeding species include Great Crested Grebe and Water Rail (both resident) as well as Common Tern (which nests on rafts) and Reed and Sedge Warblers. Hobbies may fly over during the summer.

Wire Mill is generally much quieter, though Water Rail reside in the reeds and Goosander may drop in during the winter.

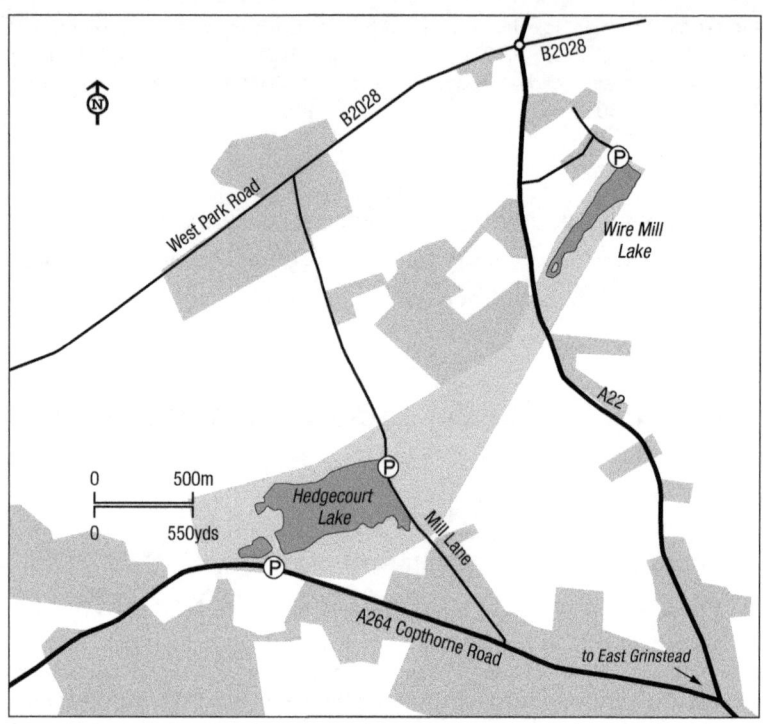

TIMING

Early morning is recommended as sailing/watersports activities on both water-bodies can disturb the birds.

ACCESS

Hedgecourt is best accessed by car. Mill Lane runs along the east side of the lake and there is a small car park where it merges with Stubpond Lane at TQ 358405. Good views of the water can be had from here and a footpath (Tandridge Border Path) runs along the north side of the lake. Alternatively, park off the A264 at TQ 350401 and take the footpath north then east to the west end of the lake.

At Wire Mill there is a small car park at the end of Wire Mill Lane at TQ 367418.

FACILITIES: The nearest facilities are in Felbridge.

CALENDAR

All year: Mandarin, Great Crested Grebe, Water Rail, Kingfisher.

October–March: Pochard, occasional Goosander, chance of Bittern.

April–September: Passage waders, Common Tern, chance of Osprey, Hobby, Reed and Sedge Warblers.

OTHER SITES IN SOUTH SURREY

C1 BLACKHEATH

This relatively compact heathland site is particularly good for Nightjar and Woodcock. Other breeders include Cuckoo, Hobby, Woodlark, Dartford Warbler, Spotted Flycatcher and Tree Pipit. Crossbill is sometimes present too. The main car park is at the end of Blackheath Lane at TQ 035462.

C2 RIVER TILLINGBOURNE: CHILWORTH TO ABINGER HAMMER

A few spots along the Tillingbourne are of note, including the meadows and paddocks east of Chilworth (TQ 037476) for Little Egret and Wheatear, Waterloo and Postford Ponds (TQ 039479) for Mandarin, Kingfisher and Firecrest, and Abinger Cress Beds (TQ 098472) for Green Sandpiper. An Osprey will occasionally pitch up in the valley in late summer, sometimes lingering for a few days.

C3 NEWLANDS CORNER AND PEWLEY DOWN

These two sites along the North Downs east of Guildford have an interesting range of chalkland flora, with open-country bird species including Skylark and Lesser Whitethroat. The slopes are good for migrant chats (especially Whinchat) and, occasionally, Ring Ouzel. There is a car park at TQ 043492 for Newlands Corner (off the A25); park at the end of Pewley Hill (TQ 005490) for Pewley Down.

C4 CLANDON WOOD NATURAL BURIAL GROUND

This small reserve comprises a wildflower meadow, hedgerows and a couple of ponds. It's a good place for vis-mig including Wheatear, Whinchat and Black Redstart. Skylark and Yellowhammer may be found year-round. It has an interesting butterfly population including Brown and White-letter Hairstreaks. There's a car park off the A246 at TQ 048512.

C5 GREAT RIDINGS WOOD AND PENNYMEAD LAKE

Pennymead Lake (TQ 098535) has a flock of Goosander in most winters as well as regular Mandarin. The adjacent Great Ridings Wood (TQ 102539) is good for Firecrest; Hawfinch is irregularly reported.

C6 HOLMBURY HILL

The smallest of the big four Surrey peaks, the woodland here holds a variety of common species of the region including Marsh Tit. There are a few Nightjars in the open areas. The viewpoint at TQ 104429 is a good place to look for raptors, including Goshawk.

C7 CRANLEIGH AREA

Cranleigh Sewage Farm (TQ 041393) has a few wintering Chiffchaffs and one or two Nightingales in the summer. Stretches of the Wey–Arun canal have suitable Lesser Spotted Woodpecker habitat. The open countryside surrounding Cranleigh – all the way to Ewhurst – has Barn Owl, Skylark and Yellowhammer.

C8 VANN LAKE AND CANDY'S COPSE

This peaceful Surrey Wildlife Trust (TQ 156395) site has Kingfisher and Marsh Tit as residents. Hawfinch is occasional. Lesser Spotted Woodpecker may be found in wetter stands of woodland close to the lake.

C9 CAPEL AND NEWDIGATE

At Capel, Nightingales breed at the private Old Stores Meadow Nature Reserve but can be heard from footpaths around (TQ 176402). Little Ringed Plover often breed at the pits off Rusper Road (TQ 177387). Newdigate Brickworks (TQ 204425) has Kingfisher and good numbers of wintering wildfowl, while Greens Farm (TQ 194411) is a good spot for Barn Owl.

C10 EARLSWOOD LAKES

These two waterbodies (TQ 269484) support impressive numbers of Mute Swan, which have attracted Whooper Swan before. Good numbers of other common wildfowl are also present, sometimes drawing in rarer species, and Kingfisher is resident.

C11 BURSTOW PARK FARM

This area of open, arable farmland (TQ 319473) north of Outwood is one of the better sites in Surrey for species like Lapwing and Yellowhammer. Grey Partridge may even hang on. In the winter, mixed finch and bunting flocks may have a few Bramblings. Chats pass through on passage.

C12 BLINDLEY HEATH

One of the best-known examples of relict damp grassland on Weald clay in Surrey, this Surrey Wildlife Trust site (TQ 367448) occasionally has Nightingales, though the population isn't stable. Cuckoo, Garden Warbler and Lesser Whitethroat breed.

C13 BRITISH WILDLIFE CENTRE

This small zoo for native British species, north of Newchapel (TQ 365433), has a wetland area with a boardwalk, offering views of a private reserve. There is a heronry here and the odd wader passes through on passage, with Wood Sandpiper among the species recorded.

C14 LITTLELAKE FARM

North-east of Horley, this farm (TQ 298448) has a similar species assemblage to Burstow Park Farm (site C11). Green Sandpiper favours damper areas in the winter. Scarcities have included White-fronted Goose, Avocet and Great Grey Shrike.

C15 HURST GREEN SEWAGE FARM

This small sewage works south-west of Hurst Green (TQ 398501) is viewable from a public footpath on the north side. A few Chiffchaffs winter and Grey Wagtail can often be seen on the filter beds.

SOUTH-WEST SURREY

MAIN SITES
20 Hindhead Common
21 Frensham Common and Ponds
22 Crooksbury Common
23 Cutt Mill Ponds and Puttenham Common
24 Shackleford Farmland
25 Thursley Common
26 Milford and Witley Commons
27 River Wey: Eashing to Shalford
28 Chiddingfold Forest

OTHER SITES
D1 Wrecclesham Water Meadows
D2 Waverley Abbey
D3 Farnham Heath RSPB
D4 Hankley Common
D5 Thundry Meadows
D6 Bagmoor and Royal Commons
D7 Eashing Fields
D8 Thorncombe Street
D9 Winkworth Arboretum
D10 Hascombe Hill, Hydon's Ball and The Hurtwood
D11 Painshill Farm
D12 Frillinghurst Wood

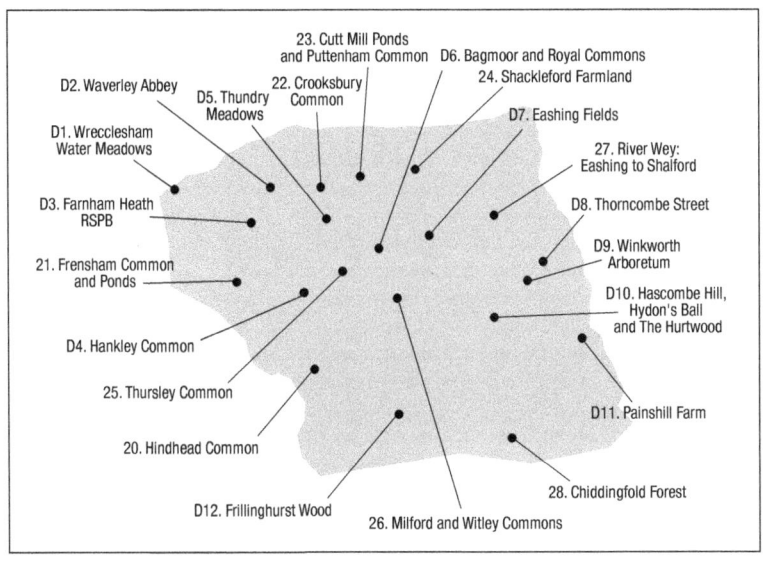

The South-West Surrey region is the area of the vice-county lying south of the A31 (Guildford to the county border at Hampshire) and west of the A281 (Guildford to the county border at West Sussex).

South-West Surrey is the smallest of the Surrey Bird Club's recording areas but makes up for this with a varied range of habitats. Indeed, South-West Surrey can almost be seen as a microcosm of the county as a whole, incorporating lowland heath, extensive woodland, river valleys, lakes, hills and farmland. Much of the region is still resisting the urban growth found in the north of the county, but a new development is never far around the corner in this part of the world. That said, there is a fairly rural feel to much of South-West Surrey and this is to some extent reflected in its avifauna (for instance, Ring-necked Parakeet remains very localised in the region at the time of writing, and there is little obvious presence of large gulls).

The River Wey winds from the far south-west of the county at Frensham north-east to Shalford, which is in the north-east corner of this region. Much of this stretch of the Wey has been spared most of the river improvements commonly applied by water authorities and, as a result, there is a wide mix of habitat along it, ranging from alder carr to wet meadows. It is a flyway for many migrant species and is also the most significant wetland habitat in the region, along with Frensham Ponds (which are collectively the waterbodies of greatest importance in South-West Surrey). Other sizeable lakes include those at Broadwater and Enton, but they are not especially renowned for their birds.

There is also significant heathland at Frensham which forms part of a much wider patchwork of commons. The South-West Surrey heaths consist mainly of dry, sandy soil with heather, birch and pine dominating. Thursley Common is an exception, with extensive boggy areas enough to see it listed as a Ramsar wetland. The heaths are home to a huge variety of flora and fauna, but their dryness leaves them susceptible to fires, which is cause for concern in our warming world.

South-West Surrey is heavily wooded in parts, especially in the south where the Low Weald woodlands span across the bottom-most periphery of the region. Many of these are deciduous and home to scarce butterfly species, including Wood White and Purple Emperor. Farmland is fairly fragmented in South-West Surrey, with important areas from a birding perspective at Shackleford/Puttenham and around Dunsfold.

The Greensand Ridge juts across the region from east to west and includes Gibbet Hill at Hindhead Common, which is the second highest point in Surrey. The hills along the Greensand are mainly wooded, though there are patches of heath at Hindhead and The Hurtwood. The North Downs in this region run from Guildford to Farnham (known as the Hog's Back), but there is little of ornithological interest along this stretch, unlike other parts of the downs in Surrey.

The variety of habitats lends itself well to the birder, even if most of the key sites are the heaths. Excellent populations of species like Woodlark and Dartford Warbler are found across such sites and the widespread positioning of the commons means these species often 'spill' out onto other habitats – for example you may find a wintering Dartford Warbler in a river meadow or a Woodlark holding territory over a young conifer plantation. In this part of Surrey, the waterbodies, while productive, don't hold significant waterbird populations, especially compared with the other regions in the county. Woodland species tend to thrive, though – Firecrest is common, Marsh Tit is locally common and Lesser Spotted Woodpecker is patchily distributed, with various raptor species also doing well.

20 HINDHEAD COMMON

OS Explorer 133
OS grid ref: SU 896361
Postcode: GU26 6AB

HABITAT

Hindhead Common is a National Trust-owned SSSI and AONB in the far south-west of the county. The site is mainly comprised of heathland and woodland and includes the Devil's Punch Bowl, a steep-sided combe running north towards Farnham. At the bottom of the combe there is extensive mature beech woodland, with most of the heathland along the slopes. A ride at Highcombe Copse affords excellent views to the north and east. At various points there are truly stunning views – it's not hard to see why the Victorians knew the area as 'Little Switzerland'.

The highest point of the rim of the bowl is Gibbet Hill, some 272m above sea level and the second-highest point in Surrey. It commands a panoramic view that includes, on a clear day, the skyline of London some 61km away. The A3 used to skirt the rim of the site before the Hindhead Tunnel was built in 2011 – the protected status of the site prevented above-ground redevelopment of the road. Historically it would have been a very remote, bleak landscape – this was the last site in Surrey where Black Grouse hung on, until dying out in the early 1900s.

SPECIES

Hindhead has all the classic heathland species, albeit in lower densities than most of the other South-West Surrey commons. Nonetheless, Nightjar, Woodlark, Dartford Warbler, Redstart and Tree Pipit are easy to find, along with Cuckoo, Woodcock, Willow Warbler and Stonechat. The open area at Sugar Loaf Valley, north of Highcombe Farm, is best for these species – this is a good spot for Garden Warbler, too. Spotted Flycatcher favours the woodland and a few pairs breed in most years. The deep shelter of the beech glades at the base of the bowl were the final Surrey stronghold for Wood Warbler. Unfortunately, this species is now a rare passage visitor, though there is an outside chance of a singing male holding fort for a day or two in April or May.

This site is particularly good for Firecrest – it is not hard to rack up double-figure counts of singing males in spring. Marsh Tit is another fairly common resident in the deciduous woodland. This is one of the better locales in South-West Surrey for Crossbill as well, and in good years several flocks may be dotted about the site. Despite the site's size, habitat and proximity to Black Down, Goshawk sightings are few and far between, though it's a species that must be kept in mind. Raven is a rather regular fly-over, often noisily giving its presence away.

The height of the site and its position along the Greensand Ridge means it can be productive during passage seasons. Pied Flycatcher is a rare but possible early autumn migrant. There's a chance of Ring Ouzel, too, although Hindhead can't be described as reliable for them – especially compared with nearby Black Down.

TIMING

Spring and summer are the best times to visit, though autumn can be productive. This site is very popular with mountain bikers, dog walkers and joggers, so early mornings are advised.

ACCESS

The main car park (National Trust) is just east of Hindhead village, at the end of London Road (formerly the A3) at SU 890357. It is pay and display. From here, there are various trails around the whole site, including a full circumnavigation or a short, flat walk to Gibbet Hill. There is a 300m circular, accessible route from the car park to a sandstone viewing platform. A much smaller, free car park can be found at the north of the site, off Boundless Road at SU 899377.

The nearest train station (Haslemere) is 5km away. Buses from Farnham and Haslemere stop in Hindhead – alight at Hindhead crossroads and walk the short distance east along London Road to the National Trust car park.

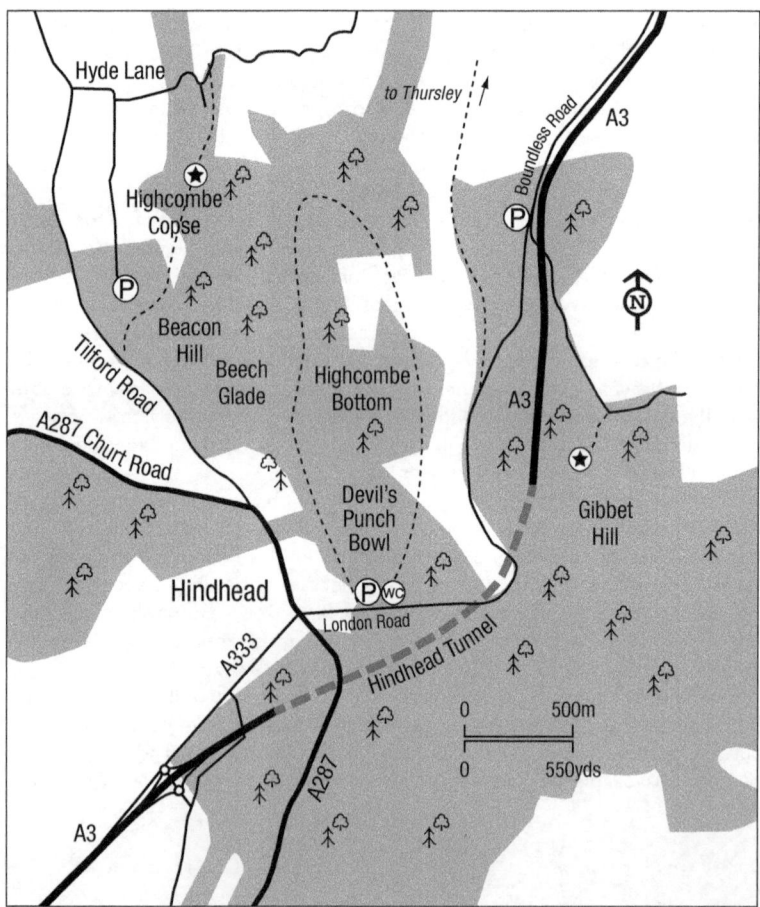

FACILITIES: The main car park has public toilets (including accessible toilets) and a café that sells a range of hot and cold drinks, snacks and lunches.

CALENDAR

All year: Raven, Marsh Tit, Firecrest, Crossbill.

April–July: Cuckoo, Woodcock, Nightjar, Woodlark, Dartford, Garden and Willow Warblers, Redstart, Spotted Flycatcher and Tree Pipit.

August–October: Chance of Pied Flycatcher in early autumn, Ring Ouzel and Hawfinch possible in October.

21 FRENSHAM COMMON AND PONDS

OS Explorer 133/145
OS grid ref: SU 853407
Postcode: GU10 2JN

HABITAT

The Surrey heaths meet their westernmost limit at Frensham Common, a 273ha heathland SSSI that includes two sizeable lakes: Frensham Great and Little Ponds. The ponds were originally created in the thirteenth century to supply fish to the Bishop of Winchester and his court when visiting Farnham Castle. Until the construction of reservoirs in the north of Surrey in the early twentieth century, the Great Pond was the largest body of water in the county. The reedbeds at both ponds are the most extensive in the region. The shores are sandy, creating 'beaches' that are popular with people in the summer. There is a sailing club at the Great Pond. A stretch of wet woodland is found where the Great Pond exits on its west side to the River Wey.

Frensham Common is typical of most South-West Surrey heaths, with heather, gorse and scattered pines. The elevation is low but undulating with high points along Kings Ridge – a small hill that dissects the common – at approximately 90m above sea level. Lowicks Pond is situated on the south-east side of the common and beyond that lies the smaller, slightly boggier Churt Common, which includes Axe Pond and the Devil's Jumps – a series of three small hills.

SPECIES

A Bittern or two winter at the ponds, but they are notoriously elusive. You may be lucky to catch a view of a bird in flight, moving from one reedbed to another, and they have a strong preference for the Little Pond – especially the reedbeds at the south end. Winter wildfowl diversity is relatively low, but there is usually a decent flock of Tufted Duck and Pochard, typically on the Great Pond. A few Goosander normally winter but they can be unpredictable, often favouring the outlet pond or Lowicks Pond. Dabbling duck are a little more erratic, though a few Shoveler, Gadwall and Teal may be found. Hard weather may produce a Goldeneye or two or, better still, a scarce grebe or diver. The only gull roost in the county away from the London reservoirs almost exclusively features Black-headed Gull, though occasionally something scarcer – perhaps a Little or Mediterranean Gull – may be picked out. Frensham is a good site for Water Rail and, although views are rare, they can often be heard at dawn or dusk, squealing from the reedbeds. Kingfisher are resident and often encountered and there will normally be a Little Egret or two around. Frensham is one of the better winter locations for Firecrest, with the holly stands at the west end of the Great Pond by the outflow especially reliable. Other winter passerines that may be encountered include flocks of Siskin, Lesser Redpoll and Crossbill or, rarely, a Great Grey Shrike out on the common. There is a Common Starling roost at the Little Pond.

As with most inland waterbodies, spring and autumn passage can produce a range of species. Hirundines are usually the first-arriving migrants and Frensham has good form for early Sand Martins in particular. Garganey occasionally drop in on one of the ponds, usually in spring, with Brent Goose, Shelduck and Pintail virtually annual. Rarer wildfowl species have included recent Ferruginous, Ring-necked and Long-tailed Duck – it might pay to check the *Aythya* flocks. The beaches will attract passage waders, though disturbance means birds rarely linger. Common Sandpiper is the most likely, but Frensham has a particularly good track record of Oystercatcher. Little Ringed Plover and Greenshank are possible, and the sandy shores have historically attracted a few Sanderlings. The ponds are also good for migrant terns, with Arctic and Black virtually annual. Along with Tice's Meadow (site 29), Frensham is probably the best site in outer Surrey for these two species. Late April/early May and late August/early September are best. Little Gulls drop by in most years as well. Great White Egret is an increasing but still rare visitor. Osprey is annual and sometimes autumn birds linger. Passerine migration is generally a little more limited, though Wheatears and Whinchats move through. Rarities are not impossible – indeed there are no fewer than three records of Great Reed Warbler, with other notable goodies down the years including Little Auk, Red-rumped Swallow and Bluethroat.

There is a diverse breeding community at Frensham, with the full suite of heath-land species on Frensham Common including Nightjar, Woodlark, Dartford Warbler, Redstart and Tree Pipit. Occasionally a Redstart pair will nest right beside the café at the great pond. Other summer visitors include Cuckoo, Garden and Willow Warblers and Spotted Flycatcher, with the latter best sought out around the Little Pond. Hobby doesn't breed but is a regular visitor to the waters, drawn in by the number of dragonflies and hirundines. The ponds themselves have plentiful breeding Reed Warbler (note that Sedge Warbler is rare on passage only), Reed Bunting, Common Tern on the rafts and Great Crested Grebe. Cetti's Warbler looks set to become a new colonist here. Firecrest is readily encountered in spring; Marsh Tit breeds in the wet woodland towards the outlet pond. Lesser Spotted Woodpecker has become rare but is possible in areas of alder.

TIMING

A visit can be rewarding year-round, but it's important to note that Frensham is a popular place. On warm summer days it can be incredibly busy with people using the beaches. Even nice weekends in the winter can attract lots of visitors, so early mornings are advised. Outdoor swimmers have recently started using the Great Pond early in the mornings – this can have a detrimental effect on any passage waterbird that's dropped in overnight. Sailing is mostly restricted to weekends.

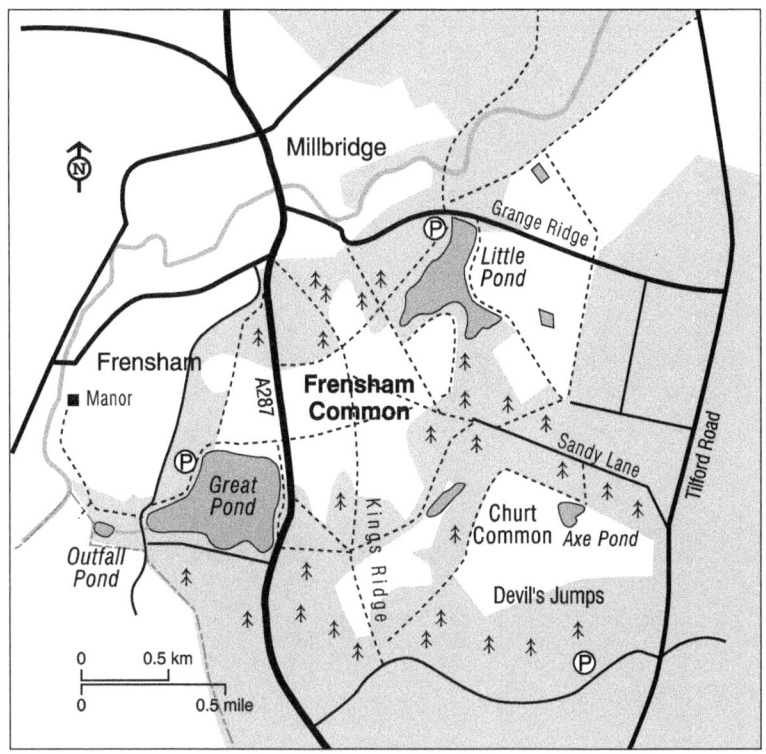

ACCESS

There are car parks at both ponds. For the Great Pond, a large car park off Bacon Lane (SU 843405) on the north side of the pond opens at 8:00 am and closes at 9:00 pm. Charges apply every day, though it's free for National Trust members. From here it is possible to circumnavigate the Great Pond – and also cross the A287 to Frensham Common and onto the Little Pond. There is sometimes room to park in the lay-by at (SU 840400) at the west end of the pond.

For the Little Pond, the National Trust car park is situated off Priory Lane at SU 858418. It is free for National Trust members but non-members must pay.

A smaller, free car park is along Grange Road at SU 863417. Another free car park, further west along Priory Lane (SU 853416), offers quick access to Frensham Common.

The 19 bus (Aldershot to Haslemere) stops in Frensham village (on the green) and along the A287 between the Great Pond and Frensham Common. The nearest train station is Farnham; the 19 bus stops there and it is roughly a 15-minute journey to Frensham. It runs an hourly service from Monday to Saturday.

Many paths around the Little Pond are smooth and level, but access for visitors with limited mobility is less practical at the Great Pond.

FACILITIES: At the Great Pond, Frensham Pond Snack Bar serves food and hot and cold drinks, and has toilets. It is open from 10:00 am until 4:30 pm daily. The Tern Café at the Little Pond serves a selection of snacks and hot and cold drinks, and has toilets. It is open from 10:00 am until 3:00 pm in the winter and 10:00 am until 5:00 pm in the summer. There is a hide by the café and a screen on the other side of the pond, but neither are much use!

CALENDAR

All year: Great Crested Grebe, Water Rail, Kingfisher, Woodlark, Dartford Warbler, Marsh Tit, Firecrest, Grey Wagtail, chance of Crossbill.

April–August: Passage wildfowl, Nightjar, Cuckoo, waders including Common Sandpiper and Oystercatcher, Hobby, Nightjar, Garden, Reed and Willow Warblers, Spotted Flycatcher, Redstart and Tree Pipit.

September–October: Passage wildfowl including a chance of Brent Goose, possible Arctic and Black Terns and Little Gull.

November–March: Pochard, Goosander, Bittern, chance of Great Grey Shrike.

---

22 CROOKSBURY COMMON

OS Explorer 145
OS grid ref: SU 889452
Postcode: GU10 1NQ

HABITAT

This small heathland site forms part of the broader Puttenham and Crooksbury Commons SSSI and lies south-east of the village of The Sands. The common has been restored from conifer plantation via a programme of felling and scrub control and is now an important site for reptiles and amphibians, supporting all six native reptile species as well as Natterjack Toad. The open heath comprises stands of mature Common Heather, interspersed with Bell Heather. Mature pine trees are dotted about, along with patches of birch and gorse scrub. Crooksbury has patches of Wild Strawberry – uncommon on heathland sites – and some boggier areas too.

SPECIES

Crooskbury is an excellent site for Nightjar, with good views possible here without walking too far from the car park. There are normally six or seven males on territory, which is good going for this relatively small site. Roding Woodcock should also be encountered on spring and summer evenings. With no major road nearby, a fine aural experience can be had at Crooksbury after dark in the summer, with churring Nightjar and displaying Woodcock over the heath, and hooting Tawny Owl in the flanking woodland.

Woodlark, Dartford Warbler, Redstart and Tree Pipit are all readily found here, along with Cuckoo, Stonechat and Willow Warbler. Hobby is a regular flyover and breeds nearby. Crooksbury is also a good site for Spotted Flycatcher, especially towards Culverswell Hill and around the paddocks north-west of the car park. They can be especially evident in August, when the congregation of family groups can

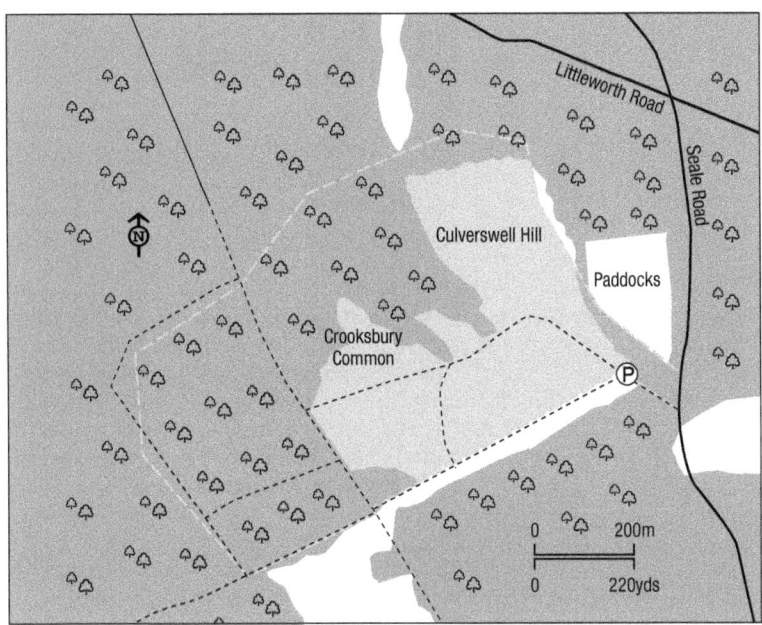

sometimes lead to double-figure counts. Unsurprisingly, Pied Flycatcher has been recorded on passage more than once.

Winter is typically quiet, although Crooksbury is a particularly productive spot for Crossbill, and sometimes large flocks form, once drawing in a Two-barred Crossbill. Bramblings also have a preference for this site. Unfortunately, a regular roost site near Culverswell Hill has been cleared, but there are still normally a few birds around.

TIMING

Spring and summer are the optimum times to visit. Early mornings are best, with dusk until after dark optimum for Nightjars.

ACCESS

There is a small car park off Seale Road at SU 895453. Footpaths north and west lead onto the common. This is quite a remote site and travelling by public transport is not practical.

FACILITIES: The nearest facilities are in The Sands.

CALENDAR
April–August: Nightjar, Cuckoo, Woodcock, Hobby, Woodlark, Dartford Warbler, Spotted Flycatcher, Redstart and Tree Pipit.

November–March: Crossbill, Brambling.

23 CUTT MILL PONDS AND PUTTENHAM COMMON

OS Explorer 145
OS grid ref: SU 911457
Postcode: GU10 1JH

HABITAT

Puttenham is the most wooded of the South-West Surrey commons, with large areas of birch, scrub oak and bracken on sloping ground. Two large areas of heathland are still to be found on the Upper Common, where a hillfort at Hillbury (which probably dates back to the Iron Age) affords far-reaching views. Puttenham Lower Common is damper and more heavily wooded. A series of tree-lined ponds run from the slopes of the Hog's Back to Cutt Mill House, bisecting the Upper and Lower commons. Most are owned by the Hampton Estate and run as fisheries, including the largest, The Tarn. Areas of wet woodland flank the ponds.

SPECIES

Cutt Mill is probably the best site in Surrey for Goosander, with a flock of birds present from mid-November until mid-March. The house pond is usually the best spot but birds can get on The Tarn and Warren Pond too. Other winter wildfowl include a few Shoveler and impressive numbers of Mandarin – both species favour the house pond. Cutt Mill is an excellent site for Kingfisher, too, and you can get good views of the species around The Tarn. The ponds usually attract an Osprey on passage, with Long Pond (stocked with trout) a favoured spot. Great Crested Grebes breed, and Common Terns and Common Sandpipers pass through on passage.

The full suite of heathland species is found on the Upper Common, although there are only a few Nightjars and Dartford Warblers. The Top Car Park is a great spot to watching roding Woodcock in the spring and summer. Woodlarks, Redstarts

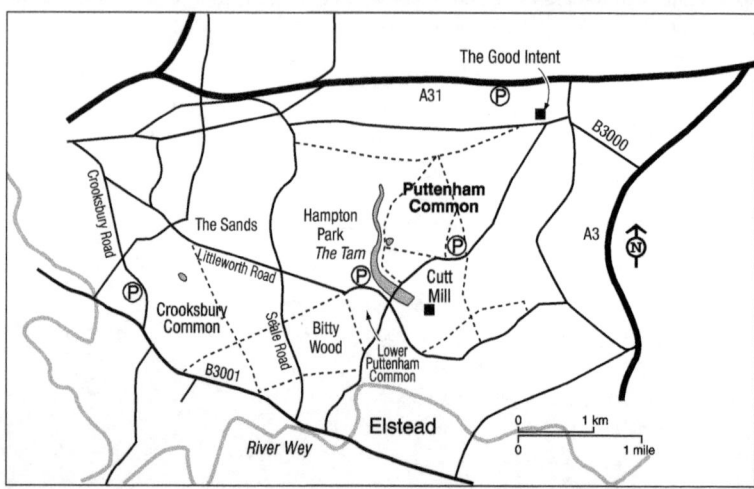

and Tree Pipits are found on both the Lower and Upper Commons – Puttenham is a good site for the latter, with the area around Hillbury Fort productive. Cuckoo and Garden Warbler are regular in the breeding season too.

The wetter woodland around the ponds has a few Marsh Tits, which are rather localised in South-West Surrey north of the Greensand Ridge. Lesser Spotted Woodpecker is resident, although it can be very elusive. Early spring is best – concentrate on the stands of damp woodland. Firecrest is fairly common and Spotted Flycatcher breeds as well – the latter can be very evident in areas of open woodland in late spring. Lesser Redpolls and Siskins are present in winter, the latter occasionally breeding. Passage possibilities include Wood Warbler and Pied Flycatcher.

TIMING

Spring and summer are the best times to visit, though the winter months are when Goosander are present. Early mornings are advisable – the Upper Common is very popular, especially at weekends.

ACCESS

The Upper Common is best accessed from Top Car Park, off Suffield Lane at SU 919461. For Cutt Mill, use either Middle Car Park (SU 911458) off Suffield Road or Tarn Car Park, off Littleworth Road at SU 909455. The public footpath down Cutt Mill Lane is the best for viewing the house pond. Britty Hill car park (SU 904456) provides access to the Lower Common.

The site is generally not suitable for wheelchairs or pushchairs, though the foot-path along the house pond is on a paved road.

FACILITIES: The nearest facilities are in Elstead.

CALENDAR

All year: Mandarin, Great Crested Grebe, Kingfisher, Lesser Spotted Woodpecker, Marsh Tit, Dartford Warbler, Firecrest.

April–August: Nightjar, Cuckoo, Woodcock, Common Sandpiper and Osprey on passage, Woodlark, Garden Warbler, Spotted Flycatcher, Redstart, Tree Pipit.

November–March: Shoveler, Goosander, Lesser Redpoll.

24 SHACKLEFORD FARMLAND

OS Explorer 145
OS grid ref: SU 936462
Postcode: GU8 6AP

HABITAT

To the north of the village of Shackleford is an area of mixed farmland, centred around Lydling Farm. The Hog's Back looms to the north but this site is open and flat, which is relatively unusual in this wooded, hilly part of Surrey. Lydling is home to a famous herd of cattle and the current management of the site has allowed wildlife to prosper, with farmer Angus Stovold creating different habitats via traditional rotational farming. Fertiliser isn't used and the rotated crops are flanked by scrubby margins and hedgerows. Winter crops are planted and areas of thistles left uncleared. Further initiatives to conserve nature include the

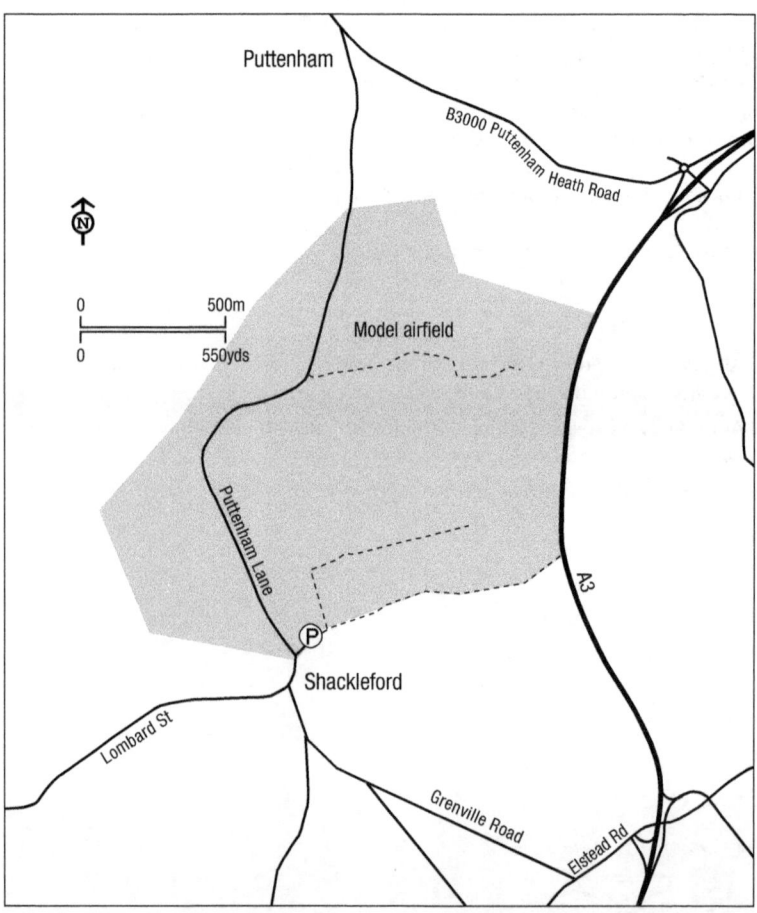

installation of various bat and bird boxes, and even the installation of a dog-proof fence around the main fields.

SPECIES

There are sometimes impressive numbers of passerines in the winter, including flocks of thrushes, Skylarks and Meadow Pipits, as well as finches and Reed Buntings. Yellowhammer is uncommon but may increase; the recent wintering of three Corn Buntings was a very notable modern-day Surrey record. A few Stonechats winter, and sometimes they have a Dartford Warbler in tow. Peregrine often visit and Raven is a regular flyover. A small flock of Lapwings may be present; Golden Plover is annual but usually flying over. Woodcock fly into the fields to feed at dusk.

The arable fields are home to good numbers of breeding Skylark in the summer, with a thriving population of Whitethroats in the hedgerows. There have been recent records of Grey Partridge – a real Surrey rarity these days – but Red-legged is far more likely. Little Owl is resident but can be elusive, and is usually heard and not seen.

Shackleford has proven productive during passage seasons, with Yellow Wagtail, Wheatear and Whinchat regular, as well as multiple records of Black Redstart and White Wagtail. Both Hen and Marsh Harrier have been recorded on migration too. Scarcities in recent times have included Wryneck and Short-toed Lark.

TIMING

A visit at any time of year can be rewarding, though high summer can be a little quiet. Model aircraft activity (usually mid-morning onwards) around the small airfield can sometimes flush birds.

ACCESS

Access is best from Chalk Lane in Shackleford village (SU 935457), where there is usually room to park a car. The 46 bus (Aldershot to Guildford) stops at Shackleford post office.

It is not a suitable site for those with limited mobility.

FACILITIES: There is a small shop in Shackleford village, as well as a pub (The Cyder House Inn).

CALENDAR

All year: Red-legged Partridge, Little Owl, Skylark.

October–March: Lapwing, Woodcock, chance of Peregrine, finches.

April–September: Whitethroat, passage passerines including Wheatear, Whinchat and Yellow Wagtail.

25 THURSLEY COMMON

OS Explorer 133/145
OS grid ref: SU 907413
Postcode: GU8 6LW

HABITAT

The Thursley group of commons, including Ockley and Elstead Commons, comprise one of the finest pieces of lowland heath in Britain. This beautiful SSSI and NNR is a real jewel in the crown not just of Surrey, but the whole of the South-East. The central part of the reserve is a peat bog – the largest mire in southern England – but boardwalks provide access over the wettest areas, including Pudmore, which is the most open pool on the common. Isolated stands of pines and birches scatter the site, which is mainly heather moor, with some elevated points. The higher ground has more extensive patches of gorse. There is an area of deciduous woodland at Parish Field/Will Reeds and Truxford Corner. A series of small ponds run along the eastern edge of the site, flanking the A3.

The acidic bog pools, ditches and ponds support several rare invertebrates, including Raft Spider. Thursley is famous for its Odonata populations – no fewer than 26 species of dragonfly and damselfly have been recorded here, including Black Darter, Keeled Skimmer, Small Red Damselfly and Southern Hawker. Butterflies includes Grayling and Silver-studded Blue, and Emperor Moth can be found. In all, a mighty 10,000 invertebrate species reside at Thursley. There are all six species of native reptile and Common Lizards can be conspicuous on warm summer days along the boardwalks.

In 2006, an out-of-control wildfire burned 60 per cent of the common, having a devastating impact on the Dartford Warbler population. The heather took a few years to fully recover. Another fire in 2020 decimated about a third of the site.

SPECIES

Spring and summer are when Thursley is at its finest, with a wonderful selection of breeding species. This includes good numbers of the heathland 'classics': Nightjar, Woodlark, Dartford Warbler, Redstart and Tree Pipit all breed, though the latter is declining. Woodlark is rather widespread across the site, with the dead trees around Shrike Hill a good place to look, as well as Parish Field. This area, along with Redstart Corner, Woodpigeon Wood and east of the Moat are particularly good for Redstart. Dartford Warbler can be found anywhere, though the south-east section of the site holds a particularly dense population. Woodcock, Skylark, Whitethroat, Garden and Willow Warblers, Firecrest, Spotted Flycatcher, Stonechat and Linnet also breed, along with Cuckoo – Thursley has become famous for this species, with Parish Field sometimes home to particularly tame birds. A pair of Hobbies breed in most years and will usually be encountered, though this species is better sought out in May, when passage birds (often second-years) will loiter around Pudmore and the veritable feast of dragonflies. At times brilliant views can be had from the boardwalk.

Another famous Thursley breeder is Curlew. The common has long been the last site in Surrey where this species breeds (and indeed one of few places in the South-East) and, from their return in early March until about late May, they can be a noisy presence, roaming around Pudmore and Ockley Common. Birds may become trickier to spot after that, as they either tend to their young or move on if they have failed to breed, but there are normally Curlew on site until early July. Lapwing occasionally attempts to nest but is becoming far less regular. Unfortunately, the presence of dogs, especially around Pudmore, has increased significantly at Thursley and both these wader species face extra breeding pressures as a result. Snipe tend to fare better and, in most years, one or two pairs nest – a notable occurrence in Surrey these days. It is quite special to be out at Thursley at dawn or dusk in the spring and hear drumming Snipe, calling Curlews and churring Nightjars. Another somewhat surprising breeder is Teal, with a pair successful in most years. Water Rail also breeds but is far more likely to be heard than seen.

Given that it is a landlocked heath, Thursley can be surprisingly productive during the passage seasons. Pudmore normally pulls a few waders in – typically Green Sandpipers, but Greenshank and Redshank are both annual. The site has good form for Wood Sandpiper, too. Wheatears and Whinchats pass through annually, sometimes in good numbers, and occasionally a few Yellow Wagtails will drop

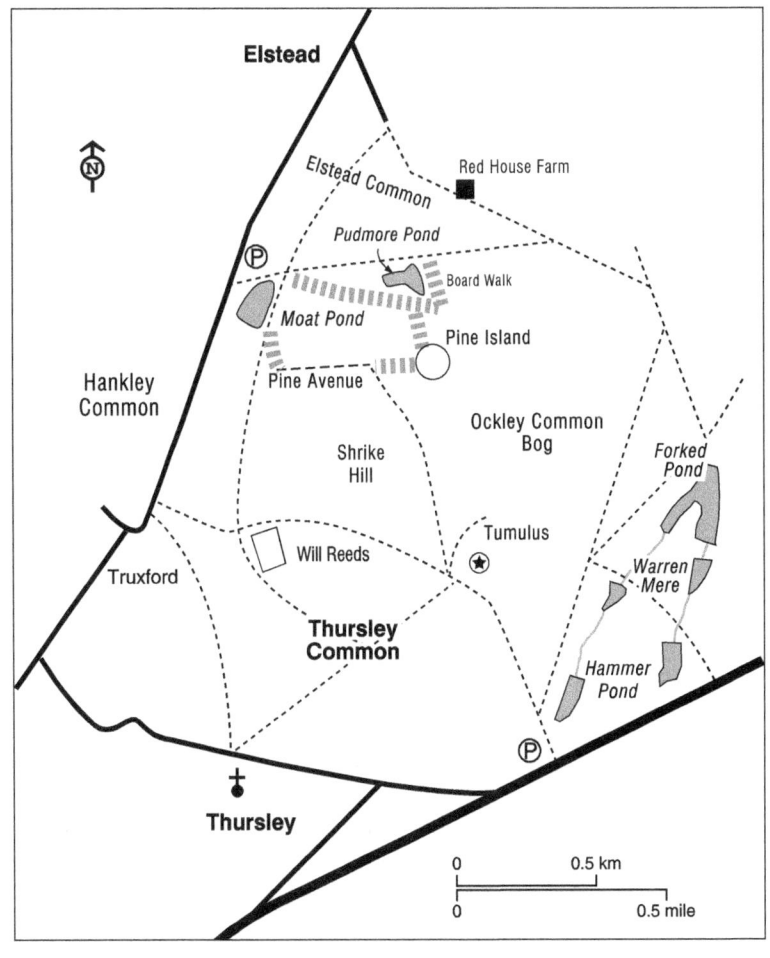

down at Pudmore. Although Spotted Flycatcher breeds, it is typically more easily seen in August. Thursley is good for passage raptors, with Marsh Harrier and Osprey annually flying over. Honey Buzzard is possible on migration too. Rarer prizes are possible and it pays to keep an open mind at Thursley – a site list that includes Fulmar, Short-toed Snake Eagle, Roller, Woodchat Shrike and Little and Rustic Buntings shows why!

Winter can be very quiet indeed, especially as two of the classic wintering species – Great Grey Shrike and Merlin – have become rarer. The former is erratic in its appearances these days but is still a possibility. Merlin seems to be more of a passage visitor now, in tandem with the decline in numbers of wintering Meadow Pipits and Reed Buntings. Hen Harrier is also less regular these days, but occurs more often than the other two species and sometimes a bird or two will roost on Ockley Common for extended periods. Finches usually include Lesser Redpolls and Crossbills; the latter species can be numerous in some years and occasionally breeds. There is often a Fieldfare roost on Ockley. A few Goosander are occasionally present at Forked Pond during the winter, with Mandarin, Great Crested Grebe and Kingfisher found there year-round.

TIMING

Spring and summer are the best times to visit. Early mornings are advised – Thursley is a popular site with dog walkers and joggers, and the boardwalk at Pudmore can get very busy (as can the Moat car park). Weekdays are recommended too. If a showy Cuckoo is present, Parish Field can be busy between April and May.

ACCESS

Thursley is best accessed by car, with a large, free car park at Moat Pond off Thursley Road (SU 899416). This is an obvious place to start, with the Pudmore boardwalk a short walk away. Alternatively, there are two large lay-bys along Old Portsmouth Road, on the south side of the common, at SU 909397. This offers quicker access to Parish Field.

Bus service 46 from Aldershot to Guildford stops in Elstead village, which is a 10-minute walk from the north end of the common.

Wheelchair and pushchair access is possible around the boardwalk at Pudmore.

FACILITIES: The nearest facilities are in Elstead, including a shop, cafés and pubs.

CALENDAR

All year: Mandarin, Water Rail, Raven, Woodlark, Skylark, Dartford Warbler, Firecrest, Stonechat, Linnet, Crossbill in good years, Reed Bunting.

April–August: Teal, Nightjar, Cuckoo, Curlew, Snipe, Woodcock, Green Sandpiper, chance of Marsh Harrier and Osprey on passage, Hobby, Garden and Willow Warblers, Redstart, Spotted Flycatcher, Wheatear and Whinchat on passage, Tree Pipit.

September–March: Osprey and Marsh Harrier on passage, chance of Hen Harrier and Merlin, possible Great Grey Shrike, Wheatear and Whinchat in September, Lesser Redpoll.

26 MILFORD AND WITLEY COMMONS

OS Explorer 133/145
OS grid ref: SU 932409
Postcodes: GU8 5QA and GU8 5QJ

HABITAT

Situated south of Milford, between the A246 and A3 and adjacent to Thursley Common, Witley and Milford Commons are owned by the National Trust and consist of woodland, heathland and scrub. The commons have a broad range of both deciduous and evergreen trees, with birch, oak and pine dominating. Parts of Witley are boggy. Milford is far scrubbier and excellent restoration work has improved areas of blackthorn and hawthorn, as well as opened up old patches of heather. Some 30 species of butterfly have been recorded across both commons and reptiles present include Adder.

Beyond the southern end of Witley Common and across Lea Coach Road is a small Amphibian and Reptile Conservation reserve (confusingly also called Witley Common!) The heathland habitat here is excellent. Both Sand Lizard and Smooth Snake have been successfully reintroduced. Rodborough Common is just the other side of the A3. It is largely wooded, with patches of wet woodland and small areas of heath. To the south-east of Witley Common, across the A246, is Mare Hill, a small heathland site that, like Milford, has benefited greatly from restoration work.

SPECIES

Milford Common is one of the best sites in Surrey for Nightingale. It has long been a haunt of this wonderful songster, but numbers crashed in the 2000s to the point that the species deserted the site. However, habitat restoration – especially the management of hawthorn scrub – has enticed birds back and several males are present each spring. They can often be heard from the car park at Milford Cemetery and a wander around the common should produce a few birds, as well as good numbers of Garden Warbler and Bullfinch. Firecrest and Spotted Flycatcher may be found at both commons.

Most of the heathland specialists are found on Witley Common. A few Dartford Warblers, Woodlarks, Redstarts and Tree Pipits breed, and it's a good site for Nightjar. For greater numbers of these species, though, it's worth visiting the ARC reserve on the other side of Lea Coach Road. Stonechat, Cuckoo and Woodcock also breed. Witley is sadly the only South-West Surrey common that still has breeding Yellowhammer, but the species has declined here and may not persist for long.

Rodborough Common has a few Marsh Tits and, possibly, Lesser Spotted Woodpecker in the wetter parts. The work at Mare Hill has brought back Dartford Warblers and Nightjars.

TIMING

Spring and summer are the best times to visit. Early mornings are best for Nightingales, especially before the A3 gets too noisy. Both sites are popular with dog walkers. Try to avoid weekend afternoons.

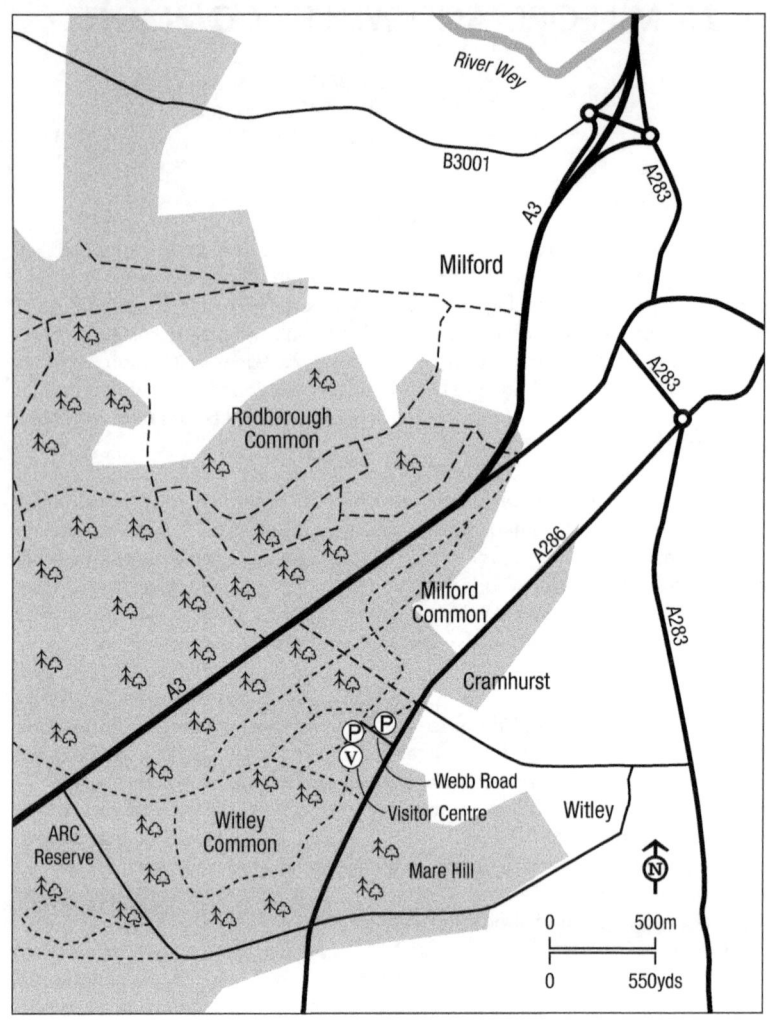

ACCESS

For Milford Common (and Nightingales), there's a car park off the A246 next to Milford Cemetery at SU 940413. For Witley Common, there are car parks at Webb Road (SU 933409) and Lea Coach Road (SU 927398). Lay-bys further along Lea Coach Road (SU 924399) offer access onto the ARC reserve, as well as the southern (and most productive) end of Witley Common.

There is a small car park on Mare Hill at SU 935399. The Rodborough Common car park is off Portsmouth Road (passing under the A3) at SU 937418.

The 70 Guildford–Midhurst and 71 Guildford–Haslemere buses alight on Petworth Road in Witley; from The Star pub it is a 15-minute walk west along Wheeler and Gasden Lanes to Witley Common.

Many paths are smooth and fairly level.

FACILITIES: The nearest facilities are in Milford.

CALENDAR

All year: Chance of Lesser Spotted Woodpecker, Dartford Warbler, Marsh Tit at Rodborough Common, Firecrest, Bullfinch.

April–July: Nightjar, Cuckoo, Woodcock, Woodlark, Stonechat, Garden and Willow Warblers, Nightingales at Milford Common, Redstart, Spotted Flycatcher, Tree Pipit, Yellowhammer.

October–March: Lesser Redpoll, possible Brambling and Crossbill.

---

27 RIVER WEY: EASHING TO SHALFORD

OS Explorer 145
OS grid refs: SU 971441 to SU 996467
Postcodes: GU7 2QE to GU3 1LQ

HABITAT

Between the hamlet of Eashing, south of Godalming, and Shalford village near Guildford, the River Wey winds through a floodplain of water meadows, willow and alder copses. The River Wey Navigation – a shallow canal – runs in parallel with part of it and is used by a small number of pleasure craft in the summer, especially between Godalming and Shalford. Various species of dragonfly and damselfly occur and, in the summer, riverside flora includes Agrimony, Common Valerian and Ragged-robin.

The stretch between Eashing and Godalming is the most wooded, with extensive alder carr in parts. The Lammas Lands is a series of marshy meadows between Godalming and Farncombe, with ditches and a few pools. The next stretch, between Farncombe and Peasmarsh, is known as Unstead Water Meadows, which has working farmland, alder and willow and more wet meadows. Broadwater Lake sits the other side of the Wey here. There is a sewage farm at Unstead which has a few lagoons, meadows, areas of reedbed, scrapes and open filter beds. The final part of the river at Shalford is perhaps the most dynamic in terms of habitat, with meadows, marshes, pools, alder woodland and scrub.

SPECIES

The part of the river between Eashing and Peperharow Road in Godalming is wooded and has a small population of Marsh Tit. Lesser Spotted Woodpecker is uncommon now, but Firecrest can be found in Milton Wood and Spotted Flycatcher in the alder carr towards Eashing Bridge. Garden Warbler is present in decent numbers during the summer. Along the Wey itself, the high riverbanks mean this stretch is good for Kingfisher. Grey Wagtail and Mandarin are also readily found. In hard weather, Goosander may appear.

The two key Lammas Lands meadows are Catteshall and Overgone. Both meadows have a similar array of species, which include up to 40 wintering Snipe (there is normally a Jack or two among them). Occasionally other waders, such as Little

Ringed Plover and Oystercatcher, will drop into the pools or fly over. Stonechat and Reed Bunting breed, with an impressive population of the latter species on Catteshall Meadow. Reed and Sedge Warblers are erratic nesting birds. In the winter, Dartford Warbler may be found tailing the Stonechats. Whinchats and Wheatears pass through on passage, as do – rarely – Grasshopper Warblers and Water Pipits. Barn Owl is an elusive visitor that may be seen at dawn or dusk. Borough Road bridge, at the west end of Overgone Meadow, can be an excellent place to watch the Kingfishers that occasionally nest in the riverbanks here.

The Unstead Water Meadows stretch has a reedbed at Tannachie that supports both Reed and Sedge Warblers. Cetti's Warbler is occasional. In the summer a Hobby may hawk overhead. As on the Lammas Lands, Stonechat and Reed Bunting breed, with Whinchat and Yellow Wagtail dropping in on passage. Winter flooding here can be dramatic and, when the site is underwater, impressive numbers of geese, ducks and gulls can be sifted through – there are recent records of

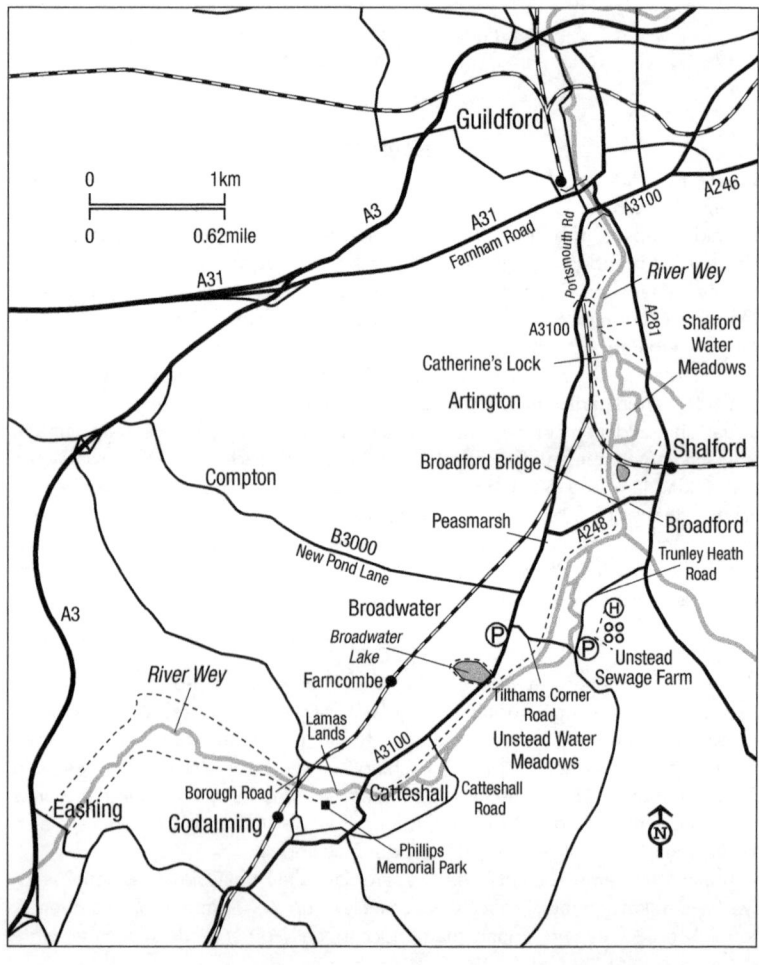

White-fronted Goose and Mediterranean Gull. Water Rail is a noisy presence in the marshier areas and in the stands of wet woodland, while a few Woodcock feed in the damp meadows after dark. The adjacent Broadwater Lake has breeding Great Crested Grebe and is frequently visited by Kingfishers (and occasionally Common Tern). Pochard sometimes appear in winter.

Unstead Sewage Farm has a rich birding history including a fine list of scarcities and rarities but, sadly, the habitat declined greatly in the 2010s. Restoration work has now given the site a new lease of life. It is an excellent place for Water Rail in the winter, which can often show well at the south end of Dry Lagoon, viewable from the public footpath. Unstead is a traditional site for Green Sandpiper, too, and the creation of new habitat should benefit this species, as well as Snipe and Lapwing, which were formerly regular winterers. Cetti's, Reed and Sedge Warblers breed, as does a small colony of Black-headed Gull. There are normally several wintering Chiffchaffs around the works, and they should be sieved through for Siberian Chiffchaff, which has occurred here several times before. Chats pass through the meadows on passage and impressive numbers of wagtails and hirund-ines loiter around and over the open beds. Wintering wildfowl include Shoveler, Teal and Gadwall. Among the more exceptional species on the site list are Purple Heron, Red-necked Phalarope and Red-rumped Swallow.

Shalford Water Meadows is essentially the section from Broadford Bridge to St Catherine's Lock. It has a healthy wintering population of Snipe and a few Jack Snipe. The pool at St Catherine's Lock often has gatherings of wildfowl, including Shoveler and Teal. Garganey and Pintail have been recorded too. This area has breeding Sedge Warbler. Shalford will occasionally attract a reeling Grasshopper Warbler for a day or two; this species can turn up anywhere along this stretch of the Wey on passage, but is rare. Grey Wagtail and Reed Bunting breed, but sadly Lesser Spotted Woodpecker has become scarce. Barn Owl is a sporadic visitor, sometimes breeding. Whinchat and Yellow Wagtail pass through annually.

TIMING

Any time of year can be productive. Periods of heavy winter rain can render the Lammas Lands inaccessible, but equally can attract many more birds to Unstead Water Meadows. Early mornings are best at the Lammas Lands and the Shalford stretch, both of which are popular with dog walkers.

ACCESS

For the Eashing stretch, there is a small car park at Eashing Bridge (SU 946438) with a footpath running from Greenways Farm to Godalming.

The Lammas Lands are best accessed from Borough Road (limited roadside parking) at SU 967441 for Overgone Meadow or, for Catteshall Meadow, from Catteshall Road (SU 980445); limited parking) or the Godalming United Church (SU 973441). Much of Overgone can be viewed from the Phillips Memorial Park in Godalming. Godalming train station is a five-minute walk to Overgone Meadow; Farncombe train station is roughly a 10-minute walk to Catteshall Meadow.

The towpath from Catteshall Road to Tilthams Corner Road affords views of Unstead Water Meadows, which can also be accessed by parking at Broadwater Lake (SU 985453), crossing the road and taking the footpath east.

At Unstead Sewage Farm, there is limited parking beside the works (SU 994455), which are situated off Trunley Heath Road. Most of the site is viewable from public footpaths and there are viewing screens at the lagoons. A hide at North Meadow is members-only and may be locked. For information about further access visit the Unstead Wetland Nature Reserve website..

For Shalford, it's best to park either at Broadford Bridge (SU 996467) or the Park and Ride car park at Artington (SU 993475). Shalford train station is a 10-minute walk from the east side of the site.

FACILITIES: The nearest facilities are in Godalming (supermarkets, shops, public toilets) and Shalford (a café).

CALENDAR

All year: Mandarin Duck, Barn Owl, Kingfisher, possible Lesser Spotted Wood-pecker, Marsh Tit at Eashing, Cetti's Warbler at Unstead Sewage Farm, Stonechat, Grey Wagtail, Reed Bunting.

October–March: Teal, Shoveler, Water Rail, Snipe, Jack Snipe at the Lammas Lands and Shalford, possible Green Sandpiper, Chiffchaff at Unstead Sewage Farm (Siberian Chiffchaff aka *tristis* is possible).

April–September: Hobby, hirundines, Reed and Sedge Warblers, Spotted Flycatcher at Eashing, Wheatear, Whinchat and Yellow Wagtail on passage.

---

28 CHIDDINGFOLD FOREST

OS Explorer 133
OS grid ref: SU 982337
Postcode: GU8 4PG

HABITAT

Chiddingfold Forest covers various blocks of Low Weald woodland spanning some 840ha in the far south of the county and crossing over the Sussex border. The woods contain a large proportion of broadleaved species including signifi-cant numbers of oak, birch and ash. There are also areas of ancient woodland and conifer plantations, as well as grassy rides and clearings. A large proportion of the complex is designated as an SSSI. Areas of note from a naturalist's perspec-tive include Botany Bay and Fisherlane, Oaken, Tugley and Sidney Woods. The latter is disjunct from the main forest and is a couple of miles to the east.

The 12ha Oaken Wood is a Butterfly Conservation reserve. Indeed, Chiddingfold Forest is well known for its butterflies and is a national stronghold for the scarce Wood White. Grizzled Skipper, Dingy Skipper and Green Hairstreak are present in spring while, during the summer, Purple Emperor can be seen in good numbers, along with White Admiral, Silver-washed Fritillary and White-letter and Purple Hairstreaks. Important moth species include Drab Looper and Broad-bordered Bee Hawk-moth. The forest also supports populations of Barbastelle and Bechstein's Bat.

SPECIES

Chiddingfold Forest is, at the time of writing, the last refuge of Turtle Dove in Surrey. Sadly it is now tricky to connect with (the core of this population is across the county border in the Sussex side of the forest). Tugley Wood is the most reliable area but in some years no birds will be holding territory. Nightingale is faring better, however, and this is one of the best places in Surrey and Sussex to enjoy the species, which dominates the dawn chorus from mid-April through to late May. Other species in voice at this time of year include Cuckoo and Willow Warbler. There is also a good population of Garden Warbler and Firecrest, and Spotted Flycatcher breeds.

Marsh Tit is resident and readily encountered. Lesser Spotted Woodpecker is elusive but may well be chanced upon; the areas of damper woodland in Botany Bay are worth checking. Another shy resident is Hawfinch. This species probably breeds most years but is better sought out at the more reliable South Surrey sites.

Winter is quiet. A few Crossbills or Lesser Redpolls may rove around or perhaps a Woodcock will be flushed from an area of bracken.

Sidney Wood has Nightingales and, rarely, Turtle Doves. It's a good site for Spotted Flycatcher with birds often present beside the car park. Hobby also breeds. The wet woodland beside the Wey–Arun canal is another place to try for Lesser Spotted Woodpecker.

TIMING

Spring and summer are the best times to visit. Early mornings are optimum for peak birdsong, with dusk also good for Nightingales.

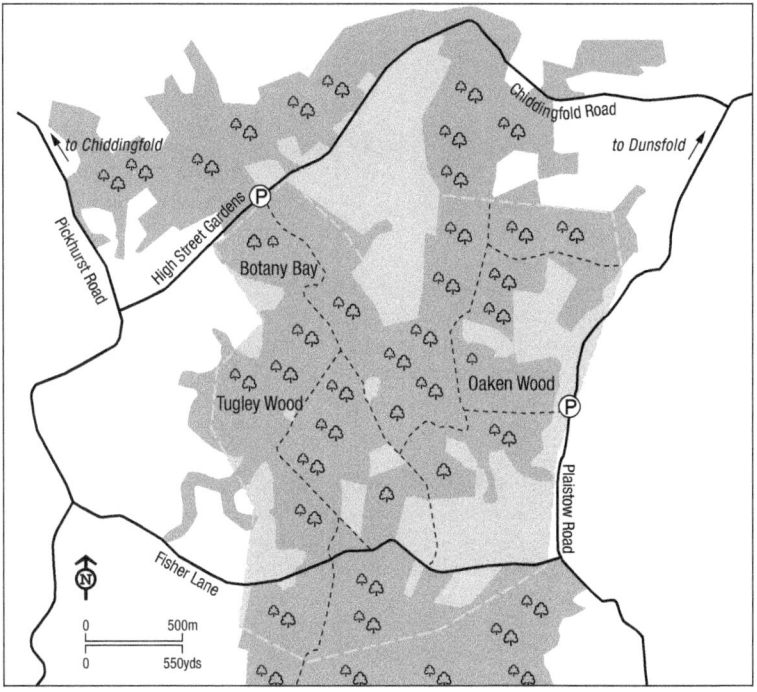

ACCESS

There is room for a few cars at the Oaken Wood butterfly reserve entrance off Plaistow Road at SU 993337. Alternatively, there are large lay-bys along Fisher Lane at SU 985331 and SU 982330. Don't block Forestry Commission gates. The Sidney Wood car park is found off Dunsfold Road at TQ 026350. This is quite a remote site and travelling by public transport is not practical. Many of the tracks are flat and (fairly) smooth.

FACILITIES: The nearest facilities are in Chiddingfold.

CALENDAR

All year: Chance of Lesser Spotted Woodpecker, Marsh Tit, Firecrest, chance of Hawfinch.

April–July: Cuckoo, chance of Turtle Dove, Hobby, Garden and Willow Warblers, Nightingale, Spotted Flycatcher.

November–March: Finches, including Crossbill.

OTHER SITES IN SOUTH-WEST SURREY

D1 WRECCLESHAM WATER MEADOWS

The River Wey south of the A31 at Wrecclesham is prone to extensive flooding, which can attract plenty of Wigeon and Little Egrets. Unusual species such as Avocet and Knot have occurred. The meadows at Wey Manor Farm (SU 821453) are best.

D2 WAVERLEY ABBEY

This English Heritage site (SU 869455) is situated on a River Wey floodplain that occasionally attracts breeding Lapwing and, during winter, wildfowl. The lake has common waterbirds including Kingfisher. Raven breeds nearby.

D3 FARNHAM HEATH RSPB

This former conifer plantation been transformed into open heathland by the RSPB (SU 861437). All the classic species can be found including Nightjar, Woodlark, Dartford Warbler, Redstart and Tree Pipit. There is usually a pair of breeding Spotted Flycatchers in the vicinity of the adjacent Rural Life Centre (SU 858433).

D4 HANKLEY COMMON

This 560ha expanse of open heath (SU 885406) is excellent for heathland species, with strong populations of Nightjar, Woodlark, Dartford Warbler, Redstart and Tree Pipit. The Nightjars are especially reliable and can sometimes be enjoyed from one of the car parks, along with Woodcock. Cuckoo and Spotted Flycatcher also breed. In winter there is an outside chance of Great Grey Shrike or Hen Harrier.

D5 THUNDRY MEADOWS

This tranquil Surrey Wildlife Trust reserve (SU 897440) sits on the banks of the River Wey and is one of the best spots in the region for Lesser Spotted Woodpecker, though they can be elusive. Spotted Flycatcher, Firecrest and Garden Warbler also breed.

D6 BAGMOOR AND ROYAL COMMONS

These two commons (SU 921425) are rather wooded and aren't great for heath-land species at present. However, Spotted Flycatcher breeds, Goosander sometimes winter on the pond at Royal Common and Lesser Spotted Woodpecker is possible.

D7 EASHING FIELDS

This Surrey Wildlife Trust managed grassland site (SU 946431) has breeding Skylarks and Linnets and usually attracts chats on passage, especially Whinchat. Its relative height means it can be productive for vis-mig.

D8 THORNCOMBE STREET

There are a few sites of interest in the Thorncombe Street area, including Snowdenham Mill Pond (TQ 001441) which is good for Mandarin and other duck, Kingfisher and occasional Red-crested Pochard; Bonhurst Farm (TQ 012428), resi-dent Little Owl and passage chats; Gatestreet Farm (TQ 012417), occasional Lesser Spotted Woodpecker; and Selhurst Common (TQ 014408), breeding Spotted Flycatcher and Firecrest.

D9 WINKWORTH ARBORETUM

This National Trust site (SU 994413) has common waterbirds including Kingfisher and Grey Wagtail, as well as breeding Firecrest, Spotted Flycatcher and Marsh Tit. Crossbill and Hawfinch are occasional.

D10 HASCOMBE HILL, HYDON'S BALL AND THE HURTWOOD

These three spots along the Greensand Ridge each have good populations of Firecrest, with Marsh Tit resident and Crossbills fairly regular. Forestry clearance at The Hurtwood (SU 995388) has attracted Woodlark, Dartford Warbler and Tree Pipit. The farmland south of Hydon's Ball (SU 972390) is good for Yellowhammer.

D11 PAINSHILL FARM

This farmland site near Dunsfold (TQ 023385) has breeding Nightingale and Yellowhammer, plus Barn and Little Owls. Whinchat and Wheatear pass through on passage.

D12 FRILLINGHURST WOOD

A few pairs of Nightingales breed at this woodland site near Grayswood (SU 929343), along with Marsh Tit, Firecrest and Spotted Flycatcher. Fields at Imbhams Farm support wintering Woodlark.

NORTH-WEST SURREY

MAIN SITES
29 Tice's Meadow
30 Blackwater Valley Gravel Pits
31 Ash Ranges
32 Chobham Common
33 Ockham and Wisley Commons and Boldemere
34 Papercourt area
35 Burpham Water Meadows and Stoke Lake

OTHER SITES
E1 Normandy and Wanborough Farmland
E2 Brookwood Cemetery
E3 Pirbright Ranges
E4 Bagshot Heath and Olddean Common
E5 Virginia Water
E6 Horsell Common and Fairoaks Airfield
E7 Wisley Sewage Farm
E8 Old Woking Sewage Farm
E9 Whitmoor Common
E10 Farnham Park

The North-West Surrey region defined in this book is the area of the vice-county west of the A3 (Cobham to Guildford) and north of the A31 (Guildford to the Hampshire border) that's outside the LNHS boundary.

There are two rivers that flow through this area: the Blackwater, which defines the border between Surrey and Hampshire, and the River Wey, which runs from Wisley down to Guildford. This part of the county is relatively built-up along the Blackwater, with various towns and villages all the way to the Berkshire border. Another feature

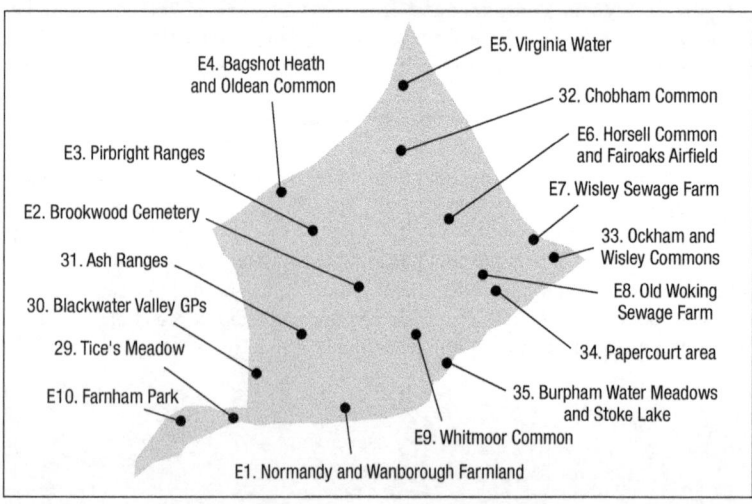

of this valley is a number of gravel and sand extraction pits – many of which are no longer in use. These include Tice's Meadow, at the bottom of the valley, which is undoubtedly the key site in the area – and the most dynamic birding site in outer Surrey.

The other wetland area of note is the stretch of the river Wey from Guildford to Pyrford. At Guildford, the Wey passes through a narrow gap in the North Downs and then broadens into the Burpham/Stoke area, where the habitat is comprised of water meadows, marsh, open water and woodland. The habitat is similar but more extensive up towards Papercourt, where there is also a series of old gravel pits. Other significant waterbodies include Virginia Water, at the Berkshire border, and Boldermere, in the Wisley Common area.

Another key feature of North-West Surrey is the heathy commons, all of which form part of the Thames Basin Heaths Special Protection Area (SPA). Access to some of them is limited or, at best, variable, due to an active MOD presence. These sites have a more rugged feel to other commons in the county and lie on the sandy Bagshot Beds, which are not found elsewhere in Surrey. These heaths, as with many in the county, hold an interesting array of other fauna and flora; Sand Lizard and Smooth Snake are among the reptile species found, while butterflies include Silver-studded Blue and Grayling.

This is the least-wooded region of outer Surrey with notably limited areas of mature, deciduous woodland compared with much of the county. The most extensive tracts of farmland are in the south, on the northern plateau of the North Downs. The further north you go in this region, the more built-up the landscape becomes, especially towards Spelthorne.

The most appealing areas of North-West Surrey to the birder are the wetland sites and the heaths. Tice's Meadow is perhaps the most popular birding site in the county and is well watched. It is the focal point for many waterbird species that are not numerous elsewhere in the region and is easily the most productive locale for waders and gulls. Of the entire Wey Valley in Surrey, the Papercourt area is probably the best stretch for birding and has a wide variety of interesting species. Red Kite and Little Egret are now rather numerous across these sites and Peregrine is well established.

The heaths in North-West Surrey hold some of the greatest densities of heathland breeders in the county. The wildness of the Ash and Pirbright Ranges, for example, holds superb numbers of Nightjar and Dartford Warbler. The previously mentioned lack of woodland does render some classic Surrey species thin on the ground in North-West Surrey. For instance, this is the only region in the county without a regular site for Marsh Tit, while Goshawk is yet to secure a foothold and Hawfinch is a rare visitor.

29 TICE'S MEADOW

OS Explorer 145
OS grid ref: SU 874485
Postcode: GU9 9LY

HABITAT

One of the major birding destinations in the county, Tice's Meadow is a relatively new nature reserve and the most significant area of wetland in outer Surrey. This 60ha site is flanked to the west by Badshot Lea, to the south by the A31, to the east by Tongham Gravel Pits and to the north by the River Blackwater, a housing estate and the border with Hampshire. Formerly the site of Hanson's Farnham Quarry, Guildford Borough Council adopted Tice's Meadow as a Site of Nature Conservation Interest (SNCI) following a recommendation from the Surrey Bird Club in 2009. Gravel extraction finished in 2010 and the majority of the site was restored by 2011, with restoration work finally completed in 2018.

The open water, scrapes and wet grassland – known as The Workings and The Meadow – are the main attraction to birders but plenty of other habitats are present, including reedbeds, scrub, woodland and ponds. A feeding station and small wildflower meadow are also found on site. Horton's Mound overlooks the wetland and is the focal point for most birders. Tice's Meadow also benefits from sitting at the bottom of the Blackwater Valley and the western end of the North Downs – this, coupled with the open nature of the site, helps attract migrating species that use these linear features.

Much of the excellent habitat management is undertaken voluntarily by the Tice's Meadow Bird Group, who are also behind the construction and installation of a hide, two Sand Martin banks, a Swift tower, some 100 nest boxes and several tern rafts. There is free open access to the site through a network of permissive footpaths, with a public footpath (the Blackwater Valley Path) running along the northern perimeter.

SPECIES

Large numbers of common wildfowl are present in the winter, with Wigeon and Teal particularly conspicuous. Occasionally a Pintail may visit, but rarer diving ducks are not regular. A flock of a few hundred Lapwing use the site; Golden Plover is occasional, though sometimes in impressive numbers. Snipe are present in decent numbers, with one or two Jack Snipe among them, though the latter species rarely gives itself up. Usually one or two Green Sandpiper winter, and may even be flushed from the River Blackwater. The gulls are worth sifting through at this time of year – Tice's Meadow is the best site in outer Surrey for Yellow-legged Gull, with winter the best period to search for one. Water Rail are more likely to be heard than seen, with the same said for Cetti's Warbler, which is a recent colonist here. Little Egret is regular and Great White Egret has increased as a visitor – it's still scarce and sporadic, but could be found year-round. Occasionally, mass panic among the wildfowl and waders will indicate the presence of a Peregrine, which breeds nearby and often visits. The feeding station is well worth a visit in the winter and affords great views of common species; appearances by Brambling are possible.

Spring passage begins in March and Tice's Meadow usually scores the first

Surrey hirundine of the year, typically Sand Martin. Little Ringed Plover breeds in most years and are normally back on site by the end of the month. April and May can be exciting. Not only are summer breeders, such as Garden, Reed and Sedge Warblers, back on site, but waterbird passage is in full swing. Garganey is near-annual here but can be surprisingly elusive. Black-necked Grebe will occasionally drop in between late March and May. Osprey and Marsh Harrier are best looked for on days with cloud cover and a bit of a headwind – there are a few records per year of both species, with autumn better for the latter. A few pairs of Common Tern breed. Arctic, Black and Sandwich Terns and Little Gull are recorded in most years, with late April and early May best for these species. Little Tern and Kittiwake are rare, but possible. Tice's is one of the better sites for Mediterranean Gull in Surrey and it seems possible that this species will one day breed here, especially given the numbers of Black-headed Gull that nest on site.

Wader passage is more spread out. Oystercatcher, Common Sandpiper, Ringed Plover, Dunlin, Greenshank and Whimbrel are all regular visitors, with Grey Plover, Turnstone, Sanderling and Wood Sandpiper as outside chances. Tice's Meadow is not particularly famous for passerine migration but you have as good a chance of a spring Wheatear, Whinchat and Yellow Wagtail here as anywhere else in the county. Spring scarcities down the years have included Black-winged Stilt, Spoonbill, Wryneck and Red-rumped Swallow.

The site is often a hive of activity in midsummer, with breeding Common Terns and Black-headed Gulls and, in good years, one or two pairs of Little Ringed Plovers. A Hobby may hawk overhead and the sound of chattering *Acrocephalus* warblers emanates from the reeds. Return wader passage begins as early as late June, and the period from July to September is doubtless the best. Flocks of Black-tailed Godwit drop in, and several Common and Green Sandpipers will be on site from mid-July. The aforementioned wader species are all possible during the autumn, with Wood Sandpiper particularly reliable here. It is possible to visit in August or September and record 10 or more wader species – good going for

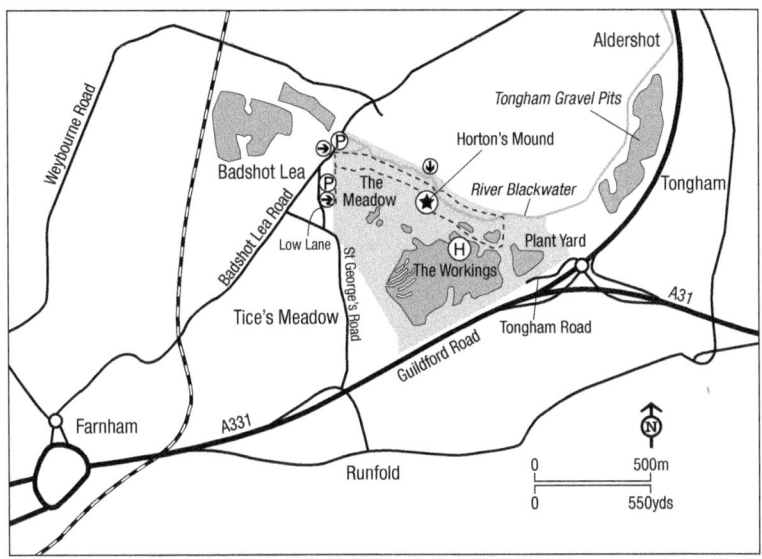

Surrey! Rarer species that have visited before include Little Stint, Spotted Redshank and Pectoral Sandpiper. It is surely only a matter of time before a true showstopping wader is found at Tice's Meadow. Wheatear, Whinchat and Yellow Wagtails are regular in August and September.

As September moves into October, wader numbers drop off and wildfowl numbers increase. Surrey's only Crag Martin was found in October 2006. Short-eared Owl is a possibility, with birds occasionally lingering. Barn Owl is resident but surprisingly hard to see.

It is worth highlighting the importance of a telescope at this site. Birds can sometimes be incredibly distant from Horton's Mound and waders in particular can be hard to see, let alone identify, if they are on the far southern shore.

TIMING

Any time of day can be productive and the site rarely gets too busy, though weekdays are quieter. Due to the south-facing nature of Horton's Mound it's advisable to visit early on sunny, clear days. During passage seasons, wet or misty mornings are good for grounding waders.

ACCESS

The nearest train station is Aldershot, with regular South Western Railway services from Guildford and London Waterloo. Stagecoach Bus run regular services from Aldershot, Guildford and Farnham with stops near the site (routes 15 & 46). The best area to park is on Badshot Lea Road (B3208) near Pea Bridge (at SU 867490), though there is also limited parking around Overton Close (SU 873488) from where you can cross the River Blackwater to the site.

The Meadow and The Workings are best viewed from Horton's Mound (SU 871487), although there is also a viewing point to the west of Horton's Mound, on the southern footpath, where a gap in the bushes allows The Meadow to be viewed. There is no public access into The Meadow, The Workings and Plant Yard. The footpaths and Horton's Mound can become very muddy following wet weather, so wellies are recommended.

Hanson's Hide is situated in the reedbed overlooking The Workings. It offers closer views of many birds but is often underwater and thus inaccessible – summer and early autumn are the best times. A shelter is also situated atop Horton's Mound but note that it is open-faced. There are also some benches here.

The unimproved nature of the site means that it is currently not suitable for wheelchair or pushchair users. However, gates with RADAR locks have been installed on Badshot Lea Road (at SU 867489) and from the housing estate (at SU 872488) to enable easier access for disabled visitors. Tice's Meadow Bird Group hope to complete an all-weather accessible path from the Estate Entrance to Horton's Mound soon.

Further information about the site, including dates for volunteer work parties arranged by the Tice's Meadow Bird Group, can be found on the Tice's Meadow website.

FACILITIES: The nearest facilities are in Badshot Lea and Aldershot.

CALENDAR

All year: Lapwing, Little Egret, chance of Great White Egret, Barn Owl, Peregrine, Cetti's Warbler.

October–February: Wildfowl, Water Rail, Golden Plover, Jack Snipe, Green Sandpiper, Yellow-legged Gull, Brambling.

April–June: Chance of Garganey and Black-necked Grebe, Osprey, Little Ringed Plover, wader passage including Oystercatcher, Common Sandpiper, Ringed Plover, Dunlin, Greenshank, Whimbrel and scarcer species, Arctic, Black, Common and Sandwich Terns, Little and Mediterranean Gulls, hirundines, Wheatear, Whinchat, *Acrocephalus* warblers, Yellow Wagtail.

July–September: Marsh Harrier, wader passage including Black-tailed Godwit, Hobby, Wheatear, Whinchat, Yellow Wagtail.

30 BLACKWATER VALLEY GRAVEL PITS

OS Explorer 145
OS grid refs: SU 890515, SU 882491 and SU 892543
Postcodes: GU12 4AL, GU16 6DS and GU14 8AG

HABITAT

Along the Blackwater Valley and flanking the county border with Hampshire is a series of gravel pits, from Badshot Lea downstream to Frimley. Badshot Lea Gravel Pit is a medium-sized waterbody, used for fishing and surrounded by trees with some small islands and patches of reed. Tongham Gravel Pit is a landscaped pit tucked beside the A331 with grassy, reedy and wood-fringed shores. The collection of pits between Ash and Ash Vale are mainly private fisheries similar to Badshot Lea, though there is public access at Lakeside Nature Reserve. Mytchett Gravel Pit, inaccessible for many years, has recently been opened up with the creation of the Water's Edge housing development. At Frimley, Quay Lake (also known as Coleford Bridge Road Pit) is used for watersports and open-water swimming, while the northernmost pits are more private and wood-lined fisheries.

SPECIES

Tongham is the most productive of the pits, with a lack of angling disturbance and the relatively open nature of the site no doubt benefiting waterbirds. In the winter, decent numbers of Wigeon, Gadwall and Tufted Duck can be found, with smaller numbers of Pochard and Shoveler. Outer Surrey scarcities such as Scaup and Black-necked Grebe have turned up on more than one occasion. Great Crested Grebe is a breeding resident. The reedbeds hold Reed and, occasionally, Sedge Warblers in the summer; Cetti's Warbler has visited and may colonise in time. A light passage of Common Sandpipers occurs but none of these pits is productive for waders.

Although the biggest pit in the area, Quay Lake attracts surprisingly few wildfowl, probably due to the recreational activities that take place at the site. Large gulls will loaf in the day and are worth scanning through for scarcer species. Frimley

Pits Fishery is made up of the four northernmost pits, including Hatches, which is the best for wildfowl, including Pochard and Gadwall in the winter. Water Rail can be found at this time of year too. Scarcities are not unheard of and recent examples include Black-throated Diver and Grey Phalarope. Mytchett has recently held substantial flocks of Tufted Duck and also attracts Wigeon and Shoveler.

Great Crested Grebe breeds at Badshot Lea, which normally holds a small *Aythya* flock in the winter, when Goosander will occasionally visit. Kingfisher is frequently seen, as they are at Ash Vale; at the latter site, Lakeside is productive for good views of this species. Reed Warbler breeds at Lakeside as well, though duck numbers are lowest here. All of these pits can receive visits from scarcities that are lingering at nearby Tice's Meadow – Great White Egret, Black Tern and Little Gull are some examples.

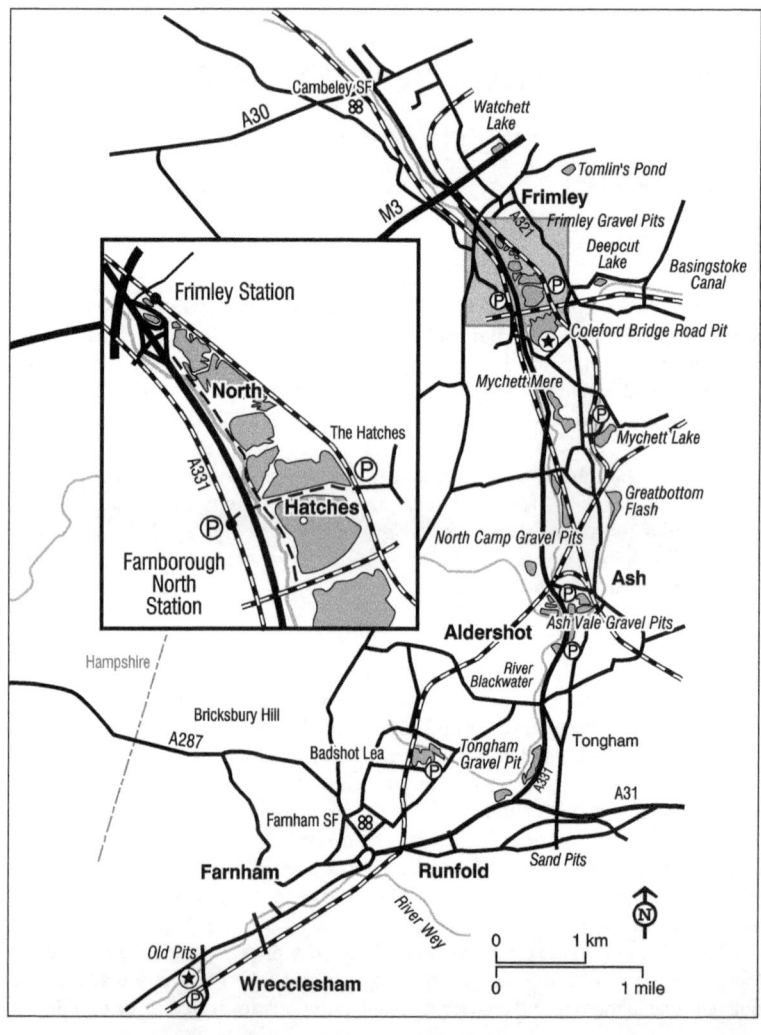

TIMING

Winter is the best time to visit due to the presence of good numbers of ducks, though passage periods could yield scarcer waterbirds. At the angling pits, an early arrival is recommended. All of these pits will freeze in cold weather.

ACCESS

For all of these sites, access is limited to public footpaths, from which the quality of viewing varies, as outlined below.

Parking for Tongham is best on Tongham Road (SU 884495) or at Aldershot Park (SU 877493). A footpath runs along the west side of the pit; viewing is rather limited but there are adequate vantages. For Quay Lake, a small area of parking at SU 881559 off Coleford Bridge Road overlooks the water and is a good place to scan from. Frimley Pits Fishery is best accessed from Farnborough North Station (where there is parking) via a footpath over the level crossing and between Hatches and the northern pits. This is the best place to view. A separate footpath runs along the west side of the pits but viewing is challenging. Badshot Lea is viewed through a chainlink fence along a footpath at SU 862489; park sensibly along nearby residential roads. There is a free car park at Lakeside (SU 888518); it's possible to view the private Ash Vale fisheries from the Basingstoke Canal at SU 887514, accessed via the canal from either Ash Lock (SU 881517) or Ash Vale (SU 893516). At Mytchett, a network of paths has been created as part of the Water's Edge development and parking is available (SU 886546).

Bus service 3 from Aldershot to Yateley stops at Tongham, service 34 from Camberley to Guildford stops at Frimley Green and bus service 12 from Aldershot to Farnham stops at Badshot Lea. Farnborough North station is less than five minutes from Frimley Pits, Aldershot station is a 30-minute walk from Tongham and a 20-minute walk from Badshot Lea; Ash Vale station is a 25-minute walk from Lakeside.

FACILITIES: Access for disabled visitors isn't good, though the north side of the main lake at Lakeside can be reached with a wheelchair from the car park. Other facilities can be found in nearby Aldershot, Ash, Camberley and Frimley.

CALENDAR

All year: Great Crested Grebe, Kingfisher.

April–September: Reed and Sedge Warblers, Common Tern, passage Common Sandpiper.

October–March: Wintering wildfowl including occasional rarities, Water Rail.

31 ASH RANGES

OS Explorer 145
OS grid ref: SU 917541
Postcode: GU3 2AQ

HABITAT

This rugged expanse of open heathland – some 1,576ha – lies on the sandy Bagshot Beds and is part of the Thames Basin Heaths Special Protection Area. Comprising Ash, Cleygate, Pirbright and Wyke Commons, this area of MOD-owned land comprises three live-firing ranges and is also used for military exercises: access is limited and caution is advised with regard to unfamiliar looking objects.

Vast tracts of heather and gorse cover the site, with scattered pines. The land is undulating and wet in parts. Fine views can be had from the area's higher ground. Birdlife probably benefits from the comparative lack of human disturbance – the ranges hold some of the densest populations of heathland species in the South-East. Henley Park Lake is a wood-fringed waterbody on the east of the site.

SPECIES

This is a quality site for heathland species and in recent years up to 251 Dartford Warbler territories, 49 singing Woodlarks and 80 churring Nightjars have been counted. It is almost certainly the most important single site in England for Dartford Warbler. Tree Pipit, Redstart and Stonechat also breed and these species, along with Raven and Hobby, are readily encountered in the summer. Good numbers of singing Willow Warbler can be found – a species declining in the region – and a spring visit will almost certainly produce vocal Cuckoos. Siskin may breed on the fringes; Crossbill is far more irregular but can be found here in most years, even if they aren't breeding.

Autumn and winter are typically quiet, with usually little more than the odd hardy resident and overflying Raven of note. Great Grey Shrike is less regular as a wintering bird in Surrey these days but the Ash Ranges are one of the most reliable locations. Hen Harrier and Merlin formerly occurred annually but now only pass through on rare occasions at these times of year, while Henley Park Lake occasionally attracts Goosander.

TIMING

Spring and summer are easily the best time to visit, with early mornings best, though dusk and beyond is ideal for Nightjars. It's important to check access availability before any visit – it is not permissible to walk the area when red flags are flying or red beacons are lit, or when access gates are locked and signage detailing site maintenance is installed. At all other times the ranges are open to the public for access on foot. The site is usually open over bank holiday weekends and a two-week period in July/August, but for an up-to-date list of firing times visit the South East Training Estate firing times on the GOV.UK website.

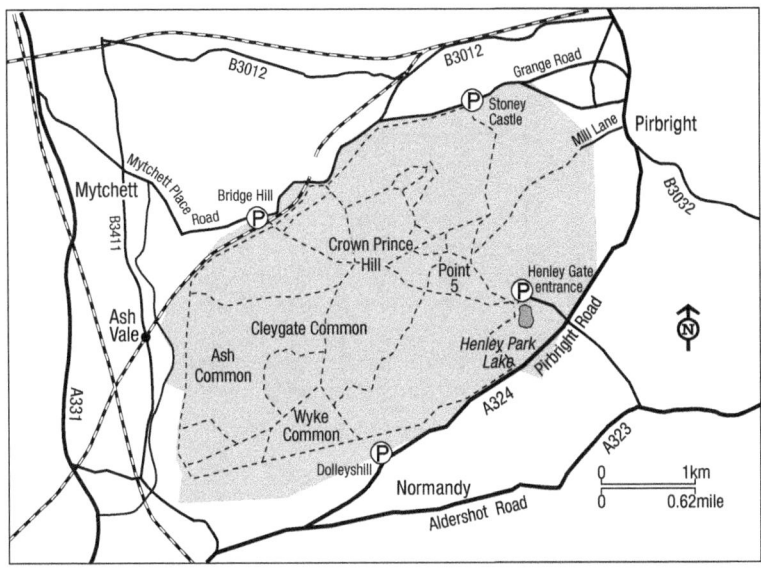

ACCESS

On the north side, there is room for limited off-road parking at Bridge Hill at (SU 904546), 1.6km east of Mytchett. The Henley Gate entrance on the west side is accessed via the track north of the A324 at SU 934538. Park sensibly in a lay-by near the gate. A nice, looped walk can be done from this point, heading west out to Crown Prince Hill and passing Henley Park Lake on the way back. Other parking options are at Dolleyshill (SU 918521) and Stoney Castle (SU 928558).

Bus service 28 from Guildford to Woking stops at Pirbright village, where the ranges can be accessed from the end of Mill Lane. Ash Vale station is situated close to the west side of the ranges – it's about a 15-minute walk.

FACILITIES: Most of the paths are sandy and not suitable for wheelchair or push-chair users. All of the parking spots mentioned are free. Other facilities can be found in Ash and Pirbright.

CALENDAR

All year: Woodlark, Dartford Warbler, Stonechat, Crossbill in some years.

April–August: Cuckoo, Nightjar, Hobby, Redstart, Willow Warbler, Tree Pipit.

October–March: Great Grey Shrike and occasional Hen Harrier.

32 CHOBHAM COMMON

OS Explorer 160
OS grid refs: SU 980643 and SU 969658
Postcode: KT16 0ED

HABITAT

At some 655ha, Chobham Common – an SSSI – is an impressive and important area of lowland heathland. Most of the site, which is bisected by the M3, comprises open heath with scattered pine and birch. The highest part is along Staple Hill, with the lower-lying Gracious Pond area boggy in parts. Most of Chobham Common is managed by Surrey Wildlife Trust and has a particularly impressive array of invertebrate fauna – including Silver-studded Blue and Grayling butterflies.

SPECIES

Chobham is a popular spot for Nightjar, with more than 50 churring males recorded in recent summers. The typical suite of heathland breeders is also found here, including good numbers of Woodcock, Dartford Warbler, Stonechat and Woodlark. Unlike on the south-west Surrey commons, Tree Pipit is uncommon, while Redstart is absent in most years. Cuckoo is present in the summer along with, occasionally, Hobby. Chobham is the most reliable Surrey site for Meadow Pipit in the summer, which is now a rare breeder in the county.

Winter is typically quiet, though Great Grey Shrike and Hen Harrier are not out of the question. Crossbill and Brambling are possible in good years. Chats move through on passage. Down the years there have been reports of Long-eared Owl from the southern half of the common and, although this county rarity is very unlikely to be encountered, it's worth bearing in mind should you be visiting after dark.

TIMING

Spring and summer are by far the best time to visit. It's worth trying to get to Chobham early if you can – this site can get very busy indeed with dog walkers, drone and model aircraft fliers all using the area. The M3 can create a noisy backdrop, too, so avoiding rush hour times is recommended.

ACCESS

Three car parks along Staple Hill provide access to the southern half of the common, including the main one at SU 973648. There is also a lay-by near Gracious Pond with room for several cars (SU 988639). For the northern half use one of the two car parks along Chobham Road, including at Ship Hill (SU 964654).

Longcross station is only a few minutes' walk north-east from the northern half of the common. Bus service 73 from Woking stops at Bowling Green Road in Chobham.

FACILITIES: Most of the paths are sandy and not suitable for wheelchair or pushchair users. All of the parking spots mentioned are free.

CALENDAR

All year: Woodlark, Dartford Warbler, Stonechat, Crossbill in some years.

April–August: Cuckoo, Nightjar, Hobby, Tree Pipit.

October–March: Occasional Hen Harrier and Great Grey Shrike.

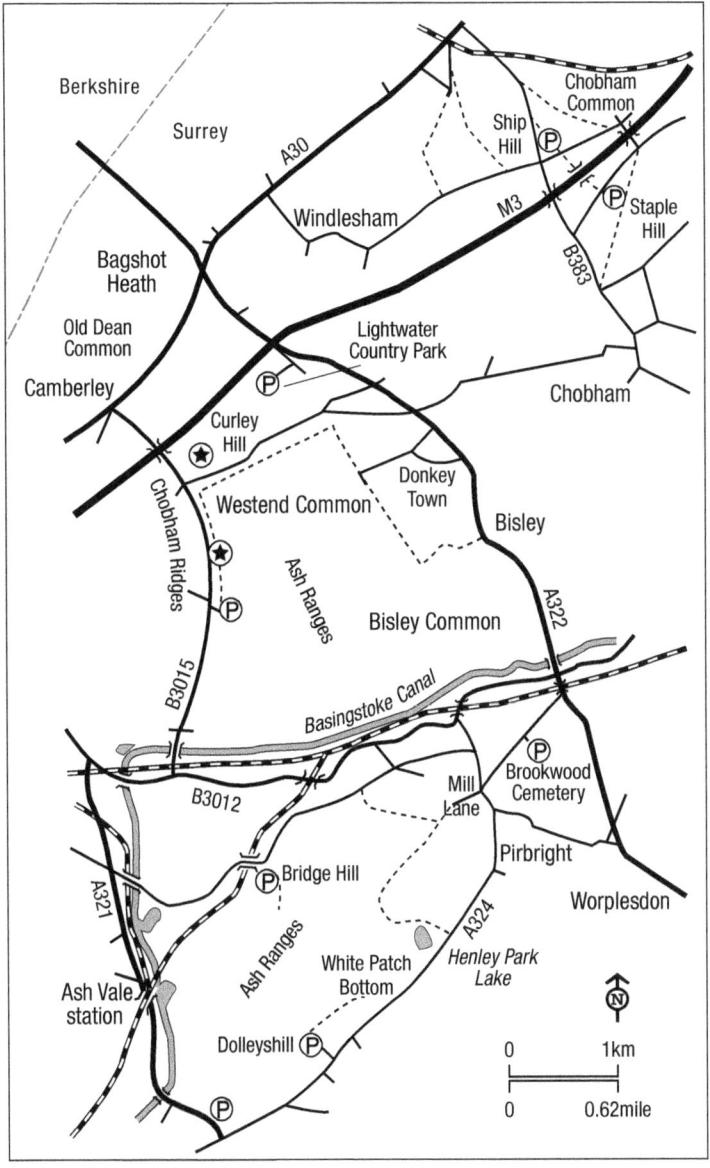

33 OCKHAM AND WISLEY COMMONS AND BOLDERMERE

OS Explorer 145
OS grid refs: TQ 082588, TQ 080590 and TQ 076583
Postcodes: GU23 6QS and GU23 6PY

HABITAT

Ockham and Wisley Commons once formed one site before they were split into four pieces by the M25/A3 junction. The Surrey Wildlife Trust has done an excellent job of restoring both commons to their full lowland heathland value, using grazing and clearance to ensure Scots Pine and birch don't take over the site. There are some marshy areas on Wisley Common. Boldermere is a large lake situated to the north-east of Ripley on the east side of the A3 to the south of junction 10 of the M25. Fishing was once allowed, but it is now a nature reserve. This is an excellent site for dragonflies and damselflies, with 20 species recorded. South of Boldermere lies Wisley Airfield, the site of a wartime airstrip which is now disused.

SPECIES

The two commons are best visited in spring and summer, when breeding species include Redstart, Dartford Warbler, Stonechat, Woodcock and Nightjar. The latter species has increased steadily here and is often encountered only a short walk from the car parks. Spotted Flycatcher breeds in most summers, with Ockham Common the most likely bet – suitable-looking habitat near Semaphore Tower and Currie's Clump is recommended as a good place to check. Winter is quiet, although the odd flock of Crossbills may be encountered.

Common Terns nest on rafts at Boldermere. Great Crested Grebe and Kingfisher are also present, while Hobby is a regular sight in the summer, with birds attracted to the plentiful dragonflies. Decent numbers of Coot and Wigeon are present in the winter.

Wisley Airfield has a decent population of Skylarks and is a good site for Wheatear and Whinchat on passage. It has attracted Stone-curlew before.

TIMING

Wisley Common can be busy at weekends. Nightjars usually become active after sunset. All sites are best visited in the early morning – the noise from the A3 can be loud.

ACCESS

The nearest train station (West Byfleet) is 6.5km away. However, the 715 Kingston to Guildford Stagecoach bus stops along the pedestrian walkway of the A3, between the two commons. It runs hourly, Monday to Saturday.

If coming to Ockham Common and Boldermere by car, there are two car parks along Old Lane (accessible off the A3): Boldermere Car Park (TQ 078586) and Pond Car Park (TQ 079583). Wisley Common is best accessed from Wren's Nest Car Park (TQ 065589).

None of these sites are suitable for wheelchair users or those with limited mobility.

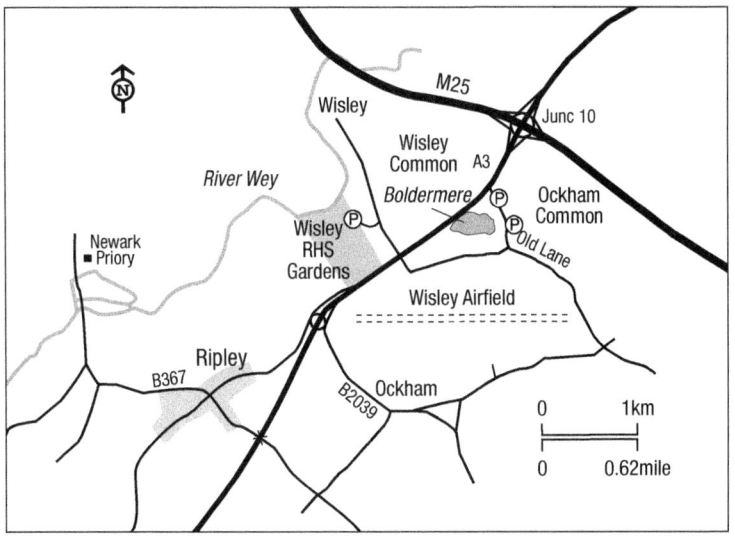

FACILITIES: Baby-changing facilities and toilets, including accessible toilets, are found at Boldermere Car Park, where Ockham Bites café is open daily from 8:00 am until 4:00 pm.

CALENDAR

All year: Kingfisher, Stonechat, Dartford Warbler.

October–March: Wildfowl on Boldermere, Crossbill.

April–September: Woodcock, Nightjar, Hobby, Redstart, Wheatear and Whinchat on passage.

34 PAPERCOURT AREA

OS Explorer 145
OS grid ref: TQ 035569
Postcodes: GU23 6JX and GU23 7ES

HABITAT

Situated to the south-east of Woking along the Wey Valley, the Papercourt area can be split into four different sectors: Papercourt Gravel Pits, Papercourt Water Meadows, Papercourt Marshes and Prews Farms. There are several old gravel pits in the area but the most productive are the two immediately north of Send Marsh village. The largest and deepest one is used for sailing and is the best for birds; a smaller, undisturbed pit lies to the south next to the sailing club building. Papercourt Water Meadows is an area of wet meadows and grassland between Broadmead and Papercourt Lock, flanked by Broadmead Cut and the River Wey.

Papercourt Marshes is run by the Surrey Wildlife Trust and access is restricted to permit holders. The habitats found within the reserve consist of a mixture of open standing water, shallow scrapes, reedbeds, marsh and wet woodland. Finally, some decent arable farmland remains north of Send village at Prews Farm.

SPECIES

Papercourt holds interest year-round, though it is perhaps the winter that offers the most excitement. The gravel pits are at their busiest with birds at this time with *Aythya* gatherings including decent numbers of Pochard. Goosander is less regular these days and prefers the quieter, smaller pit. Cold weather or periods of wildfowl passage have produced Ferruginous Duck and Smew in recent years and, going further back, the pits have rich form for rare ducks. Dabbling ducks prefer Papercourt Marshes, where Water Rail and Cetti's Warbler can be encountered. The water meadows are well known as the 'owl capital' of Surrey, with Barn and Little Owls both resident. Papercourt is probably the best site in the county for the former species, which often quarter over the meadows at dusk. Scanning the area from the banks of the Wey is a good tactic. Little Owl are more likely to be heard calling, but may be viewable in a favoured oak. The star attraction is Short-eared Owl, but the presence of this popular species varies winter on winter. In some years as many as six birds have been present, sometimes affording wonderful views. Conversely, this species can be absent for a few years in a row – it's worth checking the Surrey Bird Club sightings page or BirdGuides for the latest news. Depending on the water levels, Jack Snipe may be encountered as well. Peregrine nests in Woking town centre and visits frequently, often perching on the pylons for long periods. Scarce – but possible – visitors include Hen Harrier, Merlin and Great Grey Shrike. Winter stubble at the farmland along

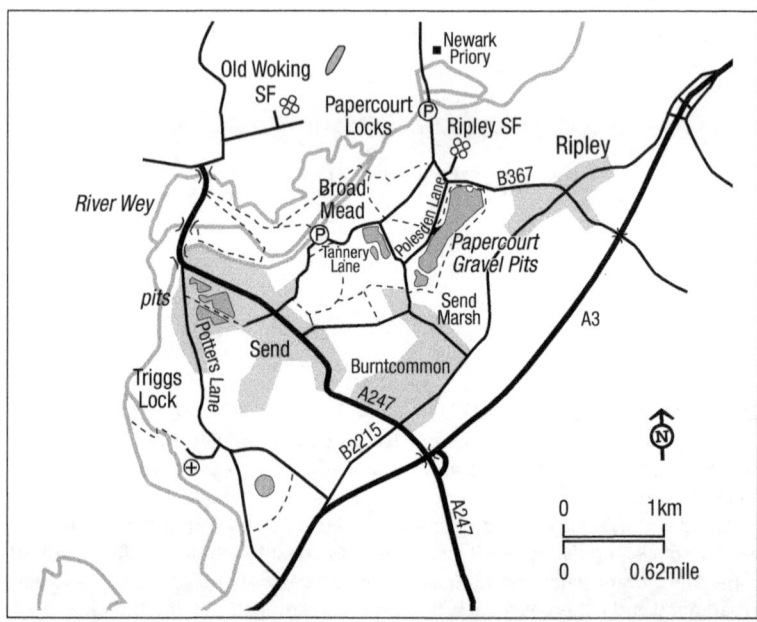

Tannery Lane can be good for seed-eaters including Brambling. Occasionally White-fronted Geese have occurred on the fields and at Papercourt Farm.

Among the breeding species in the area are Cetti's, Garden, Reed and Sedge Warblers, Stonechat and Reed Bunting. There is suitable habitat for Lesser Spotted Woodpecker along this stretch of the Wey, but few recent records. Occasionally Little Ringed Plover or Lapwing will attempt to breed, with mixed results. Great Crested Grebe nest on the pits, where Common Sandpiper is by far the most likely wader to drop in on passage; Green Sandpiper is more likely on the meadows. Passerine migration can be good out on the water meadows, with Wheatear, Spotted Flycatcher, Whinchat and Yellow Wagtail relatively regular, especially in the autumn. There's an outside chance of reeling Grasshopper Warbler in spring, though birds rarely linger. Black Redstart may be encountered around some of the many derelict buildings in the area as well. Marsh Harrier will sometimes pass through the meadows on passage and the pits can tempt down a migrant Osprey. Papercourt has form for rarities, too, with Pallid Harrier at the water meadows and Caspian Tern at the pits being the stand-out records.

TIMING

There is little sailing activity during the winter and on weekdays, which are the best time to visit the gravel pits. The water meadows can get fairly busy – especially along the towpath – at weekends and with dog walkers on weekday mornings. For owls, a dusk visit between November and March is best.

ACCESS

Access by public transport is difficult with Woking train station the closest at 3km away. Parking for the gravel pits is fairly limited along Polesden Lane, either at the small lay-by at TQ 039563 or a few metres up near the gate (without blocking it) at TQ 040563. From there, obvious paths circumnavigate the main pit, passing the smaller one as well.

For Papercourt Water Meadows, there is a car park at TQ 030563, by the small business park along Tannery Lane. From the car park, take the footpath to the north-west (TQ 030563) over two footbridges and onto Broad Mead. The entire area can be explored from here.

Papercourt Marshes requires a permit, which should be sought from the Surrey Wildlife Trust. Limited viewing can be achieved along Tannery Lane. For access to the farmland, use the Papercourt Water Meadows car park and take footpaths south into the fields, or scan from Tannery Lane.

FACILITIES: The nearest facilities are in Ripley and Send.

CALENDAR

All year: Kingfisher, Barn and Little Owls, Peregrine, Cetti's Warbler, Stonechat, Reed Bunting.

October–March: Wildfowl including occasional scarcities, Water Rail, Jack Snipe, Short-eared Owl, Brambling.

April–September: Little Ringed Plover, *Acrocephalus* warblers, Wheatear and Whinchat on passage.

35 BURPHAM WATER MEADOWS AND STOKE LAKE

OS Explorer 145
OS grid refs: TQ 005517 and TQ 006527
Postcodes: GU1 1QA and GU4 7NA

HABITAT

These two sites sit side-by-side in the north-east corner of Guildford, between Slyfield and Jacobs Well. Here, having passed through a narrow gap in the North Downs, the River Wey has spilled out into an area of water meadows and marshland. Stoke Lake was formed in the 1970s when a gravel pit was dug out for road-making materials used for the rerouting of the A3, which now serves as a boundary flanking the south-west side of the lake. The Stoke area is known as Riverside Nature Reserve by Guildford Borough Council, who have installed a short boardwalk and bird screens in the marsh on the north-west side of the lake. Sadly these are sometimes partially closed due to vandalism. Burpham Water Meadows, including Burpham Court Farm, is the area north of Stoke between the Wey and Clay Lane.

SPECIES

Stoke Lake suffers from considerable disturbance, especially from dog walkers, and the selection of waterbirds found here is not what it once was. Great Crested Grebe and Kingfisher should be expected, with the reedbeds holding Cetti's, Sedge and Reed Warblers and Reed Bunting in the summer. A raft for Common Terns is sometimes situated on the lake but breeding efforts are sporadic – and vandalism has meant rafts have been absent recently. Water Rail is present at this time of year, when Lesser Redpoll and Siskin can be found in the alders flanking the lake. A halcyon period for this site saw a spate of county rarities in the 1990s but unusual birds are not to be expected – wader passage is usually limited to Common and Green Sandpipers, and perhaps an Oystercatcher or Redshank.

Burpham Water Meadows would probably be one of the most productive stretches of the Wey in the county if access was more straightforward. Winter is the most fruitful period, when a flock of Wigeon can sometimes reach a couple of hundred birds, with a few Shoveler and plenty of Teal and geese also frequenting the site. The geese are worth sifting through for more unusual species, which have previously included Pink-footed Goose. A small Lapwing flock is usually present as well. Winter through to spring passage can attract more unusual wildfowl, such as Shelduck (which can sometimes linger well into spring) or Pintail, which will use the pools. Barn and Little Owls can be encountered year-round, but are unpredictable at best. Little Ringed Plovers pass through in most years.

Partially bisecting the two sites is Slyfield Industrial Estate, which can attract large numbers of loafing gulls in the winter. Caspian and Yellow-legged Gulls have both been recorded on the roofs of some of the buildings, but the former species is still very rare in this part of Surrey. The gulls bathe in both Stoke Lake and on Burpham Court Farm.

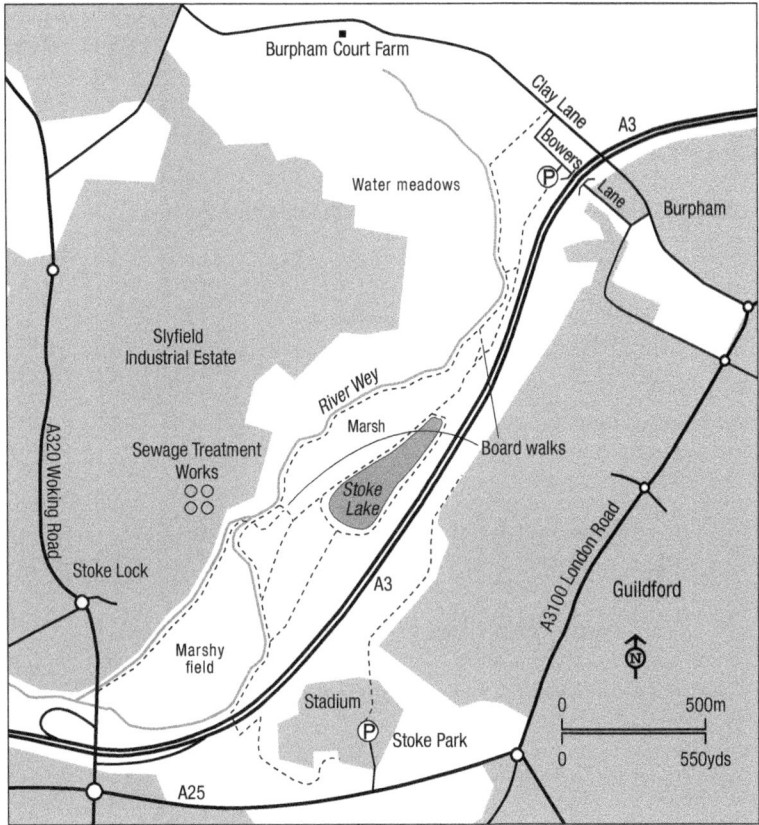

TIMING

Early mornings are important for Stoke Lake, which sees heavy footfall and a large number of dogs. Burpham Water Meadows is best early on or late in the day but can be productive at any time. Heavy winter rain and subsequent higher water levels can increase wildfowl numbers.

ACCESS

The easiest access is by car, with a free car park at the end of Bowers Lane at TQ 010526. From here, walk south-west (flanking the A3 to your left and pylons to your right) before you get to Stoke Lake. You can circumnavigate the lake. The boardwalk and bird screens are in the marsh on the north-west side of the lake (TQ 004517). The scrape in the meadows by Stoke Lock is always worth scanning in winter and at passage times.

For Burpham Water Meadows, walk north-west from the car park along Bowers Lane. You can turn left down the River Wey towpath and follow it until it bends round past some allotments. From here, you can scan part of the meadows. The best view is had from the busy Clay Lane, which is accessed by turning left out of Bowers Lane. It is not possible to access the meadows, which means plenty of birds probably go undetected.

No train stations are within realistic walking distance, but the 18 bus service runs from London Road station to Sainsbury's in Burpham, from where it is a 12-minute walk to Riverside Park car park. National cycle route 223 runs through the site and alongside Stoke Lake.

The paths and boardwalk at Stoke offer good access for visitors with limited mobility, though some areas can get muddy in the winter.

FACILITIES: There are picnic benches at Stoke Lake. Otherwise the nearest facilities are at the Sainsbury's in Burpham.

CALENDAR

All year: Great Crested Grebe, Barn and Little Owls, Kingfisher, Cetti's Warbler.

October–February: Wigeon, Lapwing, a chance of scarce gulls at Slyfield.

April–September: Occasional passage waders and ducks, Common Tern, Reed and Sedge Warblers.

OTHER SITES IN NORTH-WEST SURREY

E1 NORMANDY AND WANBOROUGH FARMLAND

The northern slopes of the North Downs here were one of the last vestiges for Corn Bunting in Surrey; Yellowhammer still persists around Manor Farm (SU 936 494) and Barn Owl breeds nearby.

E2 BROOKWOOD CEMETERY

Brookwood Cemetery (SU 956563) was originally part of a wider heathland – areas of heath remain, with much of the site parkland, grassland and open woodland. Woodcock are among the breeding species and Spotted Flycatchers pass through on passage.

E3 PIRBRIGHT RANGES

This vast MOD site (some 2,710ha), including Bisley and Westend Commons, is largely off-limits due to military training. A rewilding programme on some of the inaccessible land includes a Red Deer introduction. Brentmoor Heath (SU 938610), managed by Surrey Wildlife Trust, has public access.

E4 BAGSHOT HEATH AND OLDDEAN COMMON

The heathland at Olddean Common (SU 893625) is on the Berkshire border and includes Wishmoor Bottom, the site of a Parrot Crossbill flock in the 2017–18 winter. Bagshot Heath (SU 892632) has the typical range of Surrey heathland species. Lightwater Country Park (SU 914621) is an extension of Bagshot Heath on lower ground to the south, with nature trails and an information centre.

E5 VIRGINIA WATER

A large waterbody, half of which is in Surrey, situated in Windsor Park. Decent numbers of winter wildfowl can include scarcities, with Common Scoter, Ring-necked Duck and Smew recorded before. Osprey and Common Sandpiper pass through on passage. Park at SU 980689.

E6 HORSELL COMMON AND FAIROAKS AIRFIELD

Horsell Common (TQ 005605) is another good site for Dartford Warbler and Nightjar. Fairoaks Airfield has a healthy population of Skylarks and form for migrant chats.

E7 WISLEY SEWAGE FARM

It's possible to circumnavigate the sewage works at Wisley (TQ 059596) via a public footpath, which overlooks the open beds (Grey Wagtail and occasional Green Sandpiper) and passes the Wey (good for Kingfisher). A few Chiffchaffs winter at the works.

E8 OLD WOKING SEWAGE FARM

Situated on the north side of the River Wey in the Papercourt area (TQ 029573), this large sewage farm has a certain wildness to it. Waders and wildfowl are unlikely these days, but owls are a real possibility and insectivores are attracted to the works. Access from Carters Lane.

E9 WHITMOOR COMMON

Another Surrey Wildlife Trust site (SU 984536) near Guildford that holds breeding Dartford Warbler and Nightjar. Britten's Pond (SU 990531) is home to common waterbirds and Firecrest can be found in the holly around the water.

E10 FARNHAM PARK

This medieval deer park on the outskirts of Farnham (SU 841480) has rolling grassland, parkland and ponds. Skylarks and Little Owls breed, while Kingfisher visit the waters. Brambling sometimes visits in the winter, as does, rarely, Hawfinch.

NORTH EAST SUSSEX

MAIN SITES

36 Weir Wood Reservoir
37 Ashdown Forest
38 Broadwater Warren RSPB
39 Bewl Water
40 Brede High Woods and Powdermill Reservoir
41 Dallington Forest and Penhurst Lane
42 Sheffield Park and Garden
43 Chailey Common

OTHER SITES

F1 Birchden Wood
F2 Hargate Forest
F3 Eridge Park and Whitehill Wood
F4 Wadhurst Park
F5 Cedar Farm
F6 Flatropers Wood
F7 Darwell Reservoir
F8 Selwyns Wood
F9 Uckfield Sewage Works
F10 Fletching Mill Farm
F11 Brickfield Meadow

North East Sussex, as defined in this book, covers the area north of a line from Uckfield east to Battle, and then north-east to Peasmarsh. As a result, all the sites included here are a minimum of 10km inland. Because of the diamond shape of East Sussex, most of the sites covered in this region are nestled in the most northerly corner of the county, fairly close to the border with West Sussex. As a result, some sites that are geographically relatively nearby to one another have been divided between two regions for the purposes of this guide.

East Sussex has almost twice as much coastline as West Sussex, so it is hard to ignore its dominance from a birding perspective. That said, this more northerly, inland section of the county feels far removed from the sea in terms of habitat. Despite two major rivers (the Rother and the Medway) flowing through the area – and both the Ouse and the Medway beginning their journeys in or near Ashdown Forest – there is no real wetland habitat to speak of. There are still places to enjoy some open water birding, though, with Weir Wood Reservoir and Bewl Water attracting impressive numbers of wildfowl and gulls, as well as passing waders and lingering Ospreys.

Generally though, the landscape of North East Sussex is dominated by the mixed woodlands, rolling hills, valleys and a patchwork of farmland characteristic of the High Weald. There are also significant areas of lowland heath – most notably the great swathe of Ashdown Forest, which is the largest single tract of heathland, semi-natural woodland and valley bog remaining in south-east England. The sandstone ridges that run east to west across this part of the county are cut through with myriad streams known as ghylls, some of which go on to feed the Cuckmere and Ouse rivers. The dominance of the Wealden landscape in this region is reflected in

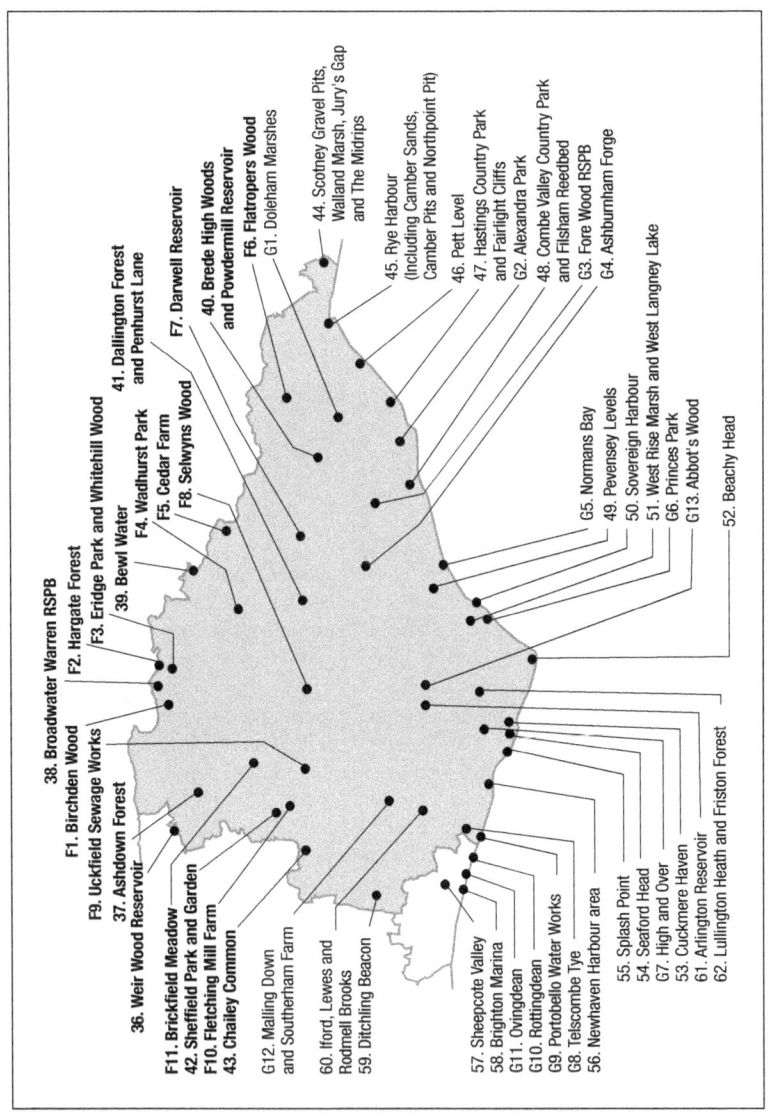

38. Broadwater Warren RSPB
F1. Birchden Wood
F9. Uckfield Sewage Works
37. Ashdown Forest
36. Weir Wood Reservoir

F2. Hargate Forest
F3. Eridge Park and Whitehill Wood
39. Bewl Water
F4. Wadhurst Park
F5. Cedar Farm
F8. Selwyns Wood

41. Dallington Forest and Penhurst Lane
40. Brede High Woods and Powdermill Reservoir
F6. Flatropers Wood
F7. Darwell Reservoir
G1. Doleham Marshes

44. Scotney Gravel Pits, Walland Marsh, Jury's Gap and The Midrips
45. Rye Harbour (Including Camber Sands, Camber Pits and Northpoint Pit)
46. Pett Level
47. Hastings Country Park and Fairlight Cliffs
G2. Alexandra Park
48. Combe Valley Country Park and Filsham Reedbed
G3. Fore Wood RSPB
G4. Ashburnham Forge

F11. Brickfield Meadow
42. Sheffield Park and Garden
F10. Fletching Mill Farm
43. Chailey Common

G12. Malling Down and Southerham Farm

60. Iford, Lewes and Rodmell Brooks
59. Ditchling Beacon

G5. Normans Bay
49. Pevensey Levels
50. Sovereign Harbour
51. West Rise Marsh and West Langney Lake
G6. Princes Park
G13. Abbot's Wood

52. Beachy Head

57. Sheepcote Valley
58. Brighton Marina
G11. Ovingdean
G10. Rottingdean
G9. Portobello Water Works
G8. Telscombe Tye
56. Newhaven Harbour area

55. Splash Point
54. Seaford Head
G7. High and Over
53. Cuckmere Haven
61. Arlington Reservoir
62. Lullington Heath and Friston Forest

119

the fact that three of the main sites featured are heathlands and seven of the other sites are woodlands. There are some urban areas, though. Indeed, at its most northerly point, the top of the 'diamond' of East Sussex is only 30 miles south of London, and Gatwick Airport is just a stone's throw across the border into West Sussex.

The draw of the coast means inland East Sussex is one of the less-birded regions in Sussex. Nonetheless, there is still plenty on offer here, especially given its proximity to the conurbations of Crawley, Crowborough, East Grinstead and Royal Tunbridge Wells – the latter just across the border in Kent. Although somewhat lacking the overall diversity of species found in more coastal or wetland locales, the woodlands and heathlands in this area offer some of the best opportunities in the region for encountering scarce breeding species such as Goshawk, Honey Buzzard and Hawfinch, as well as heathland specialists like Woodlark, Nightjar and Dartford Warbler.

36 WEIR WOOD RESERVOIR

OS Explorer 135
OS grid ref: TQ 393347
Postcode: RH19 4HP

HABITAT

Situated just over 3km south of East Grinstead, Weir Wood Reservoir is a 152ha SSSI, built in 1954 and now owned by Southern Water. The 32.6ha 'tail end' at the western end is designated as a Local Nature Reserve and managed by the Friends of Weir Wood conservation group in conjunction with East Sussex County Council, who have responsibility for the care of the whole SSSI. There is a bird hide by the small free car park, accessed from Legsheath Lane, which offers good views across the western end of the reservoir, which is an LNR in its own right. The water level here can drop quite low in the summer months, which can prove very attractive to waders.

Over 250 bird species have been recorded here but also 32 species of butterfly, including Purple Emperor, and 200 plant species including the scarce Shoreweed and Orange Foxtail grass.

SPECIES

The open waters of the reservoir combined with the sheltered bays and inlets, especially at the western end, provide many opportunities for wildfowl to feed and breed. Common species – including Egyptian Goose and Mandarin – are resident, and they're joined by Shoveler, Wigeon, Gadwall and Teal outside the breeding season. Scarcer ducks sometimes occur, such as Garganey, Scaup, Common Scoter and Goldeneye. Great Crested and Little Grebe are present year-round. Common Terns breed on the rafts at the western end, while Grey Heron and Cormorant breed in the trees by the shore. Rarer passage gulls and terns can pass through too, as with any similarly sized waterbody, with Little Gull, Arctic or Black Terns all possible in spring or late summer. There have been records of Sandwich Tern and Kittiwake here too, as well as Red-throated and

Great Northern Divers. Kingfisher is always a possibility, usually in the second half of the year.

The western end is generally the most productive for waders in late summer and autumn, while the eastern end, on or near the dam, is better in spring, with birds sometimes just passing straight through. The likes of Little Ringed Plover, Whimbrel, Green and Common Sandpipers and Greenshank are regularly recorded, while species such as Black-tailed Godwit, Bar-tailed Godwit, Dunlin, Wood Sandpiper, Curlew Sandpiper, Sanderling and Little Stint also turn up from time to time. A Long-toed Stint visited for one day in 2011 and remains the only Sussex record. Snipe are present in the winter too, in varying numbers.

Away from the waterbirds, Weir Wood also offers a good selection of passerine interest in the surrounding woodland and scrub, including Marsh Tit, Goldcrest, Nuthatch and Treecreeper year-round, and Chiffchaff, Blackcap, Garden Warbler, Whitethroat and Cuckoo in spring and summer. The dam at the eastern end often holds a few Pied and Grey Wagtails, with a chance of White Wagtail, Yellow Wagtail or Wheatear here in spring and autumn. The reedbed west of Whillet's Bridge holds breeding Reed Bunting and Reed Warbler, as well as a large roost of Reed Bunting and Starling in the winter. The abundance of alders around the perimeter of the reservoir is very attractive to Siskin and Lesser Redpoll in the winter. Just west of the picnic area (TQ 398352) on the north bank is a small Hornbeam wood which sometimes attracts Hawfinch then too. Swifts, Swallows and Martins can be seen through the summer months, but especially in the spring when they have just arrived and all four species may congregate here in their hundreds. These can often attract the attention of a passing Hobby, which frequently drop in to feed in late spring.

Weir Wood is perhaps one of the best inland sites in the South-East for Osprey, with multiple spring and autumn records every year and sometimes a non-breeding bird lingering in the area through the summer. With Ashdown Forest just down the road, it is always worth bearing in mind the possibility of Honey Buzzard from May to September; scanning from the northern side of the reservoir sometimes produces the goods, but patience and a telescope is required. Likewise, Goshawk is an ever-more frequent sight here as the species continues to increase across the High Weald.

Aside from the Long-toed Stint, Weir Wood has attracted an impressive selection of scarce and rare species in recent years, including Purple Heron, Red-footed Falcon, Black-winged Stilt, Bee-eater, Red-rumped Swallow and Bonaparte's Gull.

TIMING

Winter is best for the greatest gatherings of wildfowl and late summer to autumn for the chance of unusual waders. The hide in the south-western corner faces north, so provides suitable viewing conditions of the western end of the reservoir at any time of the day.

ACCESS

The closest train station is East Grinstead some 3km to the north. Bus route 84 (Metrobus) from East Grinstead stops at Blackland Farm, just a 10–15-minute walk from Legsheath Lane. The main car park on Legsheath Lane (TQ 382341) provides the best viewing of the western end though this area can also be viewed from Whillet's Bridge (TQ 380345), where one can also scan Whillet's Pool and reedbed to the west side of the road. Accessible parking is located at TQ 385348 and the 'Millenium Walk' trail runs both east and west from here along the

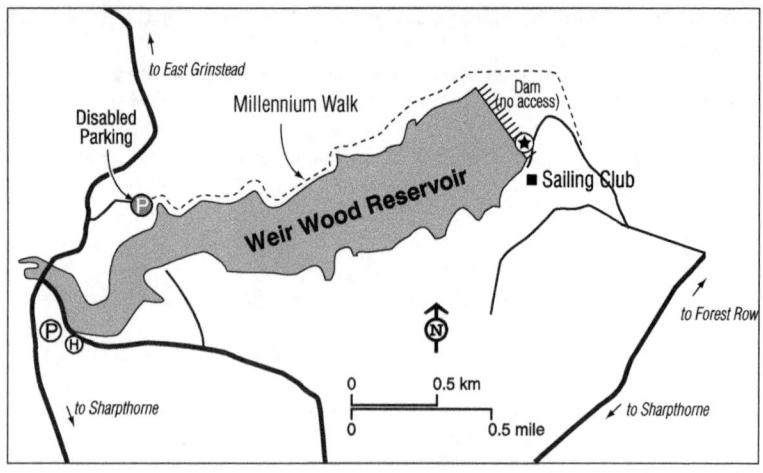

northern side of the reservoir, offering good views across the water at various points (TQ 387348 and TQ 393348, for example). Be aware, though, that this path can get very muddy in the winter and is not suitable for wheelchairs or push-chairs. Parking is also available at the Water Works (TQ 407354) and the Sailing Club (TQ 408351). The dam end can be viewed from either by the Sailing Centre or the public footpath running along the northern side near Botley Wood. Please do not attempt to access the dam itself or the shore of the reservoir, both of which have no public access, and do not walk beyond the bailiff's office. Members of the Friends of Weir Wood enjoy extended access to certain areas normally not open to the public.

FACILITIES: There is a free car park at the western end of Legsheath Lane offering easy, wheelchair-friendly access to a bird hide. Other facilities are situated in East Grinstead or Forest Row.

CALENDAR

All year: Egyptian Goose, Mandarin Duck, Pochard, Great Crested Grebe, Lapwing, Little Egret, Goshawk, Red Kite, Kingfisher, Raven, Marsh Tit, Grey Wagtail, Reed Bunting.

April–June: Garganey, Cuckoo, Little Ringed Plover, Common Sandpiper, Common Tern, rarer terns and Little Gull, Osprey, Honey Buzzard, Hobby, Sedge and Reed Warbler, hirundines, Garden and Willow Warblers.

July–September: As for April–June plus the possibility of more unusual waders, Great Black-backed Gull, Black Tern, Lesser Whitethroat and Grasshopper Warbler in scrubby areas.

October–March: Possibility of Brent Goose, Shoveler, Gadwall, Wigeon, Pintail, Teal, Water Rail, Snipe, Peregrine, Brambling, Hawfinch, Lesser Redpoll, Siskin.

37 ASHDOWN FOREST

OS Explorer 135
OS grid ref: TQ 431323
Postcode: RH18 5JP

HABITAT

At over 2,500ha, Ashdown Forest is one of the largest publicly accessible areas of land and the single largest contiguous block of heathland in South East England. It is designated as an SSSI and an EU Special Area of Conservation. Sitting atop the sandy ridge of the High Weald AONB, the climate is quite different to some other heathland sites in the south-east. Indeed, some of the higher areas can feel somewhat more reminiscent of northern or Scottish moorland sites, and visitors are often surprised to find such a wild and open landscape just 30 miles or so south of London.

The history of Ashdown Forest goes back thousands of years, with archaeological evidence of human activity here dating back to the Middle Stone Age. More recently, the forest was enclosed and protected by royal decree as a hunting ground for the King. This, combined with the advent of more recent legislation relating to Common Rights, means the forest has been protected from any cultivation, encroachment or development for centuries. It's important to note that 'forest' in this context historically meant an area designated as royal land, rather than the more modern connotation of dense tree cover.

Most famous for being the landscape that inspired A.A. Milne's Winnie-the-Pooh stories, Ashdown Forest is now a rich mosaic of dry heath, bog, wood pasture and some impressive stands of oak, birch, pine and beech woodland, all of substantial wildlife value. Aside from the birdlife, the forest supports nationally and internationally important communities of rare plants such as Marsh Gentian and Round-leaved Sundew and invertebrates including Wasp Spider and the Silver-studded Blue butterfly, as well as reptiles like Adder and Common Lizard.

SPECIES

As the largest area of unbroken heathland in the south-east of England, Ashdown supports some impressive numbers of various bird species, with all the heathland specialists present here in abundance, such as Nightjar, Woodlark, Tree Pipit, Dartford Warbler and Stonechat. The Forest is particularly good for Redstarts, especially the Sussex Wildlife Trust reserve at Old Lodge (TQ 462304) which supports several breeding pairs. Gills Lap (TQ 468319) is a particularly good area to encounter many of the trademark heathland species, and sometimes a Great Grey Shrike in the winter, although be aware it is also very popular with tourists due to the Winnie-the-Pooh connection!

From late January onwards, the somewhat bleak and barren heathland landscape begins to spring into life, with the song of Woodlarks drifting overhead and Dartford Warblers giving their hurried, scratchy warble from in the heather and gorse. Passerine migrants such as Wheatear, Whinchat and Spotted Flycatcher may be found in spring and autumn, the latter generally along the wooded edges. A dusk visit in May will usually produce the mechanical churring of Nightjars as well as roding Woodcocks patrolling their territories. Cuckoos can be heard almost anywhere in the Forest from April to June. In areas of denser vegetation and woodland are found Goldcrest, Great Spotted Woodpecker, Coal Tit and Marsh Tit throughout the year, joined by Chiffchaff, Willow Warbler and Redstart in the summer months. Lesser Spotted Woodpecker is still recorded in most years but is sadly becoming more and more elusive. Various finch species are present all year-round, such as Goldfinch, Linnet and Chaffinch, with the latter forming large flocks in the winter which can attract Brambling. Ring Ouzels often drop in on their southward migration in the autumn, with particular hotspots for the species being Old Lodge, Gills Lap and the Old Airstrip (TQ 422311). Wetter areas usually hold a few Snipe in the winter and there may sometimes be a Jack Snipe hidden away too, though these are generally very hard to find, especially without straying some distance from the footpaths.

Raptors are present in good numbers here, including now generally widespread species such as Buzzard, Red Kite, Kestrel and Sparrowhawk all year-round, Hobby in the summer and sometimes Hen Harrier and Merlin in the winter. An Osprey may be seen on occasion in spring or autumn, perhaps flying over to fish at the nearby Weir Wood Reservoir. Several pairs of Goshawk breed and there is a reasonable chance of Honey Buzzards in the summer. For both species, choose viewpoints with commanding views over areas of woodland. Two such viewpoints can be found near the Llama Park at TQ 428314, a short distance north across the A22 from the Long Car Park, and at Friends Clump (TQ 456289). Bear in mind that a telescope and patience are required to have any chance of success with both these species.

Ashdown Forest has an impressive list of scarcities and rarities to its name, most famously the Short-toed Snake Eagle which stayed in the area for a fortnight in June 2014. Somewhat less spectacular but still worth mentioning are Little Bunting, Parrot Crossbill, Wryneck and Bee-eater.

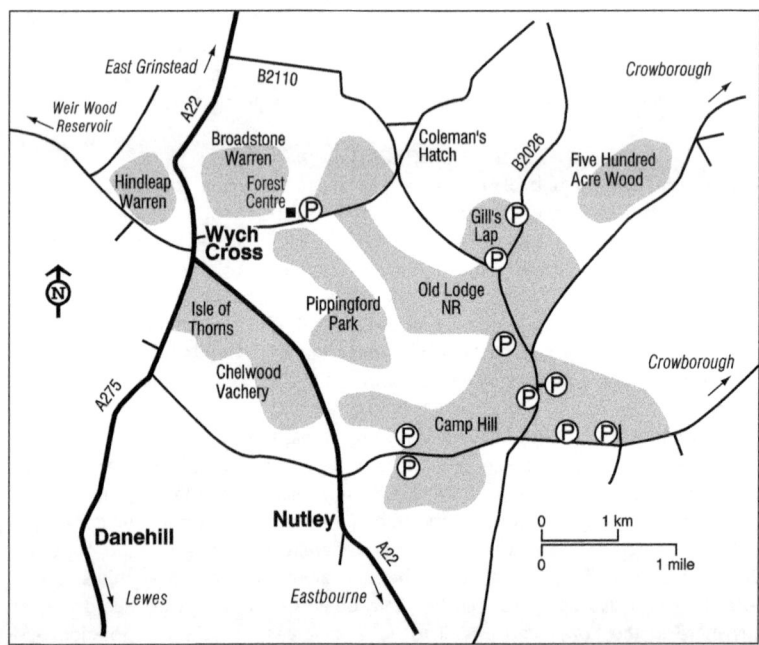

TIMING

Winter birding on any heathland can be tough going and Ashdown is no exception. While there will always be some interest all year-round, and the chance of a wintering raptor or Great Grey Shrike, for the best experience here, a spring or early summer visit is recommended. The whole forest is very popular with dog walkers and other recreational activities, so busy bank holidays are perhaps days to avoid. The sheer size of the site though means there is usually somewhere relatively quiet to explore away from the crowds.

ACCESS

There are 47 car parks dotted around the Forest, all easily accessed via the A22, A275, B2110, B2026, B2188 and various other roads which criss-cross the area. At the time of writing, these car parks have all just been made pay and display. The Ashdown Forest Centre at TQ 431323 on Colemans Hatch Road is a good place to start your explorations, as there are plenty of maps and information boards on offer here. The nearest railway station is in East Grinstead, around 7km north-west of the Forest Centre, and various bus routes operate from there to Forest Row and Wych Cross, including 261 (Compass Travel), 270 (Metrobus) and 291 (Metrobus).

FACILITIES: There are public toilets at the Forest Centre – including one with wheelchair access – and also a picnic area.

CALENDAR

All year: Woodcock, Goshawk, Red Kite, Tawny Owl, Lesser Spotted Woodpecker (easiest to find in Feb-March), Raven, Marsh Tit, Woodlark, Skylark, Firecrest, Stonechat, Grey Wagtail, Meadow Pipit, Lesser Redpoll, Crossbill, Siskin, Yellowhammer, Reed Bunting.

April–September: Cuckoo, Nightjar, roding Woodcock, Osprey, Honey Buzzard, Hobby, Willow Warbler, Garden Warbler, Spotted Flycatcher, Redstart, Whinchat, Tree Pipit.

October–March: Hen Harrier, Merlin, Great Grey Shrike, Ring Ouzel (in October), Brambling.

38 BROADWATER WARREN RSPB

OS Explorer 135
OS grid ref: TQ 554372
Postcode: TN3 9JP

HABITAT

Until the early part of the present century, 'The Warren' was a dense, monocultural conifer plantation lacking in any great wildlife value. Since 2007 though, the RSPB has been restoring the area back to its former heathland glory, by gradually removing areas of tree cover as well as breathing life back into the Decoy Pond and opening up glades in the surrounding deciduous woodland through coppicing. Half of the 182ha site has now been returned to heathland and supports breeding Woodlark for the first time in living memory, as well as increasing numbers of species such as Nightjar and Dartford Warbler. Dormice are present in the woodlands and Adders can sometimes be seen basking by the paths in spring. The ponds on site are now home to 28 species of dragonfly and damselfly including Golden-ringed Dragonfly and Black Darter. An evening visit can also sometimes produce Glow-worms on the woodland edges.

There are two signposted trails around the reserve, one short and another longer. These trails pass through all the habitats the reserve has to offer, including taking in iron-stained pools of the valley mire, which can be crossed via a boardwalk.

Just to the south lies the Sussex Wildlife Trust reserve, Eridge Rocks, with its impressive display of rocky sandstone outcrops and areas of ancient woodland.

SPECIES

Thanks to the work of the RSPB, Broadwater Warren is now home to a host of resident and migrant heathland specialist birds. From late winter through until the summer, a visit here will often produce the lilting sound of a singing Woodlark overhead, while evening visits from May should yield churring Nightjar – up to 10 males have been heard in recent years – and perhaps a roding Woodcock or hooting Tawny Owl. Tree Pipits breed too, so listen out for them performing their parachuting song flight from an exposed tree on the heath in spring, when you may also hear a Cuckoo. An abundance of insects, especially around the ponds, attract hirundines, in particular Swallow, and these in turn will often draw in a Hobby. Dartford Warblers have returned to the site, with upwards of 10 pairs breeding in recent years, and can be heard giving their scratchy warble in the heather from late winter onwards. Where there are Dartford Warblers, Stonechats won't be too far away, and this species is present here throughout the year.

The woodland areas offer a pleasing selection of passerine species including year-round Marsh Tit, Coal Tit, Goldcrest, Nuthatch, Treecreeper, Song Thrush, Chaffinch and Siskin, joined by Chiffchaff, Willow Warbler, Blackcap, Garden Warbler and Spotted Flycatcher in the summer – especially in the Northern Heath area, and sometimes Crossbill in good years for the species. Lesser Spotted Woodpeckers are still recorded here intermittently, although given the rate of decline of this species, encounters can sadly never be guaranteed. Ring Ouzel is sometimes recorded on passage.

The area around Decoy Pond, towards the south-western corner of the reserve, is good for Grey Wagtail and often attracts Snipe in the winter. Although very tricky to find, there is always the chance of a Jack Snipe among them. The abundance of invertebrate life here is also very attractive to Hobby, which has recently started breeding here, so look out for them hawking over the heath in summer. All the other expected raptor species for the region may also be encountered including Buzzard and Red Kite and, as with many other similar sites in inland Sussex, a displaying Goshawk is always possible in early spring.

TIMING
As with any heathland site, May to August is best to get the full selection of breeding migrant species, but winter can still offer interest in the form of roving flocks of passerines in the woodland areas and sometimes good numbers of finches in the pines.

ACCESS
The closest railway station is at Eridge. From there it is just over a 2km walk via Eridge Road, Park Corner Lane and Broadwater Forest Lane to reach the entrance gate to the reserve. Bus services 228/229 (Arriva) and 28/29/29A/29B (Brighton & Hove) both stop on the A26 close to the reserve. A car park with space for around 30 cars is located at TQ 558381 and can be accessed by following Broadwater Forest Lane then Fairview Lane north-west off the A26. There are also bicycle racks for 12 bikes here and an information board where you can get a map of the trails (which can also be downloaded online).

Many of the trails are uneven and can get muddy in the winter, so the full circular route is not advisable for wheelchairs and pushchairs. The boardwalk and the hard-surfaced first 200m of the short nature trail from the car park are suitable for all, and have been adapted with extra wide gates.

FACILITIES: Free car park open 9:00 am (10:00 am on Tuesdays) to 5:00 pm. The nearest toilets and other facilities are in Tunbridge Wells, around 5km away.

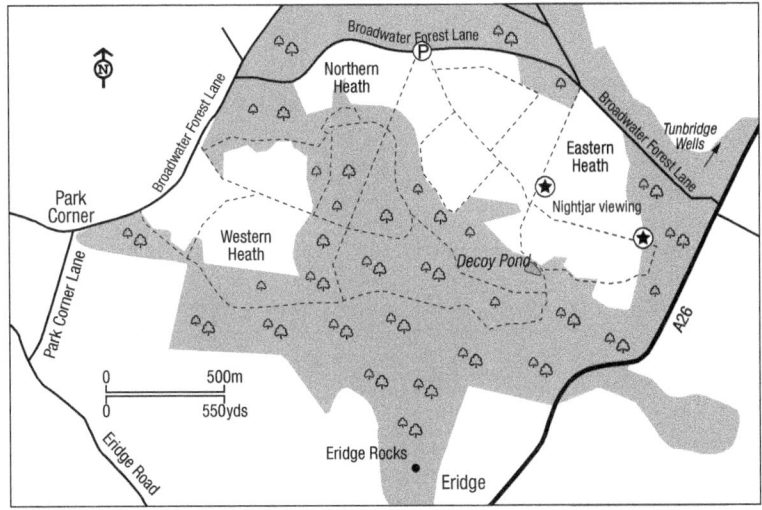

CALENDAR

All year: Woodcock, Goshawk, Tawny Owl, Lesser Spotted Woodpecker (March the best time), Raven, Marsh Tit, Woodlark, Dartford Warbler, Stonechat, Bullfinch, Lesser Redpoll, Crossbill, Siskin, Yellowhammer.

April–September: Cuckoo, Nightjar, Hobby, Swallow, House Martin, Willow Warbler, Garden Warbler, Whitethroat, Spotted Flycatcher, Tree Pipit.

39 BEWL WATER

OS Explorer 136
OS grid ref: TQ 680330
Postcode: TN3 8JH

HABITAT

Situated right on the Kent/East Sussex border, the 320ha Bewl Water is the biggest area of open water in the south-east of England, holding a colossal 31,000,000,000 litres of water when full. The reservoir's construction by Southern Water began in 1973 and was created by damming the River Bewl. Work took two years to complete and water from here now supplies households in Medway, Thanet and Hastings.

Since 2015 the reservoir has been increasingly managed as a tourist destination offering a host of leisure activities, including sailing, paddle boarding, windsurfing, swimming and fishing. The site now attracts 140,000 visitors a year. Inevitably, this causes a not insubstantial amount of disturbance to the birdlife and, coupled with the depth of the water and rather poor levels of marginal vegetation, means it perhaps isn't quite as attractive to birds as it otherwise might be. Nonetheless, the reservoir's sheer size and its many little inlets and bays do offer areas of sanctuary away from the recreational areas, and the overall site list now stands at over 200 species, helped in part by its place in the surrounding landscape.

Set in the heart of the High Weald, in an Area of Outstanding Natural Beauty, Bewl Water is surrounded by a mosaic of woodland, scrub and farmland, and part of the reservoir itself is jointly managed as a nature reserve with Sussex Wildlife Trust. Also nearby is a large conifer plantation, Bedgebury Forest.

SPECIES

Wildfowl are present in varying numbers throughout the year, including Tufted Duck and Pochard, owing to the depth of the water. Many pairs of Great Crested and Little Grebe breed and both species are also present in the winter. Winter also brings much larger numbers of dabbling ducks and geese, including many hundreds of Canada Geese, Mallard, Wigeon and Teal and smaller numbers of Shoveler and Pintail hiding around the more sheltered edges. Coot and Moorhen are also found in abundance, and the former can also congregate in the hundreds in winter. As one might expect for such a huge inland waterbody, Ospreys are a regular occurrence in passage season, especially in the autumn when one has been known to linger in the area, taking full advantage of abundance of trout and perch. Other fish-eating species are frequent visitors, particularly Cormorant and Kingfisher, but also occasionally Goosander in the winter.

Bewl Water is perhaps most famous for its gull roost, which in winter can attract tens of thousands of individuals; for example, there were 52,470 Black-headed Gulls and 21,500 Common Gulls in a recent count.

As with any waterbody, the water level at Bewl Water often drops quite substantially in dry summers, exposing muddy margins around the edges which can attract waders such as Little Ringed Plover, Green Sandpiper, Common Sandpiper, Greenshank and Dunlin. Scarcer waders sometimes occur too, such as Grey Plover, Knot, Wood Sandpiper and Curlew Sandpiper.

It's not all about the waterbirds though, as the woodland and farmland around the reservoir offer a reasonable selection of passerines too, including Marsh Tit, Goldcrest, Great Spotted Woodpecker, Nuthatch, Treecreeper and Tawny Owl all year-round, Chiffchaff, Willow Warbler, Blackcap, Garden Warbler and Whitethroat in spring and summer, and Redwing and Fieldfare in the winter. The expanse of water on offer here is also very attractive to hirundines, particularly in spring when many hundreds may congregate and feed low over the reservoir. As with any large waterbody, inclement weather especially in the spring, can also produce the likes of Little Gull or Common, Arctic or Black Tern dropping in to feed for a while.

Perhaps owing to its sheer size and/or location making it not as well watched as other sites in Sussex, Bewl Water does not have a massive list of rarities to its name, although very much worth of a mention is the first Blackpoll Warbler for Sussex in 1994 and a Black-eared Wheatear in 1988. Other somewhat more common but still

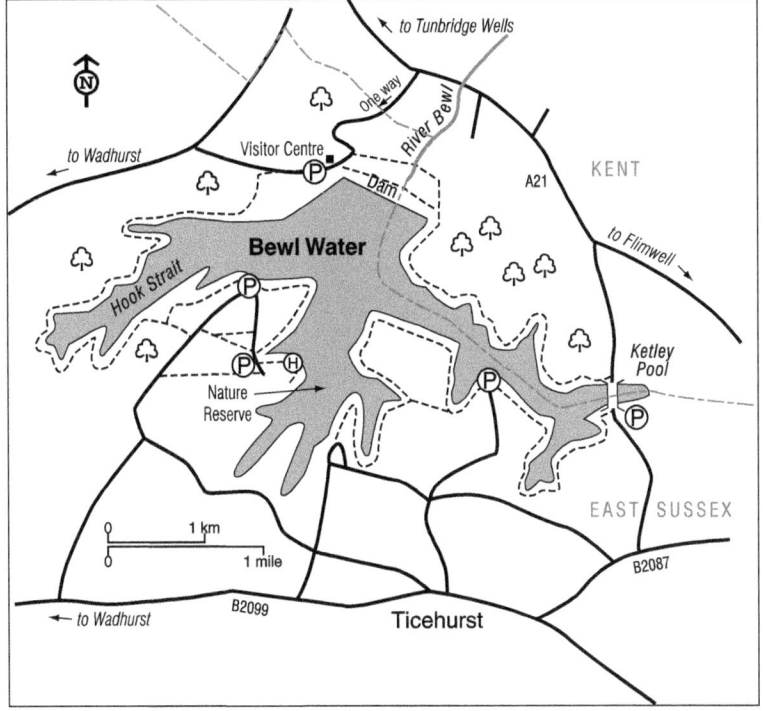

notable species have included Ring-necked Duck, Smew, Black-throated Diver, Long-tailed Duck, Glossy Ibis and Black-necked Grebe.

TIMING

Bewl Water is a very popular destination for a host of watersports and other leisure activities, so the earlier you can get there, the better – to beat the boats, swimmers and windsurfers!

ACCESS

If travelling by car, follow signage from the A21 (the main London to Hastings road) to the car park and information centre. To access the southern side of the reservoir, follow the A21 further south, then take the turning off to Ticehurst. Around 3km west of Ticehurst, turn north onto Wards Lane. There is limited parking available at the quarry along here, offering a short walk to the SWT nature reserve and bird hide. The full circuit of the reservoir takes around six hours to walk, so is not for the faint-hearted. Shorter routes are available, following the various waymarkers, although many of the trails are undulating and not well surfaced throughout so are unsuitable for wheelchairs or pushchairs. The northern shore near the dam is the easiest point from which to view the bulk of the main area of water, though there are various other viewpoints along the 20km perimeter trail.

FACILITIES: There is a café, toilets and a pay and display car park (open 8:00 am–5:00 pm) at the Aqua Park centre on by the dam on the northern side (TQ 676338). Here you can also pick up leaflets and other information on the walks and wildlife.

CALENDAR

All year: Egyptian Goose, Great Crested Grebe, Little Grebe, Little Egret.

April–May: Garganey, Teal, Common Sandpiper, Little Gull, Common Tern, Arctic Tern, Black Tern, Osprey, Hobby, Willow Warbler, Garden Warbler, Whinchat, Wheatear.

July–September: Black Tern, Green Sandpiper, Wood Sandpiper, Greenshank, scarcer waders, Osprey.

October–March: Brent Goose, Shoveler, Gadwall, Wigeon, Teal, Pochard, Goosander, Lesser Redpoll, Siskin.

40 BREDE HIGH WOODS AND POWDERMILL RESERVOIR

OS Explorer 124/125
OS grid ref: TQ 799195
Postcode: TN31 6EY

HABITAT

Brede High Woods is a 262ha site owned and managed by The Woodland Trust. The woodland, much of it ancient, contains a mixture of trees, including oak, Sweet Chestnut, Hornbeam, birch, Alder, Holly and Aspen. Modern plantations of Sitka Spruce, Scots and Corsican Pines, Douglas Fir and Larch – can also be found. Renowned for its invertebrates, Brede High Woods is the only known UK location of a very rare Flea Beetle, *Longitarsus longiseta*, previously thought to be extinct in Britain. The acid grassland also has a population of Glow-worm, a Sussex priority species, while the woodland supports a diverse butterfly population including Silver-washed Fritillary. Roe and Fallow Deer inhabit the area and there is increasing evidence of Wild Boar.

Tucked to the south of Brede High Woods is Powdermill Reservoir. Originally forming part of the Great Sanders Estate, the name is derived from the fact that there was an eighteenth-century gunpowder manufacturer on site. Nowadays it is a well-stocked trout fishery.

SPECIES

The combination of woodland and wetland birding can make for an excellent session in the field. Powdermill usually has a small range of wildfowl in the winter, including Pochard and Gadwall, plus occasional Pintail. Mandarin is resident, breeding in the adjacent woodland. Great Crested Grebe breeds and Grey Wagtail can be found around the reservoir edges. Wader passage is fairly limited, though Common Sandpiper drops in annually. An Osprey may take an interest in the numbers of trout during spring and autumn as well.

A fine array of woodland species can be found in Brede High Woods, with Marsh Tit and Firecrest relatively conspicuous residents, at least in terms of being heard. Lesser Spotted Woodpecker is elusive, but certainly present – try the woodland near the eastern of the two car parks. Goshawk is another shy and low-level resident in the woodland in this wider area. Siskin and Lesser Redpoll are present in the winter, when Crossbill, Hawfinch and Brambling may well be chanced upon.

Nightingale still breeds here, returning from mid-April. A few weeks later sees the arrival of Spotted Flycatcher, which can be encountered with some ease. Other summer migrant breeders include Cuckoo and Garden Warbler. Turtle Dove used to be regular but is elusive at best nowadays.

TIMING

The reservoir is best checked in the winter or during passage seasons. Note there can be disturbance from anglers. For Brede High Woods, early mornings from April through June are best.

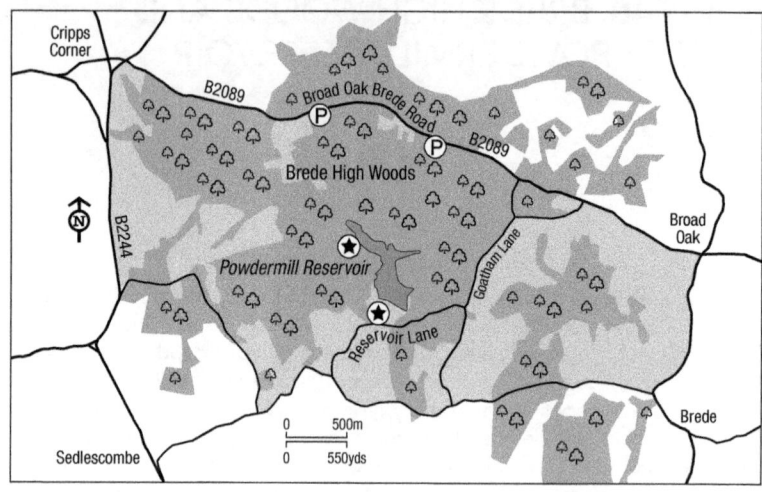

ACCESS

There are two free car parks on the south side of the B2089 at Cripps Corner/ Broad Oak Brede Road. From both there are routes into the wood via an all-ability kissing gate (RADAR key required) or steps. Three of the tracks into the wood from the B2089 are surfaced from the entrances, but beyond that most rides and paths are unsurfaced and very prone to waterlogging. Bus number 349 (Stagecoach South East) stops at Cripps Corner.

Powdermill is rather tricky to view, but tracks north of Powdermill Lane and the public footpath through Ward's Wood (TQ 796197) can take you close to the water's edge.

FACILITIES: The nearest facilities are in Broad Oak and Sedlescombe.

CALENDAR

All year: Mandarin, Great Crested Grebe, Goshawk, Raven, Lesser Spotted Woodpecker, Marsh Tit, Firecrest.

April–July: Cuckoo, Spotted Flycatcher, Nightingale, Garden Warbler.

November–March: Wildfowl on Powdermill, Crossbill, Brambling, Hawfinch.

41 DALLINGTON FOREST AND PENHURST LANE

OS Explorer 124
OS grid refs: TQ 645208 and TQ 696180
Postcodes: TN21 9JJ and TN3 9QN

HABITAT

Although geographically distinct from one another, these two sites are perhaps the most notable in the band of woodland that stretches east to west, some 25km across the southern section of the High Weald, between Battle and Uckfield.

Dallington Forest was designated as an SSSI in 1953 and part of the southern section is an SNCI. The SSSI designation was given partly due to the presence of rare plants including Hay-scented Buckler-fern and Wood Fescue. A local project, known simply as the Dallington Forest Project, aims to reconnect and enhance the wider woodland habitat, as well as educating and engaging local people and visitors to the area.

Penhurst Lane is a minor road running north to south from the hamlet of Darwell Hole to the village of Penhurst. A viewpoint along the lane offers sweeping views across the surrounding woodland, and some of the best raptor and Hawfinch watching in East Sussex.

SPECIES

Both these sites – Penhurst Lane in particular – are excellent for Hawfinch, especially in the winter, although it is likely that some stay to breed. Indeed, juveniles have been seen here in recent summers. The area was not recognised as a hotspot for the species until the influx winter of 2017–18, when good numbers were discovered here, and the confirmation of recent breeding was among the first in Sussex for two decades. Although Hawfinches can be seen here at any time and in any year, winters with bumper Hornbeam seed crops are the most productive, when counts of up to 50 birds have been recorded.

Other common woodland passerine species may be encountered here, including Marsh Tit year-round, and a chance of Brambling in winter and Crossbill in influx years.

These sites are also both excellent for raptor watching, with all the common resident species available along with Goshawk. Hobby breeds and there is a chance of Honey Buzzard in summer – although, as always, patience is required for the latter.

TIMING

For Hawfinches, especially at Penhurst Lane, the first half hour after sunrise is best. For birds of prey, mid-morning onwards on a bright late winter or early spring day should produce Goshawk, while sunny days from mid-May are best for Honey Buzzard – again from mid-morning.

ACCESS

For Penhurst Lane, turn south off the B2096 at Darwell Hole onto Penhurst Lane and stop near a metal gate just over 1km from the turn-off (TQ 696180). Scan the

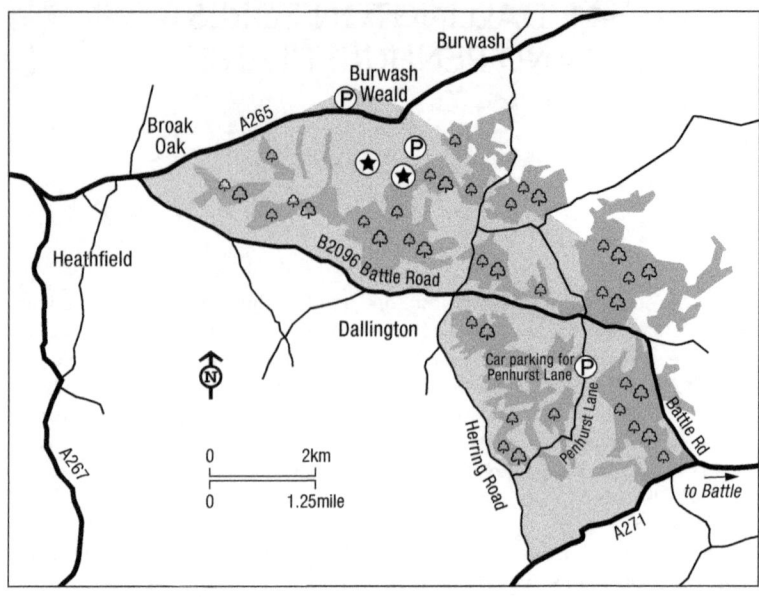

treetops on the east side of the lane. The first half hour from sunrise is the best time. Look for birds of prey soaring over the woodland between here and the A271 to the south.

For Dallington Forest, either park in the village green car park in Burwash Common or along Willingford Lane (TQ 656226). From Burwash Common, take the public footpath heading south-east for around 1km to reach a good viewpoint east of Henhurst Farm (TQ 648225). From here you can enjoy great views south across the forest and other areas of woodland. From Willingford Lane head south-west on the footpath for around 500m until you have open views towards the forest (at around TQ 654220). Access to the forest itself is rather limited, with just one public footpath running roughly west to east from Hooksdown Farm (TQ 631207) to Little Worge Farm (TQ 658213).

A telescope is recommended for scanning the distant treetops and sky.

FACILITIES: The nearest facilities are in Heathfield or Battle.

CALENDAR

All year: Goshawk, Red Kite, Raven, Marsh Tit, Firecrest, Hawfinch (though best chance in winter).

April–September: Hobby, chance of Honey Buzzard.

October–March: Brambling, chance of Crossbill.

42 SHEFFIELD PARK AND GARDEN

OS Explorer 135
OS grid ref: TQ 418238
Postcode: TN22 3QX

HABITAT

Sheffield Park is a 74ha landscape garden and parkland, designed by Lancelot 'Capability' Brown in the eighteenth century (with later developments by Humphry Repton, among others), now owned by the National Trust. Landscape features of the sites include four main lakes and several smaller waterbodies, waterfalls, areas of ancient woodland, a deer park and an arboretum holding a mix of native and exotic tree and plant species.

Although the garden, restaurant and shop are only open at set times, and the garden requires an admission fee or National Trust membership, the parkland is free to access all year-round and offers a pleasant mix of woodland and open-country bird species, as well as butterflies and dragonflies.

A recent development has been a restoration project on the stretch of the River Ouse which runs through the southern section of the parkland, reversing centuries of straightening and deepening of the river and returning it to something more akin to its natural state for the benefit of wildlife and visitors.

SPECIES

The mix of woodland, parkland and lakes at Sheffield Park offer a pleasing selection of species at any time of year, particularly for the birder keen to combine some birding with a family walk. All the expected woodland species are present, such as common tits, Goldcrest, Firecrest, Nuthatch and Treecreeper. Chiffchaff, Blackcap, Whitethroat and Garden Warbler all breed, while Spotted Flycatcher and Cuckoo are also often recorded here in the summer, and while the former is thought to still be breeding, it can prove elusive. Pied Flycatcher has passed through on occasion in late summer. Indeed, various passerine migrants are possible in spring and autumn, with recent records of Whinchat and Black Redstart. A particularly good area to check for migrants in the autumn is the higher ground in the southern section of the parkland, which also offers some good skywatching potential.

Finches of various species can be encountered at any time of year, but good winters will sometimes produce flocks of Siskin, Lesser Redpoll and a few Brambling. Hawfinch has been recorded on occasion too.

All the expected resident birds of prey are present, with Buzzard, Sparrowhawk and Red Kite the most likely to be seen throughout the year, but these are sometimes joined by a Hobby or two hawking overhead in the summer. Honey Buzzard have been recorded on occasion as well.

The main lakes, known as Upper and Lower Womans Way Ponds, Ten Foot Pond and Middle Lake all attract the expected waterbirds for the region, such as Moorhen, Coot, Great Crested Grebe, Canada Goose and, sometimes, Mandarin Duck. Kingfisher and Grey Wagtail are both very reliable here too, and there has been the odd record of Great White Egret.

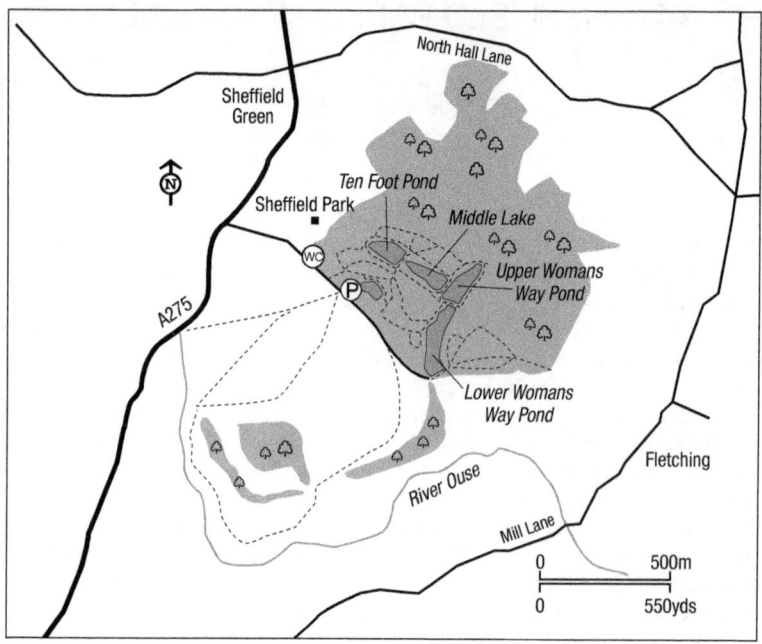

The river restoration project has created more wetland habitat towards the southern end of the parkland, with Green Sandpiper and Lapwing having been recorded here on occasion. On the Ouse itself, Goosanders are occasionally seen in the winter.

TIMING

Although the parkland is open and free to access 365 days a year, the car park can get very busy at weekends in peak season for the site (May and October) so early mornings, weekdays and quieter times of year are recommended.

ACCESS

The nearest railway station is at Uckfield. Bus route 121 (Compass Travel) from Lewes to Uckfield stops in Sheffield Park itself, but only on Saturdays. On weekdays the service only stops as close as North Chailey, roughly a half-hour walk away to the south-west. For a more novel way of travelling, the Bluebell Railway steam train from East Grinstead stops at Sheffield Park station which is just a short walk away from the parkland, and around 15 minutes to the gardens. By car, the entrance is situated 8km north-west of Uckfield on the A275. If coming south from A22, follow the brown signs for Sheffield Park and the Bluebell Railway.

The car park area and many of the main trails are level and well surfaced, so are suitable for wheelchairs and mobility vehicles. Both manual and powered wheelchairs and mobility vehicles are available for hire, and maps showing the best access routes for these can be picked up at the entrance kiosk.

FACILITIES: Café, shop and public toilets (including accessible toilets), all run by the National Trust. Free car park including accessible parking.

CALENDAR

All year: Egyptian Goose, Mandarin Duck, Great Crested Grebe, Red Kite, Kingfisher, Raven, Firecrest, Grey Wagtail.

April–September: Cuckoo, Hobby, Garden Warbler (and other common warblers), Spotted Flycatcher.

October–March: Goosander, Great White Egret, Lesser Redpoll, Siskin.

43 CHAILEY COMMON

OS Explorer 135
OS grid ref: TQ 385212
Postcode: BN8 4DY

HABITAT

Situated approximately 6km south-east of Haywards Heath, Chailey Common is a 180ha patchwork of heathland nestled to the north and west of the village of North Chailey. The site is divided into five enclosures, separated by roads, and the largest and most northerly of these is Red House Common, famous for its Grade II-listed windmill which is said to stand next to the yew tree that marks the centre of Sussex.

Although somewhat overshadowed by Ashdown Forest, Chailey Common still represents a significant portion of the remaining heathland in East Sussex and has a long and fascinating history. It was recorded in the Domesday Book in 1086 and has been used for centuries as an area for grazing animals and cutting wood and bracken for fuel.

Designated as an SSSI in 1954 and an LNR in 1966, Chailey Common supports a wealth of flora and fauna, not just birds. Scarce plants such as Marsh Gentian, Bog Asphodel, Round-leaved Sundew and Heath-spotted Orchid can be found here, as well as Adder, Common Lizard and invertebrates like Minotaur Beetle, Purse Web Spider and Black-headed Velvet Ant.

SPECIES

A walk at Chailey Common in spring and summer can deliver a rewarding experience, with a selection of heathland and woodland species on offer. As with all heathland sites, it can all appear rather devoid of life (in a birding sense anyway) in the winter months, aside from perhaps the odd flock of Siskin or Lesser Redpoll, but from early spring it begins to become a lot more lively. Listen for the song of a Woodlark overhead anytime from January onwards, while Tree Pipit can sometimes still be heard from April. March and April see the return of the various migrant warbler species that breed here, with the woodland edges of the common holding Blackcap, Garden Warbler, Chiffchaff and Willow Warbler through the summer, while Whitethroat breeds anywhere where there is scrubbier vegetation. The unmistakable song of a Cuckoo is often heard from April to early June too. Nightjars return in May, with the common supporting several breeding pairs. Visit on a warm, still evening to hear their distinctive churring song. Stonechat, Yellowhammer and Reed Bunting breed and can all be encountered here at any time of year. Dartford Warbler is not common here but always a possibility. A possibility in late summer and early autumn is a flyover Honey Buzzard, with the species' strongholds in the High Weald not far away to the north.

Red House Common, in particular, tends to be the more well-watched area and has produced unusual species such as Wryneck, Black Redstart and Siberian Chiffchaff in recent years.

TIMING
Spring and summer are recommended for the best selection of heathland-specialist migrant species, especially dusk in summer for Nightjar.

ACCESS
The nearest railway station is at Haywards Heath, from where bus service 31 (Compass Travel) to Uckfield stops at Chailey Heritage School on the A272, right on the doorstep of the Common. Car parking is available in all five areas of the Common – all with 2m height restriction barriers at the entrances (except Red House Common). For Red House Common, there is a car park at TQ 391217 on Warrs Hill Lane. For Memorial Common there is a layby on the A272 at TQ 388204. For both Pound Common and Romany Ridge Common, there are car parks on North Common Road at TQ 377207 and TQ 378207, respectively. For Lane End Common, there is a car park on Fletching Common Road at TQ 403222. There are public footpaths across all areas of the common, with Red House and Memorial Commons offering the most extensive walking routes.

Trails are unsurfaced and can get muddy in the winter, so are not terribly suited to wheelchairs or pushchairs.

FACILITIES: Aside from the car parks mentioned above, the nearest toilet and restaurant facilities are in Newick and Haywards Heath.

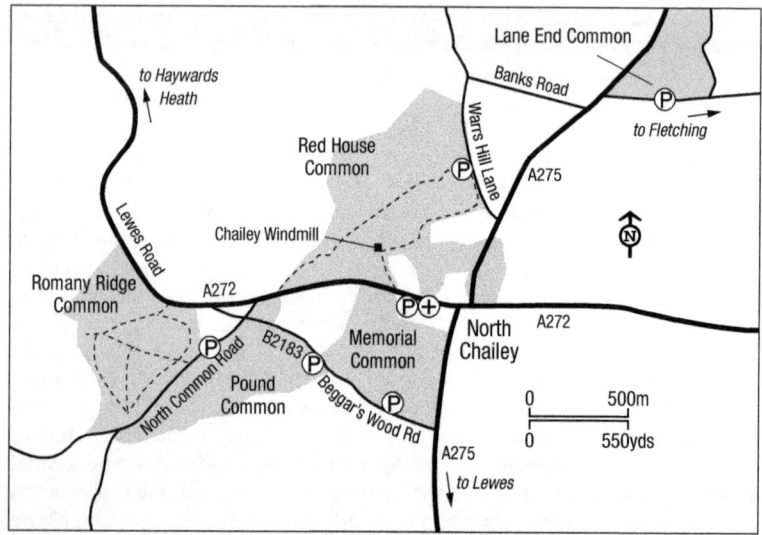

CALENDAR

All year: Red Kite, Marsh Tit, Woodlark, Stonechat, Bullfinch, Yellowhammer, Reed Bunting.

April–September: Cuckoo, Nightjar, Hobby, Willow Warbler, Garden Warbler, Tree Pipit.

October–March: Lesser Redpoll, Siskin.

OTHER SITES IN NORTH EAST SUSSEX

F1 BIRCHDEN WOOD

This 42ha Forestry Commission site incorporates mixed conifer and deciduous woodland with over a kilometre of hard-surfaced tracks. A good selection of species is on offer here including Raven, Cuckoo, Garden Warbler, Marsh Tit and Crossbill. The site is also famous for its sandstone rock outcrops known as Harrison's Rocks. A car park and toilets are located at TQ 533364.

F2 HARGATE FOREST

This Woodland Trust site (TQ 574370) is gradually being opened up from conifer plantation to a mix of heathland and deciduous woodland. Two waymarked trails wind around the site. An attractive selection of woodland birds can be found including Spotted Flycatcher, Marsh Tit, Goldcrest, Lesser Redpoll, Siskin and Crossbill. Parking is available on the road at Broadwater Down and the nearest train station is Tunbridge Wells.

F3 ERIDGE PARK AND WHITEHILL WOOD

Designated as an SSSI for its 167 species of lichen, the name Eridge derives from 'Eagle Ridge'. While eagles are far from guaranteed, the extensive woodland and parkland here (TQ 568355) support a heronry, Buzzard and Red Kite, along with a host of woodland passerines including Marsh Tit and sometimes Redstart and Lesser Spotted Woodpecker, as well as wildfowl and other waterbirds on the ponds, including the occasional Mandarin Duck. These sites are around 3km south of Tunbridge Wells West train station.

F4 WADHURST PARK

Formerly a deer park, this 850ha site is now being managed for the benefit of wildlife and is a partner of the White Stork Project, based at the Knepp Estate in West Sussex. As well as the chance of a reintroduced White Stork, Wadhurst also offers an array of wildfowl and woodland passerines, as well as Kestrel, Barn Owl and Tawny Owl. Stonegate station lies just to the east and there is limited on-road parking at TN5 7ER/TQ 656271.

F5 CEDAR FARM

This area of farmland (TN19 7QY / TQ 721296) near Flimwell is managed by a wildlife-friendly farmer and is attractive to finches in the winter. Marsh Tit can be found in the area year-round. It hosted a Serin for a couple of weeks in November 2019. Limited parking is available by the entrance gate.

F6 FLATROPERS WOOD
Just south-east of Beckley, in one of the most wooded areas of Sussex, Flatropers Wood (TQ 862234) is 38ha area of mixed woodland with open paths and rides, ponds and streams. A host of woodland birds are on offer here, including Siskin, Goldcrest and sometimes Hawfinch in the winter. The nearest train stations are at Rye and Winchelsea, and there is limited car parking in lay-bys along Bixley Lane.

F7 DARWELL RESERVOIR
This 63ha reservoir is surrounded by ancient woodland, including Darwell Wood SSSI. Viewing of the reservoir is limited although the footpath from the car park (TQ 695195) does pass close enough along the eastern and southern sides to allow some restricted viewing. Darwell Wood holds various breeding species including Marsh Tit and Nightjar, and has even hosted Serin, while the reservoir regularly holds Great White Egret and occasionally a lingering Osprey.

F8 SELWYNS WOOD
A small (11ha) area of ghyll woodland tucked away in a south-facing valley near Heathfield with some impressive old Beech trees and even a little pocket of heathland. Breeding species here including Willow Warbler, Marsh Tit and Chiffchaff, while Markly Wood just to the north-east also has produced Hawfinch on occasion. A car park is located towards the northern side at TQ 551204.

F9 UCKFIELD SEWAGE WORKS
A small sewage works (TQ 466204) on the west side of Uckfield, with viewing best from within Victoria Park. Red-rumped Swallow has occurred here before. A few Chiffchaffs winter and Yellowhammer is found in the surrounding farmland. The site is just a 15-minute walk from Uckfield train station.

F10 FLETCHING MILL FARM
South of the village of Fletching and Sheffield Park (site 42, page 135) is an area of open farmland centered around Fletching Mill Farm (TQ 423224). Both Barn and Little Owls are resident, along with Red-legged Partridge and Yellowhammer. The open nature of the site renders it attractive to passage Wheatear, Whinchat and Yellow Wagtail, with Black Redstart recorded by the farm buildings in recent years. Access via public footpaths running south from Mill Lane.

F11 BRICKFIELD MEADOW
A Sussex Wildlife Trust reserve situated on the southern periphery of Ashdown Forest (site 37), Brickfield Meadow (TQ 472265) is an unimproved, wildflower-rich meadow that is home to a range of characteristic Wealden meadow plants. Skylarks and Linnets breed, and chats are possible on passage, including Whinchat. Hobby is often seen overhead in late summer.

SOUTH EAST SUSSEX

MAIN SITES

44 Scotney Gravel Pits, Walland Marsh, Jury's Gap and the Midrips
45 Rye Harbour (including Camber Sands, Camber Pits and Northpoint Pit)
46 Pett Level
47 Hastings Country Park and Fairlight Cliffs
48 Combe Valley Countryside Park and Filsham Reedbed
49 Pevensey Levels
50 Sovereign Harbour
51 West Rise Marsh and West Langney Lake
52 Beachy Head
53 Cuckmere Haven
54 Seaford Head
55 Splash Point
56 Newhaven Harbour area
57 Sheepcote Valley
58 Brighton Marina
59 Ditchling Beacon
60 Iford, Lewes and Rodmell Brooks
61 Arlington Reservoir
62 Lullington Heath and Friston Forest

OTHER SITES

G1 Doleham Marshes
G2 Alexandra Park
G3 Fore Wood RSPB
G4 Ashburnham Forge
G5 Normans Bay
G6 Princes Park
G7 High and Over
G8 Telscombe Tye
G9 Portobello Water Works
G10 Rottingdean
G11 Ovingdean
G12 Malling Down and Southerham Farm SWT
G13 Abbot's Wood

The region referred to as South East Sussex in this book covers the 90km of coastline of East Sussex and any sites that lie within a 10km radius/boundary of the sea.

There is much diversity in this area, from the river valleys of the Ouse, the Cuckmere, the (eastern) Rother, the Brede and the Combe Haven, to the chalk headlands of Beachy Head and Seaford Head, where the South Downs reach the English Channel. Where the larger rivers flood their valleys – such as at Lewes, Iford and Rodmell Brooks and the Cuckmere Valley – a wide diversity of birds can be found, including large numbers of waterbirds in the winter. Some sites are fairly unique and exist in their own categories, such as the chalk heath of Lullington Heath and the combination of ruined industrial buildings and estuary birding at

Newhaven Tide Mills (included here as part of the Newhaven Harbour area). What links all the sites in this region is the dynamic birdlife: from coastal shorebirds to open-country passerines and woodland raptors, South East Sussex offers a bit of everything.

It will come as no surprise to anyone familiar with the area that many of the sites here are etched into birding folklore, given their illustrious history of attracting good birds. Some have earned their fame for being reliable sites for particularly desirable species (Pomarine Skuas passing Splash Point in May, for example), while others have built up an eye-watering list of rarities thanks to years of concentrated observer effort (Beachy Head and Rye Harbour, to name two).

Given its proximity to the Continent, second only to Kent in terms of its distance from the French coast, this region of Sussex serves as a kind of arrival and departure lounge for passage migrants and vagrant species. It's no coincidence that Beachy Head, for instance, is a well-known site for encountering Honey Buzzard, as it offers a prominent area of high ground jutting out into the Channel – ideal for birds needing to make the sea crossing to France. A wide range of national rarities have been unearthed along the East Sussex coast, too.

The landscape of South East Sussex is generally more open and less wooded than the rest of the county. Where blocks of woodland and scrub occur they can therefore act as magnets to a host of resident and migrant species, particularly those on or near the coast. These include sites such as Friston Forest, Sheepcote Valley and Hastings Country Park, all of which have form for producing plenty of common and scarce migrant passerines, from Redstarts to Pied Flycatchers to Hawfinches, as well as rarities.

While you will find all the famous birding locations in the region covered in the main text, we have also endeavoured to introduce other, less-known sites, as we have done throughout this book. Though perhaps less etched into the annals of birding history, the likes of West Rise Marsh and West Langney Lake, Telscombe Tye and Ashburnham Forge have rightly risen to prominence in recent years due to new generations of birders seeking out places to find birds in less expected locations close to home. Thanks to the daisy chain of large conurbations stretching along the coast, from the city of Brighton & Hove eastwards to Rye, there is an encouraging culture of birders geared up to go out looking. As a result, this region is arguably one of the best-watched areas, not just in Sussex, but in the whole of South-East England. At the same time, there are vast swathes of the area that receive poor coverage, meaning there is plenty to explore and discover.

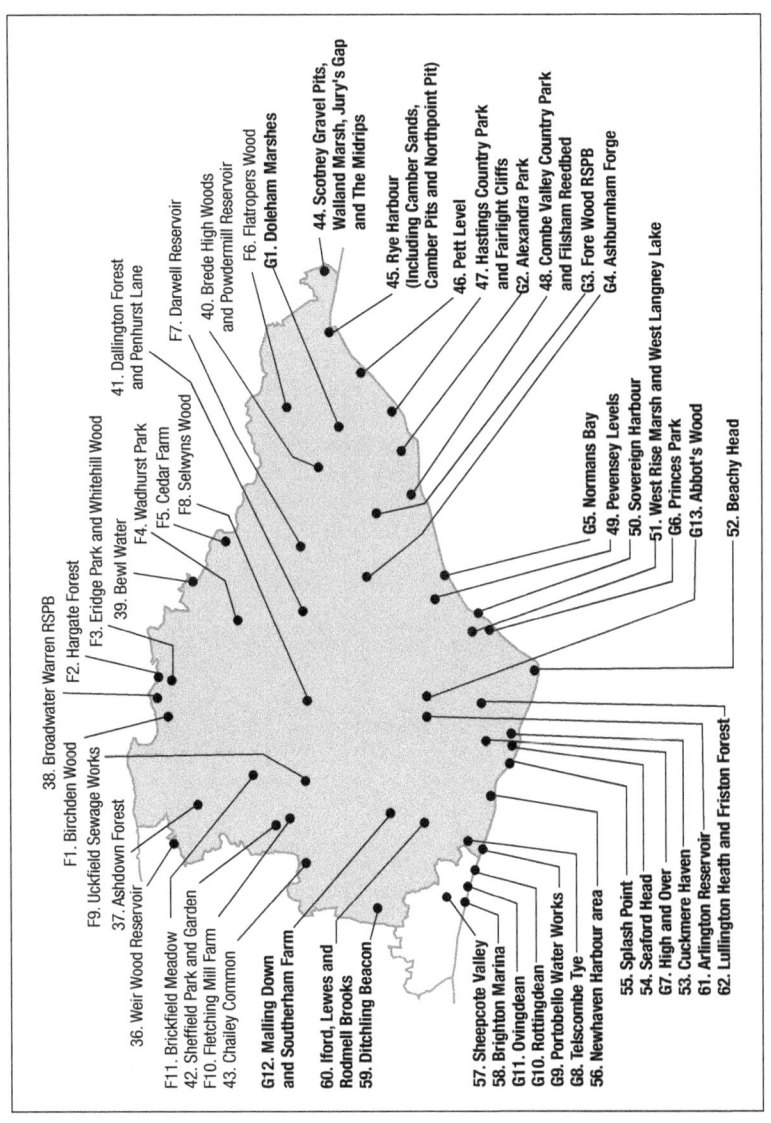

44. Scotney Gravel Pits, Walland Marsh, Jury's Gap and The Midrips
G1. Doleham Marshes
F6. Flatropers Wood
40. Brede High Woods and Powdermill Reservoir
F7. Darwell Reservoir
41. Dallington Forest and Penhurst Lane
45. Rye Harbour (Including Camber Sands, Camber Pits and Northpoint Pit)
46. Pett Level
47. Hastings Country Park and Fairlight Cliffs
G2. Alexandra Park
48. Combe Valley Country Park and Filsham Reedbed
G3. Fore Wood RSPB
G4. Ashburnham Forge

F8. Selwyns Wood
F5. Cedar Farm
F4. Wadhurst Park
39. Bewl Water
F3. Eridge Park and Whitehill Wood
F2. Hargate Forest
38. Broadwater Warren RSPB
F1. Birchden Wood
F9. Uckfield Sewage Works
37. Ashdown Forest
36. Weir Wood Reservoir
F11. Brickfield Meadow
42. Sheffield Park and Garden
F10. Fletching Mill Farm
43. Chailey Common

G5. Normans Bay
49. Pevensey Levels
50. Sovereign Harbour
51. West Rise Marsh and West Langney Lake
G6. Princes Park
G13. Abbot's Wood
52. Beachy Head

G12. Malling Down and Southerham Farm
60. Iford, Lewes and Rodmell Brooks
59. Ditchling Beacon
57. Sheepcote Valley
58. Brighton Marina
G10. Rottingdean
G11. Ovingdean
G9. Portobello Water Works
G8. Telscombe Tye
56. Newhaven Harbour area
55. Splash Point
54. Seaford Head
G7. High and Over
53. Cuckmere Haven
61. Arlington Reservoir
62. Lullington Heath and Friston Forest

44 SCOTNEY GRAVEL PITS, WALLAND MARSH, JURY'S GAP AND THE MIDRIPS

OS Explorer 125
OS grid refs: TR 008191, TR 005201, TQ 990181 and TR 001182
Postcode: TN31 7SD

HABITAT

This productive cluster of sites are located within spitting distance of the Kent border. Indeed, Scotney GPs and Walland Marsh straddle the county border, with wildfowl merrily swimming back and forth across the invisible boundary between East Sussex and Kent. As the name suggests, Scotney GPs are the remnants of gravel extraction that took place here in decades past. Now permanently flooded, the site is a magnet to a host of wildfowl throughout the year, especially in winter.

Walland Marsh to the north is a mix of arable and sheep-grazed wet grassland, while the whole area is interspersed with pockets of reedbed, scrub, drainage ditches and old farm buildings.

Jury's Gap is the stretch of shoreline between the western end of The Midrips and Camber Sands to the west, and offers a somewhat different birding experience with the rocky sea defences separating the sandy beach from the road. There are also arable fields on the north side of the road here and a sewage works at TQ 998185.

The Midrips is an expansive area of shingle with saline lagoons and saltmarsh between Jury's Gap Road and the sea. It has been used for military training for the past 150 years and, because the range is used for live firing, access is often restricted.

SPECIES

The western/East Sussex end of Scotney GPs is, naturally, of interest to the Sussex birder. During the winter impressive numbers of waterbirds congregate on the pits, with the resident Tufted Duck, Great Crested Grebe and Coot joined by Shelduck, Wigeon, Teal, Shoveler, Gadwall and Pintail. It's worth nothing that disturbance from anglers occasionally scatters the flocks. Diving duck include several Goldeneye and, traditionally, Scotney is a semi-reliable site for Scaup, though occurrences are a lot less regular these days. The same can be said for Smew – formerly present each winter, but now a rare Sussex bird. Seaduck such as Long-tailed Duck occasionally pitch up too, especially in hard weather. Little Grebes winter in good numbers and, sometimes, a scarce grebe may be found. Geese too are present in numbers at this time of year, typically Canada and Greylag, though note there is a large feral flock of Barnacle Geese in the area too. Grey geese may occur from time to time, usually White-fronted. The numerous Mute Swans – which often spread out over the farmland – can be worth sifting through for the two scarcer species (Bewick's Swans occasionally wander over from Kent, where there is usually a wintering herd near Dungeness). Hen Harrier is always possible in the winter, quartering over the arable fields, marshland or army ranges in the area. There have been occasional records of Rough-legged Buzzard here too, but this species is not to be expected. Marsh Harrier is an increasingly conspicuous presence year-round, along with Buzzard, Peregrine and Kestrel.

Large numbers of gulls may congregate at the pits in the winter to bathe and loaf. Usually a few Mediterranean Gulls can be picked out, while persistence may yield Caspian or Yellow-legged Gulls. White-wingers are possible too. Little Gull is more of a passage species. Herons and egrets are a fairly conspicuous presence – Great White Egret is increasing, and there is an upward trajectory of Glossy Ibis and Spoonbill records, too. In the winter months, the open fields of Walland Marsh can attract Curlew, Redshank, Dunlin, Ringed Plover and large flocks of Lapwing and Golden Plover. Oystercatchers and Turnstones generally favour the shoreline at Jury's Gap.

Seawatching across Rye Bay from Jury's Gap can be rewarding in the winter, when Red-throated Divers and a large Common Scoter flock will be present. The latter is worth scanning through for Velvet Scoter. Coastguards Cottages at Jury's Gap is a popular wintering spot for Black Redstart. Be sure to check the sewage works at Jury's Gap during the winter, too, as this can attract Pied and Grey Wagtails and Chiffchaff.

Passage seasons can be productive across this entire area. Waders are a theme at such times. *Tringa* species in particular favour the shorelines of the lakes and pits, especially Common Sandpiper and Greenshank in the spring and Green Sandpiper in the autumn and winter. In spring, commoner species may be joined by the likes of Avocet, Whimbrel, Bar-tailed Godwit and Little Ringed Plover, especially at Walland Marsh. Check the roosting gulls and waders on the near bank of the pit just east of Rosedale (TR 008188) for more unusual species lurking among them. In late summer and autumn, scarcer species such as Wood Sandpiper, Little Stint and Curlew Sandpiper may occur, especially on The Midrips. It's very much worth keeping an open mind here in terms of which wader species you might find, as recent years have produced such rarities as Lesser Yellowlegs, Terek Sandpiper, Buff-breasted Sandpiper, American Golden Plover, Black-winged Stilt and Dotterel. Terns also pass through during the passage seasons, especially at Scotney. Common is the most typical species, but Arctic and Black are real possibilities.

The open farmland surrounding the gravel pits and north of the road at Jury's Gap is excellent for passerines, including passage species such as Wheatear and Whinchat. Yellow Wagtail breeds in the area, though it is more obvious during the autumn. Common resident species include Linnet, Goldfinch, Meadow Pipit and Skylark, flocks of which are worth sifting through during the spring, autumn and winter. Corn Bunting still breeds and can form sizeable flocks outside the breeding season. A particularly notable resident species for Sussex birders is Tree Sparrow – a few cling on here, so do scan through any House Sparrow flocks close to farm buildings at Scotney Court (TR 015199) or Little Scotney (TR 010214). The farm buildings also host breeding Stock Dove and Kestrel, and are a good place to linger towards dusk to stand the best chance of seeing Barn Owl or Little Owl.

TIMING

Any time of year will offer some birding interest here, but for the greatest gatherings of waterbirds winter is best. Hard weather in winter will also sometimes produce the goods in terms of scarce diving ducks and grebes, while south-easterly winds in spring may deliver interesting migrant species such as Little Gull and terns. This whole area is very exposed, so be sure to dress appropriately for the harshest conditions that the winds can deliver to this stretch of the coast. Angling distur-bance can be an issue, so early visits are recommended.

ACCESS

Bus routes 102, 293 and 553 (all Stagecoach South East) all stop by Coastguards Cottages at Jury's Gap and at Rye railway station.

If travelling by car, follow the B2075 towards Lydd, following signs for Camber and Rye. Park on the roadside, in the entrance track to Scotney Farm (for Scotney) or on the outskirts of Lydd, on Tourney Road. Alternatively, you can park at the Broomhill car park between Camber Sands and Jury's Gap at TQ 980182 and walk east along the road and cycle path from there. Viewing of the Midrips is best achieved by following the beach east from Coastguards Cottages (TQ 991180) and looking north across the various pools, though note that this can only be done when the red flags are not flying, so check the South East Training Estate firing times on the GOV.UK website before visiting. Don't approach too close as to scare the waders on the pools though, and do be mindful of the tide, as it can cut you off from the return journey. For a circular walk taking in much of the habitat on offer here, park at one of the roadside pull-ins near Scotney GPs and follow the footpath north-west at TR 019195 (note that you are in Kent at this point) past the Scotney Court farm buildings. After about half a kilometre you will reach a T-junction with another footpath/farm track. Head west from here, then take the left fork which follows the northern shore of Scotney GPs (the East Sussex section) and eventually all the way to Jury's Gap. This path also offers good views of Walland Marsh to the north.

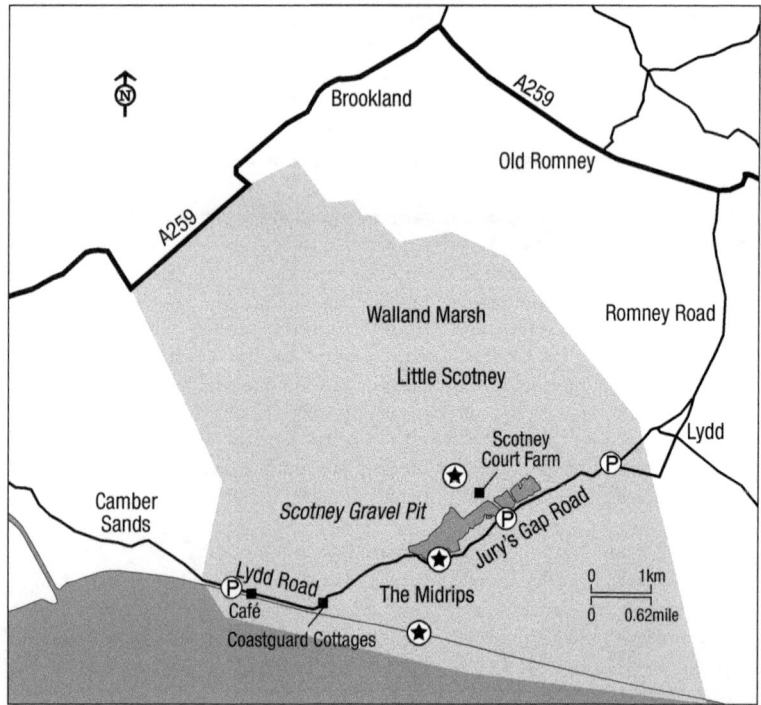

Access is not very wheelchair or pushchair friendly for the most part, although Scotney can be viewed from a parked car on the roadside. The cycle path beside Jury's Gap Road offers wheelchair access to this area.

FACILITIES: Frankie's at the Beach by the Broomhill car park offers a selection of hot and cold food and drinks. Otherwise the closest facilities are in Lydd.

CALENDAR

All year: Barnacle Goose, Egyptian Goose, Shelduck (except late summer/autumn), Great Crested Grebe, Oystercatcher, Lapwing, Curlew, Little Egret, Marsh Harrier, Little Owl, Barn Owl, Peregrine, Raven, Skylark, Cetti's Warbler, Tree Sparrow, Corn Bunting.

April–September: Avocet, Cuckoo, Little Ringed Plover, Whimbrel, Bar-tailed Godwit, Black-tailed Godwit, Common Sandpiper, Greenshank, Little Gull, Common Tern, Sandwich Tern, Hobby, common warblers, Whinchat, Wheatear, Yellow Wagtail.

October–March: White-fronted Goose, Tundra Bean Goose, Bewick's Swan, Whooper Swan, Shoveler, Gadwall, Wigeon, Pintail, Teal, Velvet Scoter, Common Scoter, Goldeneye, Goosander, Little Grebe, scarce grebes, Golden Plover, Dunlin, Redshank, Yellow-legged Gull, Red-throated Diver, Great White Egret, Hen Harrier, Kingfisher.

---

45 RYE HARBOUR (INCLUDING CAMBER SANDS, CAMBER PITS AND NORTHPOINT PIT)

OS Explorer 125
OS grid refs: TQ 932185, TQ 957183, TQ 953191, TQ 936198
Postcode: TN31 7TU

HABITAT

Just a kilometre or so west of the Kent border lies Rye Harbour, one of the richest mosaics of wetland habitats anywhere on the south coast of England. Managed by the Sussex Wildlife Trust, the 465ha of the officially titled Rye Harbour Nature Reserve stretches from Winchelsea Beach to the west all the way east to the mouth of the River Rother, with the small town of Rye just to the north. The final canalised stretch of the Rother separates the main nature reserve from Camber Sands and Northpoint Pit to the east.

Just beyond Winchelsea Beach to the west lies Pett Level, while further to the east beyond Camber, on a clear day, you can see the power station at Dungeness in Kent. Indeed, although it's divided by the county border, Rye is part of the wider Dungeness, Romney Marsh and Rye Bay Ramsar site, SPA, SSSI as well as the

Dungeness SAC. Rye Harbour Nature Reserve itself was designated as an LNR in 1970.

The reserve is divided into two parts: the Beach Reserve to the south and Castle Water to the north. Together, they form an impressive patchwork of habitats including saltmarsh, reedbed, grazing marsh, saline lagoons and vegetated shingle. Just over 300 bird species have been recorded here as well as more than 500 plant species, including the endangered Stinking Hawk's-beard and Least Lettuce. The noisy, non-native Marsh Frog is also well established and makes itself heard in some areas of the reserve. The Discovery Centre near the main entrance to the reserve has lots of information on latest sightings, as well as a café and accessible toilets.

Northpoint Pit (also called Northpoint Water or Northpoint Beach) is a former gravel pit, now primarily used for watersports, but still attractive to waterfowl given its proximity to both Rye Harbour LNR and the sea.

Camber Sands is an extensive accreting sand-dune system (meaning the dunes are gradually growing larger, as 7,500 cubic metres of sand are deposited here each year) just to the east of the river mouth; this is an unusual habitat in Sussex, with many pools and other feeding and resting opportunities available to birds at low tide. This area is also included in the same SSSI and SPA designation as Rye Harbour Nature Reserve. Just to the north of the dunes, on the other side of the Camber Road are two smaller gravel pits, known simply as Camber Pits, which also attract wildfowl in the winter.

SPECIES

The winter months see huge numbers of wildfowl congregating at Rye, especially in the event of particularly cold weather further north or on the Continent, given the site's proximity to the southeasternmost point of England. Wigeon and Teal are the most numerous ducks at this time of year, with somewhat smaller numbers of Shoveler and Pintail among them. Tufted Duck and Pochard are present year-round, especially on the larger pools and pits, but their numbers swell in the winter. Once upon a time, Northpoint Pit was one of the best sites in Sussex for Smew, with several guaranteed here, but sadly this species is rare here now. Scaup still occasionally show up from time to time, with Northpoint and Camber Pits both likely areas for one to be found. Check for scarce grebes here too – Red-necked has been recorded on Camber Pits in the past. Any of the larger pools at Rye Harbour can also produce deep-water birds in the winter, with recent records of Slavonian Grebe and Long-tailed Duck on Long Pit and Ternery Pool, respectively. Large flocks of Canada and Greylag Geese will also gather here from late summer into the autumn and winter along with smaller numbers of Brent Geese, and it is in the midst of winter that these flocks will sometimes draw in scarcer species such as White-fronted Goose, especially in influx years.

A scan of the sea in winter may deliver Red-throated Diver, Gannet, Kittiwake, Guillemot, Razorbill, Red-breasted Merganser and flocks of Common Scoter, the latter sometimes carrying a few Velvet Scoter. Sandwich Terns are increasingly a year-round feature of the coast in this area, so don't be surprised if one flies past during a midwinter seawatch. Good spots for checking the sea are at the benches on the sea wall at the Winchelsea Beach end or the viewpoint just west of the river mouth south of the Discovery Centre. Alternatively, for rather more shelter in windy conditions, tuck yourself in next to the Mary Stanford lifeboat house in front of West Beach. Another worthwhile winter highlight is scanning Castle Water from the reed-bed viewpoint on the eastern side. From here you will be able to enjoy the spectacle of Cormorants, herons and egrets flying into roost, along with an excellent chance of a Bittern – Rye Harbour is one of the few places in Sussex hosting this species. Castle Water also offers an excellent array of ducks and other waterbirds at any time of year, and is one of the few places in Sussex where one stands a chance of Black-necked Grebe, though occurrences are not regular. Another species worth keeping in mind in the winter is Caspian Gull, with several often present in the gull roosts, so always make sure to check through any gatherings of large gulls present. Passerine interest is typically rather lacking in winter, though Cetti's Warbler and Bearded Tit maintain a year-round presence in the reedbeds and ditches. Resident raptors include Buzzard, Sparrowhawk, Kestrel and Marsh Harrier. Red Kite is increasing too, while in winter Merlin may be seen dashing after Meadow Pipits or perched up on a post.

Certain waders are present in good numbers all year-round, with wintering species such as Oystercatcher, Curlew, Ringed Plover, Turnstone and Sanderling most often found scurrying around on the shingle from the river mouth west towards the Beach Reserve. Large numbers of Curlew roost nocturnally either on the New Saltmarsh or the saline lagoons near the lifeboat house. Flat Beach, best viewed from Gooders Hide just past the Discovery Centre, can be particularly good for large numbers of roosting Lapwing and Golden Plover.

While winter offers the spectacle of an abundance of wildfowl, it is during March and April that migration really starts to get into gear here, starting with the arrival of the first Wheatear, Sand Martin and perhaps a Garganey or two. Later in March the first Little Ringed Plover and Common Tern will be seen, while early warblers like Willow Warbler, Chiffchaff and Sedge Warbler may be heard singing, joining the resident breeding passerines such as Skylark, Stonechat and Meadow Pipit. A flagship species at Rye Harbour is Wheatear, with a handful of pairs of this iconic species still breeding thanks to management techniques such as pipes dug into the shingle to provide artificial nest sites. It's a real treat to hear their song in the spring (not something a birder from South-East England hears all too often) and see the fluffy juveniles hopping about on the shingle in the summer.

April is the month when migration really hots up, with numbers of arriving passerines and waders particularly of note here on good days, as well as more of the breeding Common and Little Terns. Notable arrivals of some species such as Whitethroat and Sedge Warbler can occur in the right conditions, while the diversity and numbers of waders on the pools on the reserve will markedly increase through the month and into May. Whimbrel can be particularly numerous here in late April and sometimes form impressive roost gatherings near the Ternery Pool. Other returning breeders at this time of year include Cuckoo, Reed Warbler and Sedge Warbler. The huge sky views offer ideal conditions to pick up an Osprey, Honey Buzzard or Hen Harrier drifting over at this time of year as well. A Hobby or two will often be seen hunting over the reserve in the summer, with sometimes several feeding together over Castle Water in the spring. Wader passage really hots up in April, too, with a host of species moving through the reserve including Avocet, Whimbrel, Little Stint, Curlew Sandpiper, Wood Sandpiper, Green Sandpiper, Greenshank and Spotted Redshank.

Black-headed Gull, Mediterranean Gull, Sandwich Tern and Common Tern breed on the islands in the Ternery Pool, along with a few pairs of Little Tern, Oystercatcher and Ringed Plover on the open shingle and inside the predator fence. Add to this the presence of roosting terns in late April and May and the result is an impressively raucous chorus of calls, best experienced from Crittall Hide or Parkes Hide. The areas of wet grassland, meanwhile, support around a dozen breeding pairs of Lapwing.

In spring and summer it is also worth checking the area towards the western side of the reserve near Harbour Farm, as the barns and telegraph wires here are attractive to Collared Dove which sometimes draws in a Turtle Dove, though the species sadly no longer breeds here. Another area nearby worth checking for this species is along the railway line just south-west of Rye village, near Gibbet Marsh (TQ 914198).

Despite many of the breeding species continuing on to second or even third broods, for some the return migration season begins as early as July with the first northern waders such as Green Sandpiper, Black-tailed Godwit and Little Ringed Plover beginning to build in number, sometimes joined by rarer species. Indeed, there have been July records here of Baird's Sandpiper and Wilson's Phalarope.

Into autumn proper, and the wader and wildfowl numbers will continue to increase, but passerine and hirundine numbers will also begin to build. In September, many thousands of Swallows, Sand Martins and House Martins can gather here, feeding over the pools ahead of their southward migration. Large numbers of Yellow Wagtails can also be seen moving overhead or scurrying around the place, along with Wheatears and Whinchats characteristically perched up on fenceposts and bushes like sentries on duty. Other passerines that can turn up here in varying numbers in late summer and autumn include Redstart, Spotted Flycatcher and *Sylvia* warblers.

Late autumn and early winter see the year come full circle and the return of wintering species such as Common and Jack Snipe and Rock Pipit, or perhaps a Hen Harrier or Short-eared Owl passing through. The latter species doesn't often winter here these days – nor does Long-eared Owl, which famously used to roost in some numbers.

Given the species total for Rye Harbour is now over 300, it's not surprising that the list of scarcities and rarities to have occurred here is compelling. In recent years there have been records of Ring-necked Duck, Black Kite, Collared Pratincole, Broad-billed Sandpiper, Black-winged Stilt, Terek Sandpiper, Lesser Yellowlegs, Pacific Golden Plover and White-winged Black Tern.

TIMING

Rye Harbour LNR and the other sites mentioned offer a rewarding visit at any time of year, but as with any wetland perhaps the most exciting times are spring and autumn, when almost anything can occur. In winter, the exposed nature of the landscape here can be unforgiving, so remember to dress appropriately for very cold weather – then add another layer! It is worth it, though, for the sight of thousands of wildfowl and waders beautifully lit in the low winter sunlight. The area is also very popular with dog walkers and beachgoers, so do bear this in mind if you plan to visit on sunny weekends, bank holidays or during the school holidays. Certain areas tend to be less busy, however, with crowds more easily avoided around the centre of the reserve such as Barn Pool and Harbour Farm.

ACCESS

The nearest railway station is in Rye town, just a few minutes' walk from the northern part of the reserve, with the footpath access directly from the town via the Saxon Shore Way, which crosses Brede Lock then continues south to the west of Camber Castle before joining the various footpaths that criss-cross the reserve once you reach Castle Farm (TQ 919176). Bus route 313 (Stagecoach South East) runs from Rye to Rye Harbour village every couple of hours. To get to the Rye Harbour Discovery Centre, from where most of the main trails begin, is around an hour's walk from the train station, but there are many birds to enjoy on the way.

The main car park (donation requested) at TQ 941189 can be reached by following the Rye Harbour Road off the A259 (TQ 918198), while there is also a small car park at the end of Dogs Hill Road on the Winchelsea Beach side (TQ 917160). If entering from the main entrance towards the Discovery Centre, you will find there are several different signposted trails around the reserve, varying between 2km and 10km, depending on your enthusiasm and energy levels.

The short trail (taking in the New Saltmarsh and Flat Beach) and the Beach Reserve path are both fully surfaced and suitable for wheelchairs and pushchairs all year-round, while some of the other paths are seasonally passable, with some uneven and loose surfaces, but are fine for mobility scooters. For Camber Sands, parking is available at TQ 959188, with access to the dunes just a short walk across the sand from here. Some of this area can also be viewed from the western side of the river mouth at Rye Harbour. For Northpoint Pit there is a car park at TQ 936200, while for Camber Pits there are a couple of pull-ins on the north side of the road or, alternatively, park at the western end of Camber village.

For all the sites in this area a telescope really is recommended to get the most out of your day, as some of the birds are very distant on the pools and pits.

FACILITIES: Car park, accessible toilets and hides are available. Café and shop in the Discovery Centre, open from 10:00 am–4:00 pm. Wheelchair and mobility scooter access to the Discovery Centre and most of the hides.

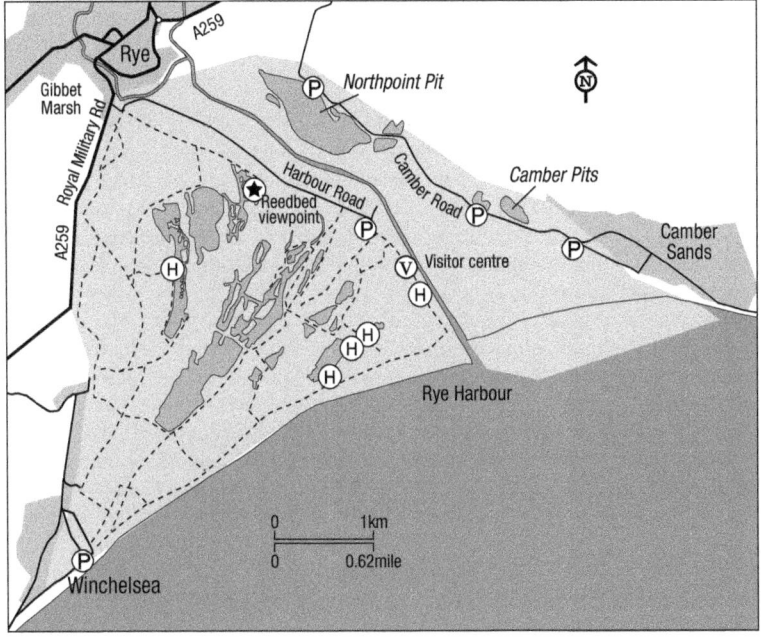

CALENDAR

All year: Egyptian Goose, Shelduck, Shoveler, Gadwall, Teal, Pochard, Common Scoter (at sea), Red-breasted Merganser (at sea), Grey and Red-legged Partridges, Little Grebe, Great Crested Grebe, Water Rail, Avocet, Oystercatcher, Grey Plover, Lapwing, Ringed Plover, Curlew, Bar-tailed Godwit, Black-tailed Godwit, Turnstone, Knot, Ruff, Sanderling, Dunlin, Common Sandpiper, Spotted Redshank, Redshank, Sandwich Tern, Gannet, Bittern, Great White Egret, Little Egret, Spoonbill, Marsh Harrier, Kingfisher, Peregrine, Raven, Skylark, Bearded Tit, Cetti's Warbler, Stonechat, Meadow Pipit, Reed Bunting.

April–June: Brent Goose, Garganey, Cuckoo, Swift, Little Ringed Plover, Whimbrel, Greenshank, Little Gull, Mediterranean Gull, Little Tern, Common Tern, Hobby, Sedge and Reed Warblers, Swallow, Sand Martin, House Martin, Wheatear.

July–September: Garganey, Black-necked Grebe, returning waders as in the spring but with an increased chance of Little Stint, Green Sandpiper, Greenshank, Wood Sandpiper, Yellow-legged Gull, terns, warblers and hirundines as in April–June, Whinchat, Yellow Wagtail, Rock Pipit.

October–March: White-fronted Goose, Brent Goose, Wigeon, Pintail, Scaup, Goldeneye, Smew, Goosander, Black-necked Grebe (and sometimes Slavonian), Jack Snipe, Snipe, Caspian Gull, Red-throated Diver, Merlin.

46 PETT LEVEL

OS Explorer 124/125
OS grid ref: TQ 903154
Postcode: TN35 4EH

HABITAT

Pett Level is tucked away behind the sea defences between Fairlight and Rye. This is an area of wet grassland, criss-crossed by ditches and, while relatively small in comparison to other 'levels' in the South-East, its sheltered location just a stone's throw from the sea makes it attractive to a host of bird species. It also offers easy viewing for the visiting birder, as most of the habitat can be easily checked from either the roadside or the sea wall. The elevated viewing position of the sea wall also offers an ideal opportunity to scan the sea, which can be productive.

Towards the western end are situated the Colonel Body Memorial Lakes (sometimes known simply as 'Pett Pools'), dug in the 1940s to produce material for the sea defences. To the north-western side is a small area of marshland, restored and managed as a private nature reserve in its own right, known as Pannel Valley, where viewing is possible from the small path which follows the Pannel Sewer, off the main Royal Military Canal footpath at TQ 895156. There are three small hides along here too, although they are not always open.

SPECIES

Impressive numbers of wildfowl gather on the sheep-grazed grassland in the winter, especially Wigeon and Teal. Geese also congregate in the many hundreds, primarily Greylag, but also Canada, Brent and Egyptian Geese, along with an increasing number of feral species. Pett Level is the only semi-reliable site in Sussex for White-fronted Goose, with varying numbers present in most winters. Rarely a Pink-footed or Tundra Bean Goose may be among them. Green Sandpiper and Water Pipit can be found in the ditches at this time of year too.

As well as the dabbling ducks grazing on the fields and swimming in the pools, the Memorial Lakes are deep enough to support diving ducks, with Tufted Duck and Pochard both present here year-round. Rare species such as Ring-necked Duck and Ferruginous Duck have occurred here on occasion too. Garganey often occur from March until late summer/early autumn, while Coot, Mallard, Gadwall, Teal, Shoveler, Wigeon and Pintail gather in the winter.

A variety of waders take advantage of the rich foraging habitat in the wet grassland, with species such as Curlew, Lapwing, Redshank, Ruff, Golden Plover and occasional Bar-tailed and Black-tailed Godwits seen just about year-round, and other species moving through in the passage months, such as Little Ringed Plover, Whimbrel, Greenshank and Common Sandpiper. In late summer, during dry years, the water levels can drop enough for the muddy margins of the Memorial Lakes to attract scarcer species such as Curlew Sandpiper, Little Stint, Wood Sandpiper and Spotted Redshank.

Little Egret and Grey Heron are a common sight throughout the year, with Great White Egret becoming an increasingly regular sight outside the breeding season. Cattle Egret, Glossy Ibis and Spoonbill are all increasing and this is as good a site as any on the south coast to encounter these species.

Marsh Harrier, Buzzard and Peregrine are a year-round sight here now hunting over the fields. Hobby is regular in summer and autumn, while Merlin and Short-eared Owl will sometimes be seen in the winter. The reedbeds around the Memorial Lakes support breeding Water Rail, Bearded Tit, Cetti's Warbler, Sedge Warbler, Reed Warbler and Reed Bunting, and sometimes a wintering Bittern.

Although seawatching in the bay is usually poor, a scan from the sea wall can be rewarding in winter, with large numbers of Great Crested Grebe, Red-throated Diver and Common Scoter sometimes seen; the latter are always worth checking through for Velvet Scoter, Surf Scoter and Long-tailed Duck in winter and spring. Common and Sandwich Terns are a common sight, patrolling up and down the shore, and Mediterranean Gull is now abundant in the area too. Turnstone, Oystercatcher, Curlew, Redshank, Grey Plover and Dunlin will invariably be scurrying about on the beach, with a good chance of a Wheatear (or several) in spring and autumn. The sea wall occasionally hosts a Snow Bunting in the winter and, even more rarely, Shore Lark or Lapland Bunting. Fulmar nests on the sandstone cliffs, in front of which Black Redstart can sometimes be found on the beach.

At the western end of the Level is an area of rough grass, reed and scrub around Toot Rock, managed by the Pett Level Preservation Trust. This can be productive in autumn for small migrants such as warblers, chats, crests and, in some years, Ring Ouzel. At that time, Toot Rock is also a good place from which to watch diurnal migration of wagtails, pipits, and finches including Crossbill.

For such a rich site for birds, it's no surprise that Pett Level has turned up an impressive array of rare and scarce species over the years including King Eider, Black-winged Stilt, Pectoral Sandpiper, Temminck's Stint, Sociable Lapwing, Squacco Heron, Common Crane, Grey and Red-necked Phalaropes, Black Kite, Red-footed Falcon, Common Rosefinch, Purple Heron and Lesser Yellowlegs.

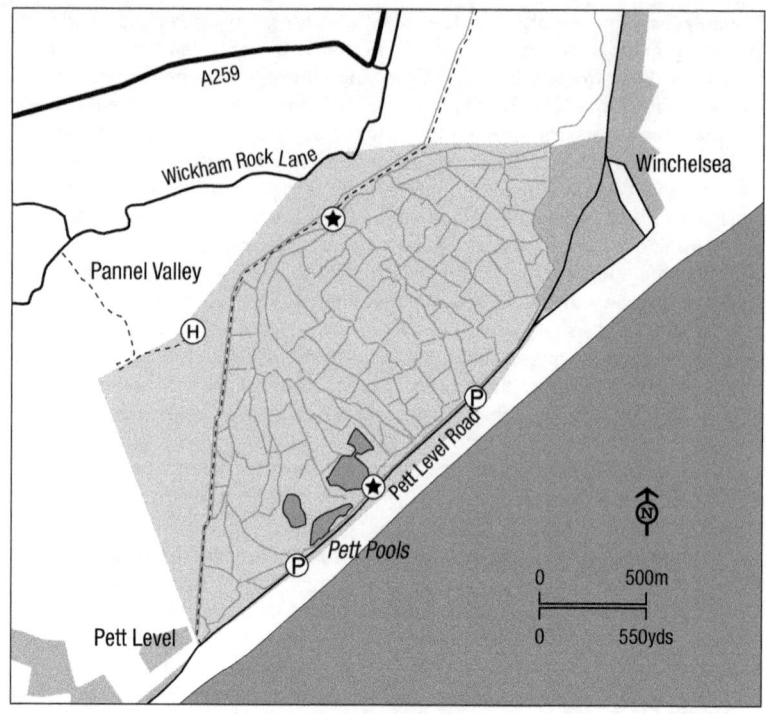

TIMING

Winter is best for the greatest assemblages of wildfowl and common waders. For the chance of scarcer species, a spring or autumn visit is recommended.

ACCESS

Public transport access to Pett Level is via train to Hastings, then a bus to Pett. Trains from Brighton stop at Hastings once an hour. Bus route 101 (Stagecoach South East) runs once an hour Monday-Saturday and once every two hours on Sunday, and stops at various points along the Pett Level Road.

By car, turn off the A259 onto Fairlight Road and follow this all the way along through Fairlight and Fairlight Cove, where it becomes Pett Level Road. This eventually winds its way to Pett Level itself, with parking available by the roadside. Although viewing much of the habitat here is possible from the sea wall, there are also various public footpaths on offer, including a handy loop following the Royal Military Canal from TQ 894138 up to the Saxon Shore Way then east and back south to the road further along from Newgate Cottages (TQ 900161) towards Winchelsea Beach.

FACILITIES: The nearest facilities are in Winchelsea.

CALENDAR

All year: Egyptian Goose, Shelduck, Shoveler, Gadwall, Teal, Pochard, Common Scoter (at sea), Little Grebe, Great Crested Grebe, Oystercatcher, Lapwing, Ringed Plover, Curlew, Black-tailed Godwit, Turnstone, Redshank, Mediterranean Gull, Fulmar, Gannet, Little Egret, Marsh Harrier, Kingfisher, Raven, Skylark, Bearded Tit, Cetti's Warbler, Reed Bunting.

April–September: Garganey, Avocet, Little Ringed Plover, Whimbrel, Common Sandpiper, Greenshank, scarcer waders, Common Tern, Sandwich Tern, Hobby, Sedge and Reed Warblers, hirundines, *Sylvia* warblers, Wheatear, Yellow Wagtail.

October–March: White-fronted Goose and other scarce grey geese, Brent Goose, Wigeon, Goosander, Red-breasted Merganser, Grey Plover, Golden Plover, Knot, Ruff, Red-throated Diver, Great White Egret.

47 HASTINGS COUNTRY PARK AND FAIRLIGHT CLIFFS

OS Explorer 124
OS grid ref: TQ 851107
Postcode: TN35 4AG

HABITAT

Nestled in the 'gap' between Hastings to the west and Fairlight Cove to the east, Hastings Country Park LNR, owned by Hastings Borough Council, is the largest area of public open space in the borough. As well as receiving its LNR designation in 2006, the reserve is also part of the wider Hastings Cliffs SAC and Hastings Cliffs to Pett Beach SSSI. A rich mix of habitats sit atop the sandstone and clay cliffs here, with glens of ancient ghyll woodland (Warren Glen, Fairlight Glen and Ecclesbourne Glen), open pasture, heathland, farmland and semi-natural grassland.

As well as being host to an array of resident and migrant birds, the park is also rich in invertebrate life, supporting species such as Wall Brown butterfly, Glow-worm, Grey Bush-cricket, and Chalk Carpet and Webb's Wainscot moths. In spring there are spectacular displays of Bluebell, Red Campion and Wild Garlic in the glens, while the heathland area at Firehills turns purple in late summer as the heathers flower. There is also archaeological interest, with much of East Hill listed as a scheduled ancient monument.

Fairlight Cliffs stretch from Hastings Country Park east towards Pett Level. The site is part of the 163-mile-long Saxon Shore Way path which stretches all the way from Hastings to Gravesend in North Kent. The cliffs at the park are managed by Hastings Borough Council, while the cliffs east of the park are looked after by the National Trust. Although very much part of the wider area in terms of its appeal for passerine migrants, the cliffs here are primarily known in the birding community for their seawatching appeal.

SPECIES

Hasting Country Park packs a rich mosaic of habitats into its 345ha and to get the most out of a visit here you should allow at least a whole morning, especially in spring and autumn when the area will be at its most bird-filled. That's not to say there isn't interest at other times of year, however, as a winter visit here may produce a pleasant selection of woodland and open-country species such as tits, Linnet, Stonechat, Meadow Pipit and Dartford Warbler, which is resident in the areas of gorse.

From March onwards though, Hastings CP really bursts into life, with the arrival of the first Wheatears and Black Redstarts heralding the start of spring migration. The country park is a very well-placed area of undeveloped coastal habitat, especially its wooded glens which are magnets for a host of passerine migrant species, from common warblers such as Chiffchaff, Blackcap, Willow Warbler, Garden Warbler, Whitethroat and Lesser Whitethroat, to somewhat scarcer passage species like Nightingale, Redstart and Black Redstart (the latter having bred at the bottom of the cliffs in recent years), Tree Pipit, Whinchat, and Spotted and Pied Flycatchers.

Any or all of these species are possible on a good day, especially in late summer and early autumn when the scrub and glens can be teeming with birds. It is in the autumn proper (September–November) that other scarce migrants may be considered more likely to be found here too, from Ring Ouzel to Yellow-browed Warbler. Being the last patch of suitable habitat along this stretch of coastline, Ring Ouzels sometimes gather here in numbers, with over a dozen having been recorded on good October days. It is also another fairly reliable site for Wryneck, with September being the best month to look. Even scarcer and rarer passerine species encountered in recent years include Serin, Pallas's Warbler and Red-flanked Bluetail. Despite this fine selection, Hastings CP is a little under-watched and thus offers any visiting birder the opportunity to find something unusual.

Perhaps the main autumn birding interest here comes from visible migration which can be spectacular. Viewing is best from Firehills early morning from late September onwards when pipits, wagtails and thrushes, among others, may be seen passing overhead in large numbers on good days.

Raptors such as Buzzard, Sparrowhawk and Peregrine may be seen here at any time of year (the latter breeds on the cliffs), with sometimes a Hobby or Merlin dashing about in summer or autumn, respectively.

Seawatching from Firehills or Rock-a-Nore at the western extreme of Fairlight Cliffs can be rewarding, though not quite so much as more notable headland sites because the birds are generally more distant. Nonetheless, good days in late winter and early spring may produce dozens of divers and hundreds of auks and wildfowl moving east, followed later in April and May by terns, more wildfowl (including Common and Velvet Scoter) and perhaps a Pomarine Skua or two. Sooty Shearwater usually puts in an appearance in the autumn, but this is far from guaranteed.

TIMING

Hastings CP is a very popular tourist destination, with hundreds of visitors on pleasant weekend days and bank holidays. An early start is highly recommended for this reason, especially in spring and autumn when the birding will be at its best. Late August and September are likely to bring the greatest abundance and diversity of migrant species, especially during periods of easterly winds, while early morning from the high ground at Firehills is best for vis-mig.

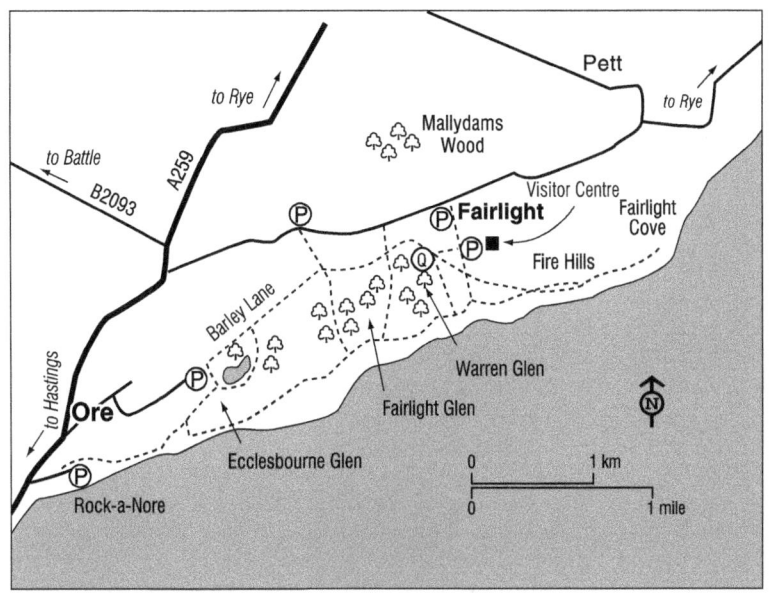

ACCESS

The closest railway station is at Ore, some 3km west of the park, but the area is well served by bus routes, including numbers 20, 100 and 101 (all Stagecoach South East) all stopping at Old Town just west along the coast. From here it's about a 2km walk east along the Saxon Shore Way to reach the Park. By car, follow Fairlight Road off the A259 to reach the car park at TQ 848117. Alternatively, there are car parks on the seafront in Old Town at TQ 828094, near Shear Barn Farm holiday park at TQ 838105 and in Fairlight village at TQ 859118 and TQ 860115. A network of public footpaths from all the car parks lead into the country park, though the steepness generally makes them unsuitable for visitors with limited mobility .

FACILITIES: Several car parks, accessible toilets at the northern and eastern car park and picnic area at the northern car park. Visitor centre at the Fairlight car park which includes a café serving drinks and light meals.

CALENDAR

All year: Common Scoter (at sea), Oystercatcher, Fulmar, Gannet, Peregrine, Raven, Marsh Tit, Skylark, Stonechat, Meadow Pipit.

April–September: Sandwich Tern, Common Tern, Hobby, Sedge, Reed, Willow and Garden Warblers, Lesser Whitethroat, Spotted Flycatcher, Redstart, Whinchat, Wheatear, Yellow Wagtail, Tree Pipit, Yellowhammer.

October–March: Brent Goose (at sea – especially in late winter into spring), Great Crested Grebe (at sea), Snipe, Merlin, Firecrest, Ring Ouzel, Grey Wagtail.

48 COMBE VALLEY COUNTRYSIDE PARK AND FILSHAM REEDBED

OS Explorer 124
OS grid refs: TQ 760103 and TQ 777097
Postcode: TN38 8AL

HABITAT

Combe Valley Countryside Park was once a landfill site but, thankfully, in the 1990s and early 2000s was developed into its current state, in a combined effort by Hastings Borough Council, Rother District Council and Sussex County Council. The Park incorporates over 6km² of land between Bexhill and Hastings, extending from Bulverhythe at its southern boundary up to the A2690 to the north, plus a finger of land beyond which extends all the way up to Crowhurst, as well as some 2.5km of shoreline from Galley Hill to Glyne Gap. The Park is criss-crossed by various watercourses including the Powdermill Stream, Combe Haven and the Watermill Stream. Combe Haven passes alongside Filsham Reedbed before it reaches the sea just east of Bulverhythe.

Filsham Reedbed, owned by Hastings Borough Council and managed by Sussex Wildlife Trust, is one of the largest reedbeds in Sussex (19ha). Designated as an LNR and SSSI along with the rest of the valley, the area is rich in all manner of invertebrate life, including many dragonfly species and the Yellow Loosestrife Bee, which thrives here thanks to an abundance of its namesake plant.

SPECIES

Combe Valley Countryside Park packs a lot of diverse habitats into its 640ha and as such hosts an impressive selection of birdlife. Sixty or more species are possible on a good morning, if you have time to cover as much of the habitat as possible.

The whole valley basin floods in the winter, sometimes turning into one giant waterbody, as it is intended to do to prevent flooding of residential areas downstream at Bulverhythe. Needless to say, this proves very attractive to wildfowl, with many hundreds of Wigeon, Teal, Pintail and Shoveler often present. Deeper-water ducks will appear in periods of flooding too and, in recent winters with very high water levels, there have even been occasional records of Whooper Swan and Scaup. Where wet areas persist into the spring, a Garganey or two is possible. There will often be large flocks of Greylag and Canada Geese, and these sometimes attract a few White-fronted Geese, particularly in influx years for the species. The lake at TQ 766091 is always worth checking for waterbirds, with all the likely species present. It's also one of the most reliable spots here for Kingfisher and even had a Penduline Tit years ago in the Reedmace around the edges.

Snipe and Lapwing gather in the winter, with a few pairs of the latter still breeding within the confines of the park. Redshank has attempted to breed in recent years but otherwise, in recent decades at least, are just a reasonably frequent passage migrant here.

Water Rail are commonly heard at any time of year as they breed in Filsham Reedbed. Grey Heron, Little Egret and, increasingly, Great White and Cattle Egrets are likely year-round, the latter out in the open areas of grazed farmland, especially

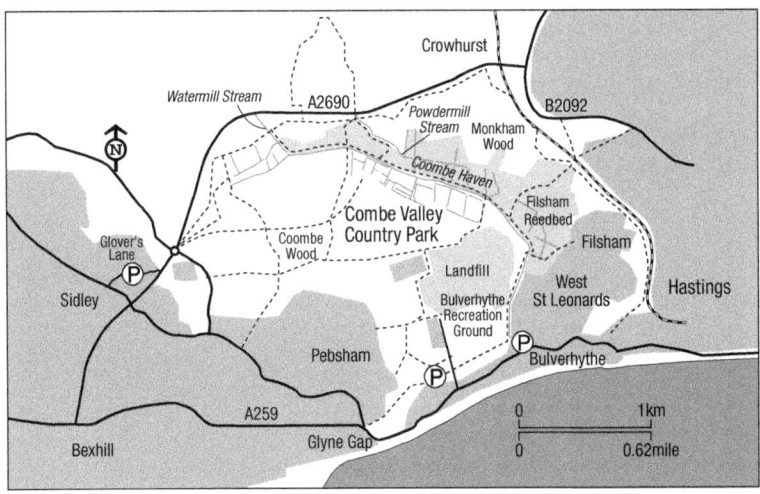

given the recent development of a small heronry, supporting five or six breeding pairs of Grey Heron. Bittern very occasionally winters at Filsham but is typically elusive.

Cetti's Warblers are abundant right across the park, with perhaps 40 or more territories in recent years. Reedbeds such as at Filsham and Pebsham hold breeding Reed Bunting, Sedge and Reed Warblers in the summer. Bearded Tits once bred too but are now more of a winter feature, particularly at Filsham. The areas of woodland hold all the expected species such as common tits, Bullfinch, Goldcrest and Treecreeper, but also Marsh Tit and even Hawfinch on occasion. A Cuckoo will often be heard in spring and early summer, along with Garden Warbler and Lesser Whitethroat around the areas of scrub and woodland edges, which will sometimes produce more unusual passerine passage migrants like Pied Flycatcher, especially in the autumn. The more open arable areas support breeding Yellowhammer.

The whole area is excellent for Water Pipit in winter, particularly some of the central sections of the park, although this is largely dependent on water levels and the birds can be very elusive at times.

TIMING

Winter is best, especially for wildfowl and Water Pipit. Some areas can become busy with dog walkers and cyclists, especially at weekends, so early morning visits are recommended.

ACCESS

The nearest railway stations are at Bexhill, Crowhurst and West St Leonards, with all of these being roughly 1.5km from the closest entrance to the park. Bus routes 98 and 99 (Stagecoach South East) all stop near Bulverhythe Recreation Ground, offering access to the southern entrances to the park. By car, there is a car park just north of the A259 at TQ 775087 with a public footpath directly north from here following the river up to Filsham Reedbed and beyond. There is also a limited amount of parking at the end of Glover's Lane in Sidley at TQ 745093. The Combe Valley Greenway path which runs east–west from Monkham Wood to

Sidley offers good access across the park all year-round, including for pushchairs and wheelchairs, but other paths may become extremely muddy or entirely impassable, especially in very wet winters.

FACILITIES: Aside from the car park on the south-eastern side, the nearest facilities are in Hastings.

CALENDAR

All year: Water Rail, Oystercatcher, Lapwing, Turnstone, Little Egret, Chiffchaff, Cetti's Warbler, Yellowhammer, Reed Bunting.

April–September: Cuckoo, Common Sandpiper, Green Sandpiper, Redshank, Hobby, Sedge and Reed Warblers, Garden Warbler, Lesser Whitethroat, Wheatear, Yellow Wagtail.

October–March: White-fronted Goose, Shoveler, Wigeon, Pintail, Teal, Snipe, Bittern, Marsh Harrier, Kingfisher, Bearded Tit, Firecrest, Stonechat, Water Pipit.

---

49 PEVENSEY LEVELS

OS Explorer 124 and OL25
OS grid refs: TQ 664058 (Pevensey Bridge Level), TQ 683069
(Hooe Level), TQ 616083 (Horse Eye and Down Levels)
Postcodes: BN24 5JW, BN24 6QG, BN27 2SD

HABITAT

Covering an impressive 14 square miles, or around 4,000ha, Pevensey Levels is the largest single area of wetland in East Sussex. In fact, it represents one of the largest lowland wet grazing systems in all of South East England. Although historically drained for agriculture, the site remains of great wildlife value, not least because of its complex and extensive ditch network, many of which are lined with reeds, adding to the available habitat for breeding birds. Much of the area is now managed as wet pasture, with grazing sheep and cattle present in varying areas and numbers throughout the year, all helping to nurture the wet grassland landscape which is so important for breeding waders, as well as poaching the edges of ditches and creating all sorts of ephemeral pools.

The Levels are divided into nine sections, with Hooe and Pevensey Bridge Levels near to Bexhill, and Horse Eye and Down Levels near Hailsham generally the best and most well-watched areas. That is to say, the most well watched in the context of the whole site which, given its size, is likely under-watched in comparison to other notable birding locations in the county.

The area boasts an impressive list of conservation protections and designations, including being listed as a Nature Conservation Review site, a Ramsar site, SSSI and SAC. Towards the southern end, the 180ha Pevensey Marshes is also designated as an NNR, with 150ha of this owned and managed by Sussex Wildlife Trust. As well as being rich in birdlife, the Levels are also home to various other rare creatures, including Shining Ram's-horn Snail and Fen Raft Spider, as well as several scarce aquatic plants including Frogbit and Sharp-leaved Pondweed.

SPECIES

Flocks of wildfowl gather in the winter, especially Wigeon and Teal, and it's worth scanning any of the many little pools for Garganey from March onwards. Mute Swans are plentiful in the winter months, with up to a hundred wintering in the area, and there is always the outside chance of a Whooper or Bewick's or two among them. Likewise, a scan of the assembled Greylag and Canada Geese could sometimes produce something rarer, such as White-fronted or Tundra Bean Goose, particularly in the second half of winter. Bitterns are sometimes seen in the winter too, although can be typically elusive. Dartford Warblers are sometimes sighted here in late autumn and occasionally overwinter in the area. Water Pipit occurs outside the breeding season, wintering in small numbers. Bearded Tit and Firecrest are both recorded fairly regularly during the winter too.

As winter gives way to spring, however, the Levels really begin to burst into life. Pevensey is one of few locales in Sussex still able to boast of hosting breeding Yellow Wagtail in the summer, albeit in small numbers. Listen for their *sweep* calls as you walk around. The reed-filled ditches are home to Sedge, Reed and Cetti's Warblers as well as Reed Bunting, while in the hedgerows and areas of scrub you can hear Whitethroat and Lesser Whitethroat, Chiffchaff, Blackcap, Linnet and sometimes Grasshopper Warbler.

In spring (and autumn), almost any of the classic migrant passerines are possible, especially Wheatear and Whinchat, but also Spotted Flycatcher and Redstart.

The wetter areas are also very attractive to waders, not just the breeding Lapwing and Redshank but also the likes of Wood Sandpiper, Greenshank, Green Sandpiper, Black-tailed Godwit, Dunlin, Snipe and Whimbrel, and on occasion rarer species such as Black-winged Stilt. Spring is generally better than autumn for waders here.

Grey Herons maintain a year-round presence (there are three heronries in the area) and all three egret species are now a routine occurrence here, with Little and Great White Egrets favouring the ditches and Cattle Egrets sometimes following the grazing animals around.

The habitat at Pevensey is ideal for both Marsh Harrier and Hen Harrier, with the former present year-round now, as it continues to increase as a breeding species in the South East of England. Hen Harrier is usually around in the winter months, as is Merlin, although often the mass flushing of wildfowl and passerines may be the first or only indication that this species is present. A Hobby (or sometimes several) will often be seen in spring and summer, taking full advantage of the plentiful insect populations and the hirundines that these attract. Peregrine, Kestrel and Buzzard are recorded all year, while there is also the chance of a fly-over Honey Buzzard or Osprey in spring and autumn. Barn Owl and Short-eared Owl can often be seen quartering at dusk in the winter, although the latter species is of course irruptive, so not guaranteed every year. For owls and Hen Harriers a good place to position yourself at dusk is near Lookers Cottage, at the junction of New Bridge Road and the public footpath that runs west towards Marshfoot Lane, Hailsham (TQ 625097). There is also parking space for a couple of cars on the roadside here.

For such a large and ecologically rich area it's no surprise that the list of scarcities and rarities to have occurred at Pevensey Levels is very impressive, including Sociable Lapwing, Broad-billed Sandpiper, Purple Heron, Marsh Sandpiper and Oriental Pratincole.

TIMING

The Levels are arguably at their best from late autumn through to the winter, especially for wildfowl, herons and egrets, raptors and owls. For passage waders and wildfowl visit in spring, while late spring and early summer are the time for breeding migrants such as Cuckoo, Hobby and Yellow Wagtail.

Any time of day can offer rewards, though for owls and wintering raptors a pre-dusk winter visit is recommended.

ACCESS

There are many kilometres of public footpaths and little lanes criss-crossing the whole area, offering plenty of access to view the various areas, although inevitably some spots are less easy to view or indeed completely without public access. To get the best out of this site, you need to allow a couple of hours at least and a telescope really is essential. Sadly, given its size and relative remoteness, much of the area is tricky to get to via public transport, but we have attempted to give as much information as we can for the main areas outlined earlier in this account.

For Hooe Level, enter from any of the public footpaths near the Star Inn in Normans Bay (TQ 686061), although be aware parking is rather limited in this area. The nearest train station is at Normans Bay.

For Pevensey Bridge Level by car, follow Sluice Lane from Normans Bay which offers various little pull-ins from which to scan the landscape, including at

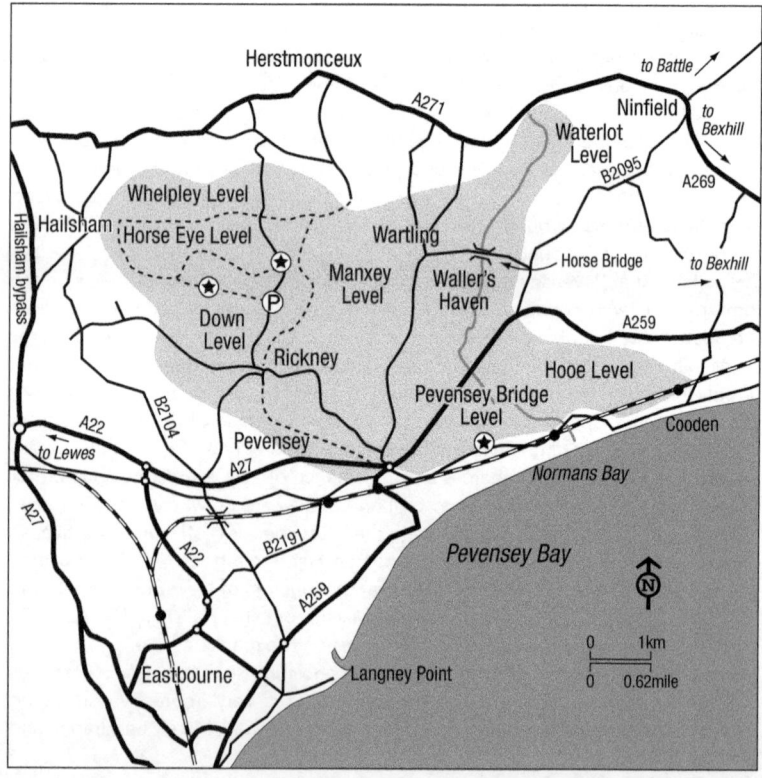

TQ 674056. The nearest railway station is Pevensey Bay to the west. Various bus routes stop in and around Pevensey and Pevensey Bay, including 8, 99 (both Stagecoach South East) and 195 (Cuckmere Buses), although some only run once a week so it's worth checking recent timetables before you travel.

For Horse Eye Level by car, follow New Bridge Road north from Rickney for just over a kilometre and park just before the bridge at TQ 627081. From here you can follow the public footpath west, with Down Level also viewable to the south. There are various opportunities to stop and scan both sides of the path, including at TQ 620083. If you're feeling energetic, continue west to White Dyke Farm, then head north-east from here to follow a 5km loop which eventually brings you back out a little further north along New Bridge Road from where you have parked. The nearest railway station is at Polegate, around 4.5km to the south-west. Bus routes 45 and H1 (both Cuckmere Buses) stop on the eastern fringes of Hailsham, less than 1km west of White Dyke Farm.

Note that in the winter some of the paths can become completely impassable due to flooding or deep mud.

FACILITIES: The nearest facilities are in Hailsham and Eastbourne.

CALENDAR

All year: Egyptian Goose, Water Rail, Lapwing, Great White Egret, Little Egret, Cattle Egret, Marsh Harrier, Kingfisher, Peregrine, Raven, Skylark, Cetti's Warbler, Meadow Pipit, Reed Bunting.

April–September: Shelduck, Garganey, Cuckoo, Whimbrel, Dunlin, Green Sandpiper, Wood Sandpiper, Redshank, rarer waders, Hobby, Sedge Warbler, Reed Warbler, Grasshopper Warbler, Swallow, House Martin, Willow Warbler, Lesser Whitethroat, Wheatear, Yellow Wagtail.

October–March: White-fronted Goose, Tundra Bean Goose, Shoveler, Wigeon, Pintail, Teal, Golden Plover, Green Sandpiper, Jack Snipe, Snipe, Common Gull, Hen Harrier, Barn Owl, Little Owl, Short-eared Owl, Merlin, Stonechat, Water Pipit.

---

50 SOVEREIGN HARBOUR

OS Explorer OL25
OS grid ref: TQ 642015
Postcode: BN23 5BJ

HABITAT

Sovereign Harbour, on the eastern side of Eastbourne, has risen to prominence as a birding destination in recent years, thanks mainly to the sheltered winter refuge it offers to divers, auks and seaduck.

Opened in 1993, it is the largest marina complex of its kind in Northern Europe, with five linked harbours – the Outer, Inner, West, North and South, as well as the Waterfront retail and restaurant development. The development of the marina was controversial, owing to the loss of the Crumbles, a former area of shingle and gravel pits that was a valuable wildlife site in its own right.

Nonetheless, Sovereign Harbour is an ideal place to drop in for half an hour's birding on a winter's day, offering easy viewing from one of the various walkways and car parks, and it can readily be combined with a trip to nearby West Rise Marsh in the same morning. In addition to the avian interest, Sovereign Harbour is also a very good place to see Common Seal, with sometimes several hauled out on the mudflats.

SPECIES

While offering the possibility of gulls, Sandwich Tern and Oystercatcher in the summer months, Sovereign Harbour is primarily a winter birding destination, especially if you are keen to find a diver, auk or scarce duck or grebe. Generally, the Outer Harbour is the main area of interest for species such as these.

Great Northern, Red-throated and Black-throated Divers have all been recorded in recent winters, with all three species present simultaneously on one occasion. The latter is rare in Sussex, though. The beauty of this site is that the birds are often close to the harbour walls, offering great views on relatively calm waters. Shag, Razorbill and Guillemot often turn up in the winter too, and Black Guillemot and

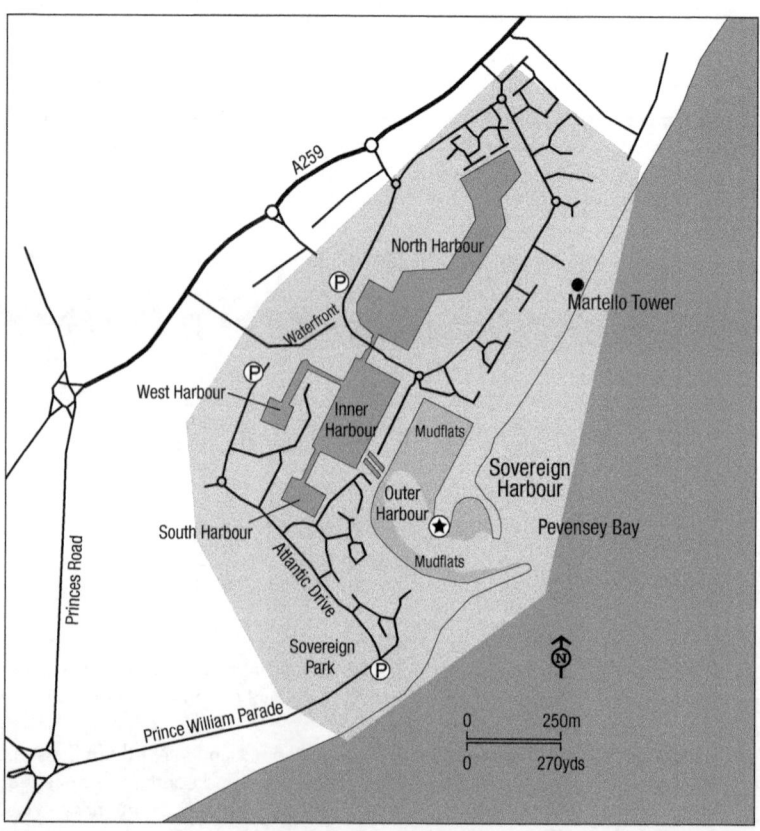

Velvet Scoter have overwintered. Waders such as Ringed Plover, Turnstone, Redshank and Dunlin may also be found on the tidal mudflats around the edge of the Outer Harbour. It's also always worth checking through any gulls you see, especially in the winter, as Sabine's, Iceland and Glaucous Gulls have all been recorded in recent years.

It's not just about the seabirds though. Black Redstarts are regularly found in the winter on the outer arms of the harbour, Shore Lark has also turned up on the beach just east of the harbour and Wheatears can sometimes be encountered hopping around here in the spring and autumn. Any of the bushes around the harbour may also produce a migrant passerine such as Chiffchaff or Firecrest.

Recently a Little Swift turned up during the winter, flying around the Martello Tower for a day – by far the rarest bird recorded here to date and proving the old adage that anything can appear anywhere at any time.

TIMING

Winter is best, both in terms of the birds on offer and the reduction in the number of tourists and the amount of sailing activity.

ACCESS

The nearest mainline train station is Eastbourne. The harbour is served by various bus services including the circular Eastbourne routes 5 and 5A (Stagecoach South East) which stop on Pacific Drive, as well as 36, 37, 195, 196 (all Cuckmere Buses) and others which stop on Pevensey Bay Road, just a short walk north of the harbour. If travelling by car, the harbour is easily reached by the A259, with parking available at the Waterfront (TQ 639021) as well as along Atlantic Drive (TQ 637018) and at Sovereign Park just to the west of the harbour entrance (TQ 641009).

FACILITIES: There are several car parks and toilet facilities at the Waterfront.

CALENDAR

All year: Oystercatcher, Great Black-backed Gull.

April–September: Sandwich Tern, House Martin.

October–March: Great Crested Grebe, Ringed Plover, Turnstone, Dunlin, Redshank, Guillemot, Razorbill, divers, Shag, Black Redstart, Rock Pipit.

51 WEST RISE MARSH AND WEST LANGNEY LAKE

OS Explorer OL25
OS grid ref: TQ 624021
Postcode: BN23 7LU

HABITAT

Surrounded on all sides by residential and industrial buildings and busy roads, West Rise Marsh and West Langney Lake (sometimes referred to jointly as West Langney Levels or West Langney Marsh), along with nearby Shinewater Lake, represent a wetland oasis in the heart of suburban Eastbourne. The lakes and wetlands here act as floodwater storage for the nearby residential areas, and were developed in the 1990s as a form of mitigation for new road building in the area. They also form part of the Eastbourne Park Local Wildlife Site.

Sometimes referred to as 'the last wilderness in Eastbourne', the marsh is just under 50ha and is leased to nearby West Rise Junior School, with the schoolchildren using it as an educational resource to learn about farming, nature studies and history. Five Water Buffalo that have been used to graze the site are the property of the school, and both the marsh and the school are situated on the site of the second-largest Bronze Age settlement in Europe.

Aside from the lake, the habitat here is a mix of marshy grassland, reedbeds and drainage ditches, all of which are very attractive to a variety of insects, especially dragonflies.

SPECIES

West Rise is perhaps best known as one of the most reliable sites for Water Pipit in East Sussex, with several birds often present in the winter. It's easy to see why, as the site can become extremely wet at this time of year, providing some excellent boggy areas in the grassland which also prove very attractive to both Common and Jack Snipe, though you would almost have to tread on the latter to discover they are there; please don't stray off the main paths. Other waders can occur here in the passage months (particularly in late summer when the water levels drop) such as Common, Green and Wood Sandpipers or, more unusually, Dunlin, Whimbrel and Black-tailed Godwit.

The resident Tufted Ducks on the lake can be joined by Pochard at any time of year and the full suite of commoner dabbling duck species in the winter: Shoveler, Wigeon, Teal, Pintail and Gadwall. A scan through the ducks in spring may also produce a Garganey. Scarcer ducks and grebes occur here from time to time such as Goldeneye, Goosander, Long-tailed Duck and Slavonian Grebe. The lake also attracts good numbers of large gulls, with Yellow-legged Gull sometimes seen from late summer into the winter months, and Caspian Gull in early winter.

The extensive reedbed here is one of the best on offer in this part of Sussex; unsurprisingly, it holds Cetti's Warbler year-round and often a few Bearded Tits in the winter. Reed and Sedge Warbler breed here in good numbers in the summer. Rarer reedbed specialists have occurred too, including the first twitchable Penduline Tits in Sussex, 'White-spotted' Bluethroat and Savi's Warbler. Bittern will sometimes winter, but may often go undetected as they are masters of camouflage and

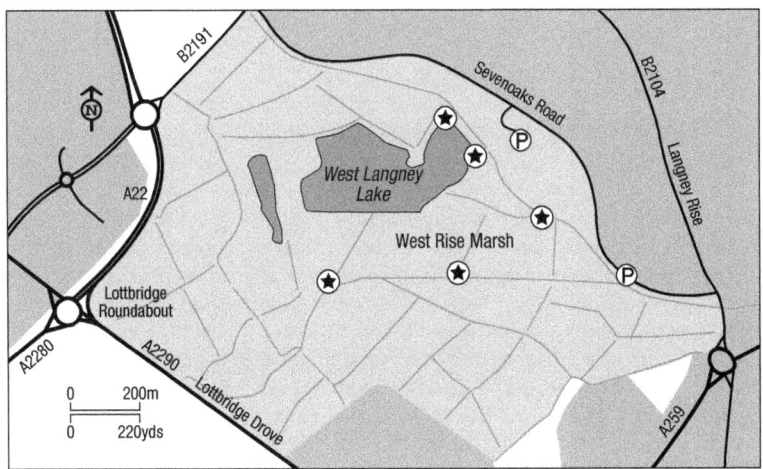

concealment. More regularly occurring heron species such as Little Egret and Grey Heron may be seen all year, while Cattle and Great White Egrets are both increasingly likely too.

The area has gained in popularity with local birders in recent years, leading to a number of scarce species being found, including Red-rumped Swallow, Spotted Crake, Ring-necked Duck and Bee-eater.

TIMING

Although good all year, the site has perhaps most to offer during the winter – although do bear in mind it can be extremely wet at this time of year and so wellies are essential.

ACCESS

The nearest railway station is Hampden Park, just over 1km to the west, while various bus services including 1, 1A and 501 (Stagecoach South East) stop on the Sevenoaks Road. There is a car park at TQ 626023, while there is also room for a few cars to park on Sevenoaks Road, near the gate through to the Langney Sewer path. There are various points from which to easily view the main lake from quite near the car park, but the paths around the wider area can become very muddy in winter so are not always suitable for wheelchairs and pushchairs.

FACILITIES: Aside from the car park, all other facilities are available in Eastbourne centre.

CALENDAR

All year: Gadwall, Pochard, Great Crested Grebe, Little Egret, Cetti's Warbler, Reed Bunting.

April–September: Swift, Hobby, Sedge Warbler, Reed Warbler, Sand Martin, Swallow, House Martin, Wheatear, Yellow Wagtail.

October–March: Shoveler, Wigeon, Teal, Pintail, Water Rail, Snipe, Marsh Harrier, Bearded Tit, Stonechat, Water Pipit.

52 BEACHY HEAD

OS Explorer 123
OS grid ref: TV 585955
Postcode: BN20 7YA

HABITAT

Beachy Head is surely one of the most iconic natural landscape features in the UK, its impressive chalk cliff towering up from the sea, more than twice the height of the highest of the Seven Sisters cliffs to the west. Indeed, at 162m, Beachy Head is the tallest chalk sea cliff in Britain. The 'Beachy' part of the name has nothing to do with beaches and is, in fact, thought to be a derivation of the French for 'beautiful headland' – 'beauchef'. It is listed in just about every tourist guide as a must-see natural wonder for overseas visitors, but it is also arguably one of the most famous and visited birding destinations in the south-east of England – and for good reason.

As one of the most southerly and elevated points along the south coast, Beachy Head is ideally placed to act as an arrival and departure point for migratory birds, especially given its proximity to the French coast, just 60 miles away. Beachy Head's prime location, combined with the ideal mix of habitats – grassland, scrub and pockets of woodland – plus some sweeping panoramic views all add up to the perfect recipe for some superb birding. The total recording area list now stands at over 300 species and, on a good day in spring or autumn, this is arguably one of the best places to be on the south coast in terms of finding a scarce or rare vagrant.

The main Beachy Head point itself is peppered with patches of scrub, which must look so inviting to any tired migrant after its Channel crossing. Spots worthy of exploring here include the Old Trapping Area (TV 583954) and the scrub near the hotel (TV 588957). Rather more sheltered patches of scrub and woodland can be found near the Belle Tout Lighthouse (TV 563955), Horseshoe Plantation (TV 561958), Shooters' Bottom (TV 574954) and Cow Gap (TV 595957).

SPECIES

In the spring, favourable winds from the south-east can produce some spectacular movements of wildfowl, including sometimes thousands of Brent Geese and many hundreds of ducks. Garganey is always on the cards on a seawatch here in March, with the day record standing at 48! In May, south-easterlies will usually yield a good passage of Pomarine Skua heading east, although there are plenty of other species that can be seen moving past in the spring in good numbers, including Bar-tailed Godwit, Whimbrel, Mediterranean Gull, Kittiwake, Sandwich Tern, Common and Arctic Tern, Gannet and Red-throated Diver. In terms of location, Birling Gap and Cow Gap are the preferred spots for seawatching, with the former best in the spring and the latter better in late summer and autumn.

Passerine movement really gets going from March into April, with the earliest Chiffchaffs, Black Redstarts and Firecrests soon giving way to the first truly sub-Saharan arrivals, with the likes of Wheatear, Willow Warbler, Redstart, Nightingale, Pied Flycatcher and Ring Ouzel all likely here in varying numbers on good days. At this time of year, light northerly winds are often the most productive in terms of falls of passerines, although do bear in mind every day is different on a headland. Given its prominent location, relatively close to the near Continent, Beachy Head is a hotspot for all of the classic spring overshoot species when warm winds blow up from the south, with Bee-eater and Serin now just about annual here, Black Kite occurring every two or three years, plus rarer historical occurrences of Hoopoe, Alpine Swift, Golden Oriole, Blue Rock Thrush, Crag Martin and River Warbler.

After a slight lull in June, return migration kicks off again in July and August, when the first southbound passerines begin to appear in the bushes, although they are sometimes a little trickier to find when they're not singing. Croaking and whistling Nightingales and lemon-yellow juvenile Willow Warblers are among the first to return to the area, with many hundreds of the latter possible on good days. These are followed by all the common warblers – Whitethroat, Lesser Whitethroat, Blackcap, Chiffchaff, Grasshopper Warbler, Reed Warbler and Sedge Warbler and other passerines such as Spotted and Pied Flycatchers, Redstart, Whinchat, Wheatear and Yellow Wagtail. From late August into September is the time to really start thinking about scarcer or rare species, with Wryneck and Red-backed Shrike perhaps the most likely to be found – but also very much in play are Melodious Warbler, Ortolan Bunting and Little Bunting, and perhaps even a rarer warbler such as Icterine, Sykes's, Booted or Sardinian.

It's not all about the passerines though, as Beachy Head is one of the best spots on the south coast to observe raptor migration. Red Kite, Common Buzzard, Peregrine, Kestrel, Hobby, Merlin, Marsh Harrier and Osprey are all regularly seen, while rarer records have included Pallid Harrier, Montagu's Harrier, Rough-legged Buzzard and even Bearded Vulture, albeit a bird from a reintroduction scheme in the Alps. It is Honey Buzzard, though, for which the area is perhaps most famous, with favourable days in late summer and early autumn sometimes producing multiple individuals. If the winds are light, these birds will often power straight through and across the Channel, but sometimes they may linger for a while, offering great views as they thermal above the headland. Perhaps unlikely to be repeated was the exceptional influx in 2000 which saw 63 Honey Buzzards recorded here in a single day.

Keeping with the theme of skywatching, the overhead passage of hirundines, Meadow and Tree Pipits and Yellow Wagtails can be very impressive in September, and this again can always offer the chance of something more unusual among the moving flocks such as Tawny Pipit, though this is not as regular as it once was.

Later into the autumn, once the bulk of the sub-Saharan migrants have departed, the last of the Chiffchaffs move through and then it's the turn of migrants and vagrants from the east to take centre stage. Yellow-browed Warbler is more or less an annual passage migrant nowadays, although Pallas's Warbler is almost as likely here. Indeed, Beachy Head is probably the best site in Sussex to find this species in late October or early November. Mid- to late October is also the best time to encounter Ring Ouzel on the headland, with dozens or even hundreds having been recorded on some days. Rarer records at this time of year have included Red-breasted Flycatcher, Radde's Warbler, Dusky Warbler, Greenish Warbler, Rustic Bunting and Black-and-white Warbler. Essentially, anything is possible here and the rarity scale is always at its highest in October. Of course, it's not all about the ultra-rarities, with more common species putting on spectacular shows in the autumn. Movements of Goldfinch and Woodpigeon can be particularly impressive overhead, often numbering many thousands of birds, while Short-eared Owls are sometimes spotted passing through or lingering for a while, and a wintering Woodcock or Dartford Warbler may be seen.

TIMING

Winter birding at Beachy Head can be tough-going and the weather unforgiving on such an exposed and elevated headland. From March right through until October the area really comes into its own, with just a slight lull in the action from late May until early July when return migration begins to get into gear. For the sheer spectacle of numbers of migrants, any time between late August and October is recommended, with an abundance of passerines in the bushes most days, and lively passage of hirundines, finches and raptors overhead.

As Beachy Head is a very popular tourist destination, it's advisable to avoid sunny bank holidays if possible, and generally try and steer clear of the busiest spots, such as the car park and café at Birling Gap. Early morning is best to avoid the crowds, although bear in mind some of the best falls of migrants and passage of species such as Honey Buzzard will often be best from mid- to late morning to mid-afternoon.

ACCESS

As a popular beauty spot, Beachy Head is easy to find by car. Follow signs from the A259 around East Dean. Parking is available at Birling Gap (TV 554960) and the Beachy Head hotel (TV 590959), as well as at various smaller car parks dotted along the Beachy Head Road – all of which are pay-and-display. All of the main areas described in the habitat section are within relatively easy walking distance of one of these small car parks. The nearest railway station is at Eastbourne and various bus services run from Eastbourne out to the Beachy Head area, although only the Sunday-only 13X (Brighton & Hove) stops along the Beachy Head Road itself. Otherwise, routes including 12, 12A, 12X and 41 stop in and around East Dean, which is just over a mile walk to Birling Gap.

Access to the beach can only be achieved at either Birling Gap or Cow Gap but, sometimes even in these areas, the paths and steps may be impassable due to erosion. Take great care along the cliff edges, as they have sheer drops and are not

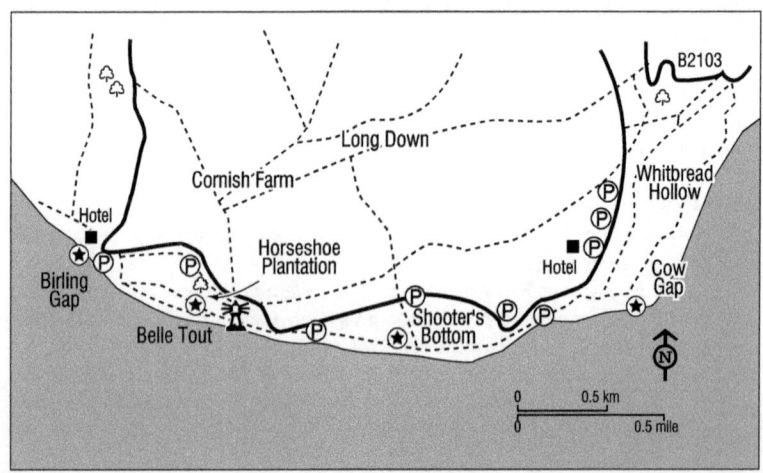

fenced. Also, do not attempt to follow the coastline from Birling Gap to Cow Gap, or vice versa, as this is only possible at the lowest of tides and there are no other ways to return to higher ground when the tide starts coming in.

Access for visitors with limited mobility is variable here, with some areas viewable from right by the car park, such as at Birling Gap or the Old Trapping Area.

FACILITIES: There is a National Trust café, shop and toilets at Birling Gap. There are also toilets at the car park by the Beachy Head hotel.

CALENDAR

All year: At sea – Common Scoter, Oystercatcher, Curlew, Kittiwake, Fulmar, Gannet, Little Egret. On shore – Raven, Skylark, Stonechat, Grey Wagtail, Meadow Pipit, Rock Pipit, Bullfinch, Greenfinch, Linnet, Goldfinch (with a peak in autumn), Corn Bunting, Yellowhammer.

March–May: At sea – Garganey, Shoveler, Pintail, Eider, Velvet Scoter, Red-breasted Merganser, Avocet, Whimbrel, Bar-tailed Godwit, Great Skua, Pomarine Skua, Arctic Skua, Little Gull, Mediterranean Gull, terns, divers and auks. On shore – Osprey, Short-eared Owl, Bee-eater, Hobby, Peregrine, Sedge Warbler, Reed Warbler, Grasshopper Warbler, Sand Martin, Swallow, House Martin, Wood Warbler, Willow Warbler, Garden Warbler, Lesser Whitethroat, Firecrest, Ring Ouzel, Spotted Flycatcher, Nightingale, Pied Flycatcher, Redstart, Black Redstart, Whinchat, Wheatear, Yellow Wagtail, Tree Pipit, Brambling, Serin, Siskin.

June–July: Quail, Yellow-legged Gull, Bee-eater.

August–October: Swift, Yellow-legged Gull, Osprey, Honey Buzzard, Marsh Harrier, Hen Harrier, Short-eared Owl, large gatherings of hirundines, all passage passerines as for March–May, but increased chance of the likes of Wryneck, Ring Ouzel, Nightingale and rarer species such as Pallas's and Radde's Warblers and Lapland Bunting.

November–February: At sea – Brent Goose, Shelduck, Great Crested Grebe. On shore – Hen Harrier, Short-eared Owl, Merlin, Peregrine, Pallas's Warbler, Dartford Warbler, Firecrest, Black Redstart.

---

53 CUCKMERE HAVEN

OS Explorer OL25
OS grid ref: TV 517988
Postcode: BN25 4AD

HABITAT

The River Cuckmere reaches the sea at the Seven Sisters Country Park between Seaford and Eastbourne. The valley is situated on the Vanguard Way which runs all the way from Croydon to Newhaven, and the area is also part of the wider SSSI which includes Beachy Head to the east and Seaford Head LNR to the west.

Part of the last stretch of the river south of Exceat Bridge has been canalised, but thankfully the original river meanders and oxbow lakes were left intact to the east of the new canal. These now form a tidal channel that is a real magnet for waders and wintering wildfowl.

The water meadows each side of the old river are grazed by sheep and cattle, keeping them at an ideal height for waders, while the shingle beach supports some interesting plant species such as Yellow-horned Poppy and Sea Kale. Other interesting species such as Round-headed Rampion, Red Star-thistle and Burnt-tip Orchid occur in the drier areas, particularly on the hillside to the east of the river. The chalk reef on this stretch of coastline is one of the best examples of the habitat in Europe and part of the Marine Conservation Zone that includes all coastal areas of the South Downs National Park.

SPECIES

A winter visit to the Cuckmere is really all about the wildfowl and gulls. Wigeon and Teal are abundant along with Canada and Greylag Geese, which may be joined by a few White-fronted Geese or perhaps even a Tundra Bean or Pink-footed Goose, the latter having become a more frequent winter visitor to Sussex in recent years. The gull roost is always worth scanning through, with both Caspian and Yellow-legged Gulls recorded here regularly in winter. The larger gulls only roost on the western side of the river, never the east.

The exposed mudflats towards the river mouth are worth checking at any time of year for waders, with the scattering of Dunlin, Oystercatcher, Grey Plover, Ringed Plover and Redshank sometimes joined by a Spotted Redshank in the winter. In the passage months, almost any wader can drop in here, either on the scrape or along the meander, with the likes of Whimbrel, Greenshank, Avocet, Little Ringed Plover, Black-tailed Godwit and Bar-tailed Godwit among the most regularly occurring, but with Wood Sandpiper, Temminck's Stint, Little Stint or rarer species appearing from time to time.

Meadow and Rock Pipits are present on the grazing marsh or shoreline year-round, and can sometimes be joined by a Water Pipit in winter or on passage in the spring, when it's also worth keeping your eyes peeled for 'Scandinavian' Rock Pipit (*littoralis*). Other passerine interest here includes year-round Skylark and Stonechat, with Wheatears, Whinchats and Yellow Wagtails moving through from March to May and returning in late summer and autumn. Scrubbier patches are always worth checking for warblers too, with Chiffchaff, Blackcap and both Whitethroat species present throughout the summer.

Another species that Cuckmere has become famous for is wintering Short-eared Owl, with multiple birds sometimes seen hunting over the fields to the east of the valley, in the Foxhole area (TV 520982). In spring and autumn, migrant raptors can be seen moving up the valley, with Marsh Harrier, Osprey and Hobby being the most commonly seen, but there have been occasional records of Hen Harrier and even Montagu's Harrier here.

Rarities in recent years have included Baird's and Semipalmated Sandpipers and Black-winged Pratincole.

TIMING

The Seven Sisters CP is an extremely popular tourist destination, often listed in travel guides for foreign visitors to the UK, so bank holidays and the summer holidays can see the area becoming very busy indeed. Luckily, the meanders of the river are set a little way away from the footpath, so birders can enjoy relatively undisturbed viewing at most times.

ACCESS
Public transport access to the site is much the same as for Seaford Head (site 54, page 174), with Seaford railway station around 5km away, on the other side of Seaford Head. Brighton & Hove bus services 12, 12A and 12B run all year between Eastbourne and Brighton, stopping at Exceat, as does the Sunday-only 13X. Bus route 40 (Cuckmere Buses) runs from Berwick Station to Seaford on Tuesdays and Fridays, also stopping at Exceat. By car, follow the A259 between Eastbourne and Seaford, which passes through Exceat. There are plenty of parking opportunities, with the Seven Sisters car park (TV 518994) being the closest, or alternatively, park just across the road by the visitor centre.

If you're feeling more energetic you can also park at South Hill Barn (TV 504980) and follow the footpath from here down into the west side of the valley. Note that it is inadvisable to try and cross the mudflats from this side, so you'll need to walk up to Exceat Bridge to cross in order to walk around and view the site from the east side.

The main path from Exceat south past the meanders is hard surfaced and wheelchair and pushchair friendly, but other paths heading up onto the Downs become very steep and some minor paths can become very muddy in winter.

FACILITIES: A visitor centre with accessible toilets is located at Seven Sisters CP car park.

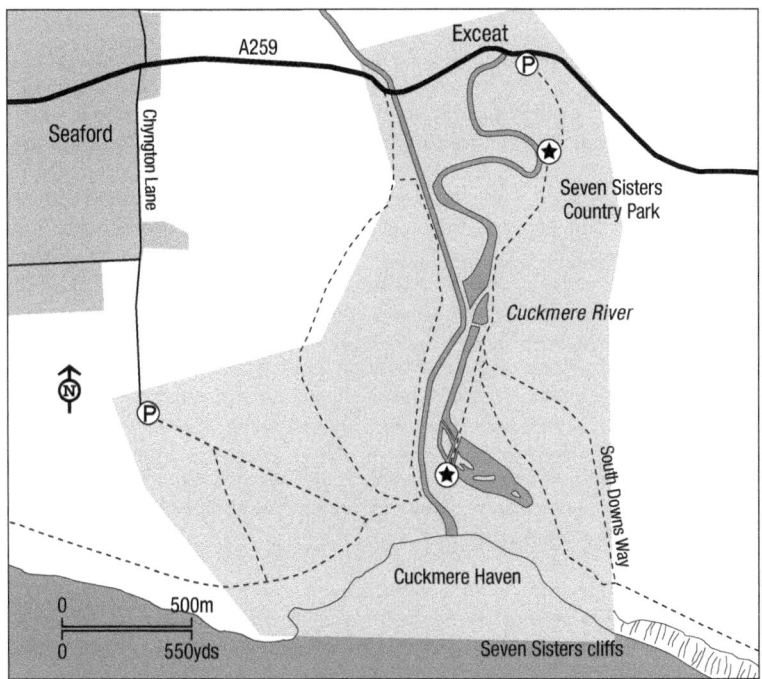

54 SEAFORD HEAD

OS Explorer OL25
OS grid ref: TV 493978
Postcode: BN25 4JQ

HABITAT

Situated in between Newhaven and the Cuckmere, Seaford Head is a 150ha LNR, owned in various parts by East Sussex County Council, Sussex Wildlife Trust and the National Trust. It is in many ways a smaller cousin of Beachy Head to the east, as a scrub- and grassland-covered chalk promontory, although it extends to less of a southerly point than Beachy. Nonetheless, the various habitats on offer here, particularly the areas of scrub, attract an enviable amount of bird species and the site has turned up an impressive list of scarcities and rarities over the years.

On the western side is a golf course, while to the eastern side is Hope Gap, a sheltered scrubby valley leading down to the beach. The area of scrub at TV 508981 known as Harry's Bush has form for turning up various unusual birds over the years. From various points on the headland, Hope Gap included, visitors can enjoy spectacular views of the Seven Sisters chalk cliffs farther east along the coast.

The site is not just famous for its birds, as the chalk grassland here supports specialised and, in some cases, rare plants like Kidney Vetch, Devil's-bit Scabious and Squinancywort. Seaford Head is one of only six sites in the UK where one can find Moon Carrot – a relative of Wild Carrot that is said to glow in the moonlight, hence the name. The rich array of flora in turn attracts butterflies such as Adonis Blue and Silver-spotted Skipper.

SPECIES

Although the likes of Short-eared Owl, Merlin or Black Redstart are possible throughout the winter months, and Brent Geese and Fulmars may be seen passing at sea, it is in the passage season that Seaford Head really comes to life.

Wheatears are more or less a constant presence from late March to May, and again from August through into September and October, hopping about on the golf course and other open areas. These will often be joined by Whinchats, Redstarts and Yellow Wagtails on good 'fall' days in spring and autumn, when the areas of scrub will also be worth checking for warblers and other migrant passerines. Late spring overshoots from the Continent must also be considered, with Hoopoe having occurred here several times in recent years, as well as Alpine Swift, Bee-eater and Black Kite.

As is the case with Beachy Head, it is in late summer and autumn that one can expect the best days of birding at Seaford, especially during periods of easterly winds. Pied Flycatcher is pretty much guaranteed in mid to late August in such conditions, while Wryneck has occurred almost annually in recent autumns. Red-backed Shrike is less expected but always a possibility. At this time of year it is worth bearing in mind the possibility of a fly-over Honey Buzzard too, or even Dotterel or Stone-curlew. Further into the autumn, the site is an excellent place to observe gatherings and movements of hirundines, finches, Meadow Pipits and wagtails, while Ring Ouzel is almost guaranteed in mid- to late October, with sometimes several counted on a good day.

As might be expected from its location and similarity to Beachy Head, Seaford Head has attracted an impressive selection of scarcities and rarities over the years including Booted Warbler, Icterine Warbler, Radde's Warbler and Tawny Pipit.

TIMING

Spring and autumn are the best times, especially if you are hoping to find something unusual. The area is popular with dog walkers, so early mornings are recommended, especially on weekends and bank holidays.

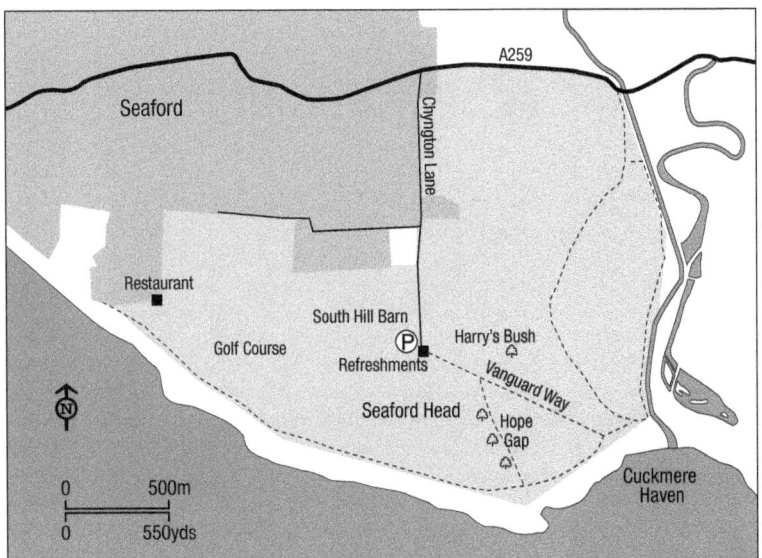

ACCESS

Seaford Head is easily accessible by public transport, with Seaford station just a 15-minute walk or so from the western edge of the reserve. Trains from Brighton run every half hour on weekdays. Various bus routes serve the area including Brighton & Hove routes 12, 12A, 12X and 13X (Sundays/bank holidays only). There is a good-sized car park near South Hill Barn at TV 504980, accessed by turning off the A259 onto Chyngton Lane and following it all the way down.

There are two main public rights of way, one from the car park heading west towards the golf course or east towards Cuckmere and the other following the cliff edge from Hope Gap around to Seaford town. These are linked by smaller paths at various points, making the area relatively easy to cover. South Downs National Park Authority has produced a 'Miles Without Stiles' leaflet for the area, showing which routes are most accessible for wheelchairs, pushchairs and mobility scooters.

FACILITIES: Brewster's Coffee trailer can be found at South Hill Barn (next to the car park) and The View restaurant is on the western side of the golf course.

CALENDAR

All year: Oystercatcher, Curlew, Fulmar, Little Egret, Peregrine, Raven, Skylark, Stonechat, Meadow Pipit, Rock Pipit.

April–October: Whimbrel, Kittiwake, Hobby, warblers, hirundines, Firecrest, Ring Ouzel (especially in October), Spotted Flycatcher, Nightingale, Pied Flycatcher, Redstart, Black Redstart, Whinchat, Wheatear, Yellow Wagtail, Tree Pipit, scarcer migrants (Wryneck, Red-backed Shrike, etc.).

November–March: Short-eared Owl, Merlin.

55 SPLASH POINT

OS Explorer OL25
OS grid ref: TV 488982
Postcode: BN25 1BW

HABITAT

Splash Point, tucked in at the western foot of Seaford Head, is regarded as the best seawatching site in Sussex and, indeed, one of the best in South East England. Rightly so, as the numbers of Pomarine Skua and the like that can be seen here on a good day in spring are mouth-watering, especially as many of the birds that pass the coast here do so relatively close to shore – and the location of the watchpoint means the observer enjoys almost eye-level views of the birds in their breeding finery.

The site also offers excellent views of the famous Kittiwake colony (one of the last in South East England) and the Seven Sisters cliffs stretching out to the east.

SPECIES

Pomarine Skua is an iconic species, especially in breeding plumage, and one that birders eagerly travel some distances to watch passing southern coastal watch-points in the spring. Splash Point offers one of the best opportunities in South East England to see this species (usually in early May), but it also offers interest at other times too.

From February onwards a seawatch from here can produce birds moving, with Brent Geese, auks and Red-throated Divers heading east in good numbers, but it is generally later in the spring when the seawatching here really gets into gear. In March, Common Scoter flocks increase and may sometimes be joined by a few Velvet Scoter. Flocks of ducks will be seen moving up-channel, and it's always worth checking through flocks of dabblers like Shoveler for a Garganey or two lurking among them. On very good days, double-figure counts of these attractive summer migrants can sometimes be seen flying past.

Skua and tern numbers increase into April, initially Sandwich Tern and Great and Arctic Skuas, followed by Common and Arctic Terns (often indistinguishable at long range, so sometimes lumped together for recording purposes as 'commic terns').

From late April into early May, favourable conditions can produce some spectac-ular movements of wildfowl, waders, skuas, gulls and terns. The results are very wind-dependent though, with a south-easterly ideal or a light north-easterly with anticyclonic conditions. Flocks of Little and Mediterranean Gulls and waders can be particularly impressive towards the end of April, especially Whimbrel and Bar-tailed Godwit with day totals of these species sometimes into the hundreds or even thou-sands. It is at this time of year that any avid seawatcher's thoughts turn to that most hallowed of species: the Pomarine Skua (or 'Pom' for short). A good day here in early May can produce tens or (rarely) even hundreds of these spoon-tailed pirates sculling east, occasionally lingering in the area for a while to chase the gulls and terns. If you are very lucky, you may even bear witness to a Long-tailed Skua moving past, but these are far from common anywhere on the South Coast. Again, it's important to check the weather forecast before you set out, with south-easterlies always the most productive. Another species to watch out for in late spring is Black Tern which passes in small flocks, sometimes mixed in among the many hundreds of 'commics' which will still be moving through. Little and Roseate Terns are both possible at this time of year too.

It's not just about the gulls, skuas and terns, mind you, as a session at Splash Point can also produce some impressive movements of hirundines, Swifts, pipits and wagtails 'in off', and sometimes raptors as well.

Return passage in late summer and autumn can produce some interest in the form of Manx and Balearic Shearwaters, but generally this is an altogether less spectacular time of year for seawatching here, and certainly lacks the large shear-waters that are a feature of more south-westerly locations.

Even on quiet days for seawatching, the Kittiwake colony on the cliff below Seaford Head always offers entertaining viewing, as the adults come and go from their nest ledges.

TIMING

March to mid-May is the peak time for seawatching, especially the first week of May. Checking the weather forecast is essential before travelling, as the 'wrong' winds can produce very little. That said, sometimes a change in the weather can be good, but the wind ideally needs to be in the east for best results.

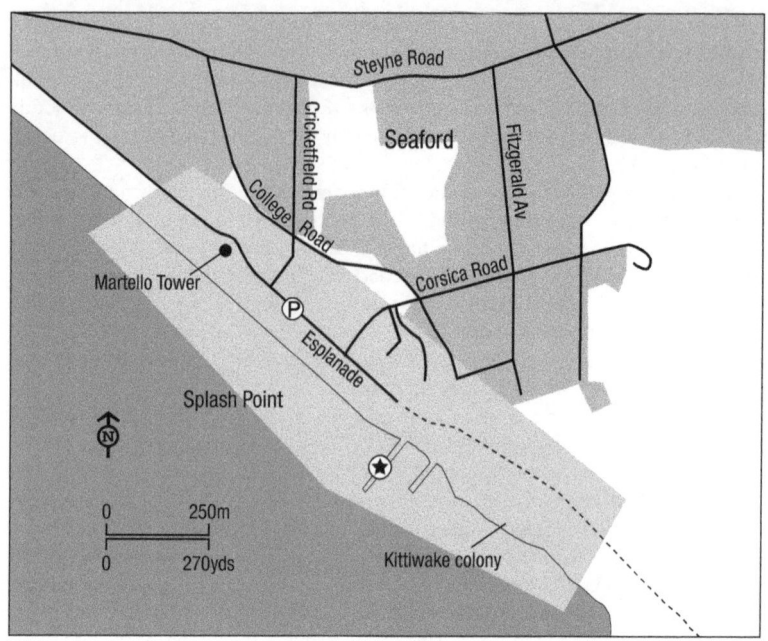

ACCESS

Seaford train station is just a 15-minute or so walk away, around 1km to the north-west. Bus services from Brighton to Eastbourne, including 12, 12A and 12B (Brighton & Hove), all stop in Seaford town centre, while the Seaford Circular 119 (Compass Travel) stops on the Esplanade, just a few minutes' walk away from the point. The site is easy to access by car too, following signs to the seafront from the A259, and ample parking available near the Martello Tower at TV 486983. In terms of where best to stand, anywhere along the beach or on or near the concrete ramp is ideal.

FACILITIES: Public toilets at the Martello Tower car park. Other facilities are located in Seaford.

CALENDAR

All year: Oystercatcher, Dunlin, Kittiwake, Lesser and Great Black-backed Gulls, Sandwich Tern, Fulmar, Gannet, Shag, Peregrine, Rock Pipit.

March–May: Brent Goose, Shelduck, Garganey, Shoveler, Gadwall, Wigeon, Pintail, Teal, Eider, Velvet Scoter, Common Scoter, Red-breasted Merganser, Great Crested Grebe, Avocet, Grey Plover, Whimbrel, Bar-tailed Godwit, Sanderling, Greenshank, Great Skua, Pomarine Skua, Arctic Skua, Little Gull, Mediterranean Gull, Little Tern, Black Tern, Common Tern, Arctic Tern, Red-throated Diver, Black-throated Diver, Manx Shearwater, Sand Martin, Swallow, House Martin, Black Redstart, Wheatear, Yellow Wagtail, Meadow Pipit.

July–September: Yellow-legged Gull, Manx Shearwater, Balearic Shearwater, Meadow Pipit.

November–February: Brent Goose, Great Crested Grebe, Guillemot, Razorbill, Red-throated Diver, Black Redstart.

56 NEWHAVEN HARBOUR AREA

OS Explorer OL11 or OL25
OS grid ref: TQ 452000
Postcode: BN9 9DS

HABITAT

The Newhaven Harbour area offers variety in terms of birding locations. The harbour itself, where the River Ouse reaches the English Channel, is typical of the stretches of industrial coastline in Sussex, with high-sided walls flanking the harbour mouth, from where ferries run to and from France. On the east side of the harbour mouth is a concrete pier, while the west side is overlooked by Castle Hill (an LNR in its own right), atop which sits a nineteenth-century fort. Running west from here is a stretch of the England Coast Path which winds its way to Brighton via Peacehaven and Telscombe Cliffs.

Tide Mills is an abandoned village situated on the River Ouse Estuary between Newhaven and Seaford. Once a tidal mill surrounded by workers' cottages, the mill operated from 1760 until around 1900. What remains of the ruined buildings are now being swallowed up by brambles and scrub and surrounded by rough grassland, making it an ideal refuge for a host of migrant passerines arriving or departing from this stretch of coastline. The Mill Creek which runs through the site is a great little sheltered spot for waders.

The Ouse Estuary NR to the north is a 10ha habitat-restoration project created in the early 2000s as mitigation for the nearby development of a road and business park. It is owned and managed by East Sussex County Council and a popular area for local dog walkers and cyclists.

SPECIES

Newhaven is one of the most reliable areas in Sussex to encounter Purple Sandpiper in the winter, with the East Pier a great place to look; sometimes double-figure counts of birds can be seen roosting here at high tide. Walk out onto the pier and look down below at the concrete struts on the left-hand side. It's worth keeping an eye out for Black Redstart anywhere in this area too, as there is often one wintering locally, sometimes joined by a passing migrant or two in early spring. The sailing club at the Bishopstone end is a particularly good area to look, as is the clifftop path towards Telscombe Cliffs. From the end of the East Pier one can also view the gull roost on the opposite side of the river, which sometimes produces Caspian or Yellow-legged Gulls. Fulmars breed on the cliffs west of the harbour, with excellent views of the birds on their ledges or in flight possible from either the beach just west of the harbour mouth, or various points along the clifftop and undercliff paths. Look out for Peregrine patrolling along the cliffs too. It's also always worth keeping one eye on the sea, as Gannets will often be fishing just offshore and divers and skuas can sometimes pass quite close; these included a stunning adult Long-tailed Skua a few years ago.

Several species of wader can be found in the winter, including good numbers of Redshank (often in the Mill Creek), Curlew, Turnstone, Dunlin and Ringed Plover. Lapwings will sometimes gather in the fields either side of Mill Drove. In spring, the resident and wintering species will sometimes be joined by the likes of Avocet, Whimbrel, Greenshank and others. Little Egrets are an ever-present feature in the Creek too. The marshier areas of the Ouse Valley NR are popular with Snipe in the winter, so it's worth being vigilant for Jack Snipe as well. Another area worth checking for waders is the wet flash at TQ 456006, which sometimes hosts species such as Little Ringed Plover and Green Sandpiper.

The areas of rough grassland at Tide Mills are excellent for Linnets and other finches, with sometimes impressive counts of Goldfinch and Greenfinch, for example. Stonechats are present year-round, sometimes joined by a Whinchat or two in spring or autumn. There are usually a few Wheatears around in the passage months too, either out in the fields or hopping about among the ruins. Migrant warblers, unsurprisingly, love the thorny scrub on offer here, with species such as Blackcap, Whitethroat and Lesser Whitethroat all breeding, and others such as Garden Warbler possible on migration. Sedge and Reed Warblers both breed in the reedbeds at the Ouse Valley NR. Rock Pipit winters along the coast here, sometimes in good numbers.

The Newhaven area has good form for turning up scarce and rare passerines, particularly in late summer and autumn. Wryneck has been almost annual at Tide Mills in recent years and the likes of Short-toed Lark, Grey Phalarope, Red-backed Shrike, Serin, Tawny Pipit, Richard's Pipit, Ortolan Bunting, Hoopoe have also occurred on occasion. Other notable species to have been encountered include White-rumped Sandpiper on the beach at Telscombe Cliffs.

TIMING

Winter is best for wildfowl and waders (especially Purple Sandpiper), and late summer and autumn for migrant passerines. The area can become busy on weekends and bank holidays.

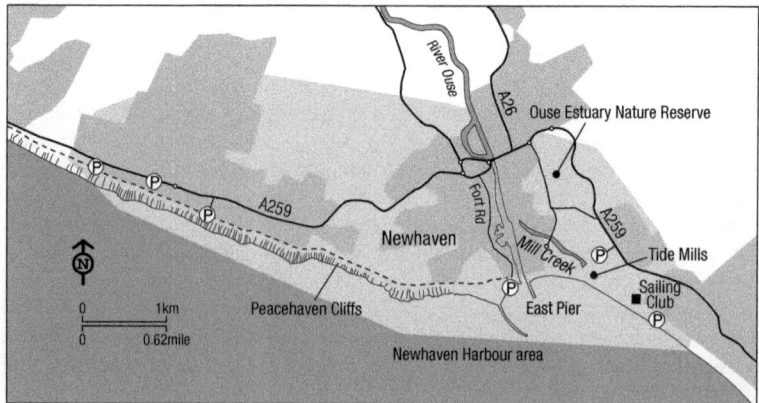

ACCESS

Newhaven town and harbour stations offer the closest access to the area by train (although only one train an hour stops at the harbour). Several bus services stop at the town station, including 12, 12A, 14 and 14C (all Brighton & Hove). All these services also stop at Denton Island, Tide Mills, Bishopstone and various points along the South Coast Road through Peacehaven and Telscombe Cliffs. If travelling by car, there is a car park right by the junction of Mill Drove and the A259 at TQ 463005, for Tide Mills and the East Pier, while for the west side of the harbour there is a car park at TQ 450000. Parking is also available at the sailing club near Bishopstone station. The Ouse Estuary NR is well served by a circular footpath from the A259 around towards Newhaven Harbour and back. For Peacehaven and Telscombe Cliffs there are car parks at TQ 411007, TQ 404011 and TQ 397012.

Some sections of paths here, such as Mill Drove (to Tide Mills) and part of the trail at Ouse Estuary NR, are hard-surfaced and suitable for wheelchairs and pushchairs, but other minor paths are not (for instance, the clifftop path) and can become very muddy in the winter. There is also a cycle path through the Ouse Estuary NR.

FACILITIES: Aside from the car parks, there are various food and drink vendors near Newhaven Fort; otherwise, all other facilities are situated in Newhaven town centre.

CALENDAR

All year: Ringed Plover, Curlew, Redshank, Fulmar, Little Egret, Peregrine.

April–September: Avocet, Whimbrel, Mediterranean Gull, Sedge Warbler, Reed Warbler, Lesser Whitethroat, Whinchat, Wheatear, Yellow Wagtail.

October–March: Shelduck, Teal, Oystercatcher, Lapwing, Curlew, Turnstone, Purple Sandpiper, Jack Snipe, Snipe, Short-eared Owl, Cetti's Warbler, Black Redstart, Rock Pipit.

57 SHEEPCOTE VALLEY

OS Explorer OL11
OS grid ref: TQ 341049
Postcode: BN2 5TS

HABITAT

Just north of Brighton Marina, right on the eastern edge of the city, lies Sheepcote Valley. Used for much of the twentieth century as a rubbish tip, after various protests by local people in the 1970s, tipping finally ceased here in the 1980s and the area was proposed as a nature reserve. In 1997, the Friends of Sheepcote conservation group was formed to help Brighton and Hove City Council's ranger team care for this new area of regenerating greenspace, which now forms a valuable habitat corridor between East Brighton Park and Blackrock Valley to the south and Whitehawk Nature Reserve to the north. The area is perhaps most famous locally for the giant 50m chalk hawk which has been cut into the hillside on the western side.

The landscape here is now an enticing mix of open chalk grassland and areas of dense scrub, all situated on a south-facing valley just 1km from the coast, providing ideal habitat for any arriving or departing migrant bird to drop in and rest and refuel. Given that the site was formed on top of a former landfill, there are several artificial tiers to explore as one descends or ascends through the reserve, interspersed with banks of scrub and hedgerows.

As well as being rich in birdlife, it is also good for butterflies, with all the chalk downland species present, including Chalkhill and Adonis Blues and Silver-spotted Skipper. Nearby Whitehawk Hill has also become a reliable site in recent years for Long-tailed Blue.

SPECIES

Although offering a good supporting cast of resident birds at all times of the year, such as common tits, Skylarks and Stonechats, the real draw of Sheepcote Valley is its potential for attracting excellent numbers of migrant passerine species in spring and autumn.

Wheatears will often be seen hopping around in open areas from mid-March through to May, and again in late summer and autumn, particularly on the higher slopes towards the northern car park. These will frequently be joined by Whinchats and perhaps a Ring Ouzel, especially in the autumn. The denser areas of scrub are teeming with breeding Whitethroats in the summer and may attract a Dartford Warbler in the winter. On passage, any of these scrubby patches can hold migrant warblers, Spotted Flycatcher, Pied Flycatcher, Nightingale, Redstart or Tree Pipit, while later still in the autumn there is the possibility of Firecrest, Yellow-browed Warbler or something even rarer.

All the resident raptor species of the region may be encountered here, with Hobby sometimes seen during the passage months. Short-eared Owls sometimes winter and Barn Owl is possible too.

The valley offers good vis-mig potential, especially in the autumn when impressive numbers of thrushes, hirundines, pipits and wagtails may be seen moving overhead.

Sheepcote is one of the most reliable locations in Sussex for Wryneck, with birds having been found here almost annually in recent years, especially in autumn. Other scarce and rare species to have occurred here include Iceland Gull, Hume's Warbler, Pallas's Warbler, Radde's Warbler, Serin, as well as fly-over Dotterel, Black Kite and Tawny Pipit.

TIMING

Spring or autumn are best for producing the greatest diversity of species, especially August and September. Bear in mind, too, that the site is fringed by lots of housing as well as sports pitches and a bike track, so recreational use can be very high on weekends and during the summer months. The earlier you can get here in the morning the better!

ACCESS

The nearest railway station is Brighton, around 3km west of Sheepcote Valley. Various bus routes including 2, 21A and 72 (all Brighton & Hove) stop at either Woodingdean or Whitehawk, with easy walking access into the reserve from any of the perimeter roads. By car, the reserve is easily accessed from either the B2066 (Roedean Road) via Wilson Avenue on the south side or Warren Road on the north side, with car parks situated at TQ 339038 and TQ 345055. There is also space for cars to park on the Sheepcote Valley access road off Wilson Avenue. If accessing from the south, follow the main path north past the football pitches and caravan park to join the network of informal paths which cover the area well. From the north, simply follow the main public footpath south to join the same network of paths. There are many routes available to the visiting birder, some only taking half an hour to 45 minutes, others lasting closer to two hours. Note that some of the paths here involve some steep uphill or downhill stretches, so it is not very suitable for visitors with limited mobility.

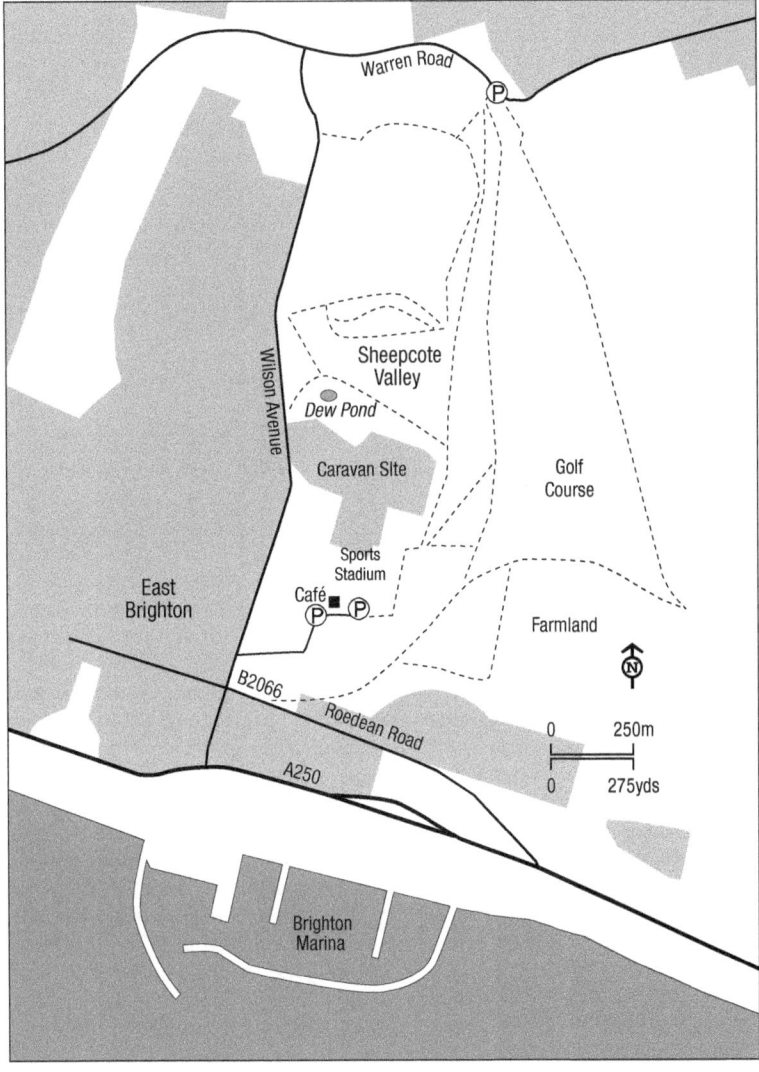

FACILITIES: The East Brighton Café (with toilets) is situated on the southern side, and there are two car parks.

CALENDAR

All year: Kestrel, Skylark, Stonechat.

April–September: Wryneck, Hobby, Willow Warbler, Reed Warbler, Garden Warbler, Lesser Whitethroat, Spotted Flycatcher, Nightingale, Pied Flycatcher, Redstart, Whinchat, Wheatear, chance of scarcer migrants, especially in autumn.

October–March: Short-eared Owl, Firecrest, Ring Ouzel (October), chance of rarer passerines in late autumn.

58 BRIGHTON MARINA

OS Explorer OL11
OS grid ref: TQ 338029
Postcode: BN2 5WA

HABITAT

At some 51ha, Brighton Marina is the largest marina in Europe. It features a working harbour and residential housing alongside a variety of leisure, retail and commercial options. Situated below a chalk cliff on the eastern periphery of Brighton city, the complex was built in the 1970s, though developments continue to this day. Concrete breakwaters flank the marina and they are usually accessible to the public.

SPECIES

This site offers a great selection of birds right on the edge of the city. Waders including Turnstone and Oystercatcher can be seen on the walls and breakwaters virtually all year-round, though it is in the winter when the main prize this site has to offer can be sought out: Purple Sandpiper. As many as 20 used to winter but numbers are much reduced these days. The concrete boulders on the seaward side of the far end of the east breakwater are the best place to look. The outer harbour may have a few auks and Red-throated Divers at this time of year, with Shag occasionally seeking refuge during more severe weather. Rock Pipit and Kingfisher are also present, with Black Redstarts often found near the cliffs.

Seawatching can be productive. If nothing else, the marina is good for Fulmar, which breeds a little to the east (see sites 56, G10 and G11) The usual array of wildfowl, gulls and terns move through at the times outlined in other site accounts in this book (such as Splash Point, site 55). Both Grey and Red-necked Phalaropes have appeared in the harbour as well, typically after rough weather.

Rarer species on the site list include Franklin's and Laughing Gulls, Shore Lark and Pallas's Warbler.

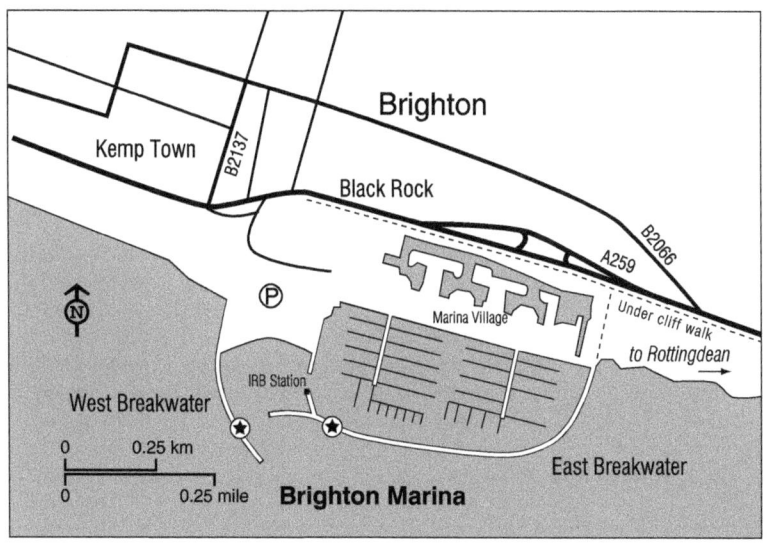

TIMING

November to March is best if you want a chance to see Purple Sandpiper, and indeed a winter visit probably offers the most variety here.

ACCESS

A multistorey car park can be found via the postcode BN2 5UT. It offers free parking with a four-hour limit. There is also customer parking at Asda for up to two hours – the undercliff path next to the supermarket hosted the aforementioned Pallas's Warbler in 1998!

If travelling via bus from central Brighton, the number 7 runs every 6 minutes during the daytime, with the night bus running twice an hour throughout the night. Services 7, 21, 21B, 25C, 47, 52 and 57.

FACILITIES: There are plentiful facilities in the marina complex, including shops and toilets.

CALENDAR

All year: Gannet, Oystercatcher, Mediterranean Gull.

November–March: Red-throated Diver, Fulmar, Shag, Turnstone, Purple Sandpiper, Guillemot, Razorbill, Kingfisher, Rock Pipit, Black Redstart.

April–October: Fulmar, Common Scoter, wader, gull, skua and tern passage.

59 DITCHLING BEACON

OS Explorer OL11
OS grid ref: TQ 338130
Postcode: BN1 9QD

HABITAT

The highest point in East Sussex (248m), Ditchling Beacon is a large chalk hill with a steep northern face, covered with grassland. Many of the fields are grazed by sheep. The Beacon commands a 360-degree view of the Weald and coastal strip, and is the third-highest point on the South Downs. It forms part of the Clayton to Offham Escarpment SSSI. A 24ha part of the beacon is a reserve managed by the Sussex Wildlife Trust. The slopes represent some of the best chalk downland in the area. A variety of plants can be found, including Musk Orchid and Marsh Fragrant-orchid.

The chalkpit on the scarp is famous for bryophytes, with over 120 different

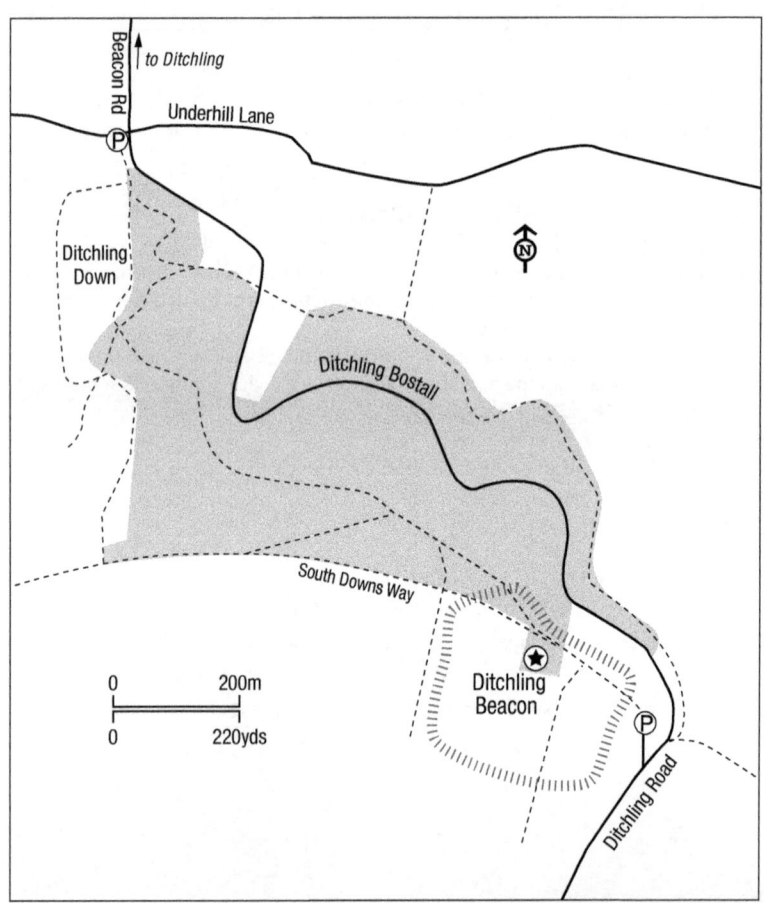

mosses and liverworts having been recorded. Chalkhill Blue and Silver-spotted Skipper are the headline butterfly species, with Green Hairstreak also present.

SPECIES

Perhaps the most obvious presence during the breeding season is Skylark, with many pairs breeding here. Ditchling Beacon is a good spot for Corn Bunting as well, with Yellowhammer another resident breeder. Grey and Red-legged Partridges and Raven are also present year-round. One or two Lesser Whitethroats can be found in the scrubbier areas during the summer, but are outnumbered by Whitethroats. Meadow Pipit also breeds. In some summers, Quail may be heard singing.

Its high position along the Downs makes Ditchling Beacon a good site for vis-mig. Wheatears and Whinchats can be particularly conspicuous on passage, especially in early autumn. Ring Ouzel is another species that moves through at such times, perhaps loitering in a sheep field in April or nervously perching in a Blackthorn bush in October. Indeed, numbers of thrushes and other passerines moving overhead can be impressive in this month. The site can boast multiple records of passage Dotterel and Stone-curlew, too.

Winter is generally very quiet. Occasionally a Hen Harrier or Merlin may move through.

TIMING

Spring, summer and autumn are best. It's worth noting that Ditchling Beacon is very popular with walkers – try to arrive early and avoid weekends.

ACCESS

There is a National Trust car park at the top of the reserve part of the site, along the Ditchling to Brighton Road at TQ 333129. A path also leads to a small car park on Underhill Lane. These other paths are steep, slippery and have some stiles.

Brighton & Hove, Brighton's main bus operator, operates a special service – the number 79 – from locations in the city centre to Ditchling Beacon, via Ditchling Road, on Sundays and bank holidays year-round and Saturdays in the summer.

FACILITIES: The nearest facilities, including accessible toilets, are found in Ditchling village.

CALENDAR

All year: Grey and Red-legged Partridges, Raven, Skylark, Meadow Pipit, Corn Bunting, Yellowhammer.

April–September: Chance of Quail, Lesser Whitethroat, Meadow Pipit, Wheatear and Whinchat on passage.

October–November: Ring Ouzel, chance of raptors including Hen Harrier.

60 IFORD, LEWES AND RODMELL BROOKS

OS Explorer OL11
OS grid ref: TQ 420077
Postcode: BN7 3HF

HABITAT

Situated on the floodplain of the River Ouse, south of Lewes, this extensive network of fields and ditches offers a variety of habitats for a host of birds and other wildlife. The north-western section of around 330ha nearest Swanborough and Kingston is designated as an SSSI, some of which is owned and managed by the RSPB.

Rodmell Brooks is located at the southern end (between the Ouse and the village of Rodmell), while the middle section is known as Iford Brooks. Although all part of the same area, the habitats of the three sites are quite different. Lewes Brooks is managed as a wet grassland so is good for waders, including breeding species such as Lapwing and Redshank. Iford Brooks is arguably the best overall birding site, with the lake attracting the highest numbers of wildfowl in the winter and passage waders in spring and autumn. Rodmell is better for owls and raptors, particularly in the winter.

The ditches here are home to a variety of aquatic plant species, some of them locally scarce including Shining Pondweed, Frogbit, Soft Hornwort and Water-violet, while some uncommon moths have also been recorded in the area, including both Silky and Webb's Wainscots.

SPECIES

Large numbers of wildfowl gather here in the winter, especially Canada Geese, Greylag Geese, Wigeon and Teal. In the second half of winter in particular, the goose flocks can sometimes attract more unusual species, most commonly White-fronted Goose but also Tundra Bean and Pink-footed on occasion, so it's always worth double checking any 'grey geese' at this time of year. The rarer geese are almost always at Iford Brooks, on the north side of the footpath running north-east from Iford village. Large gatherings of Mute Swan in the winter some-times attract the odd Bewick's too, though these are sadly present in increasingly small numbers across Sussex nowadays.

During good winters for Short-eared Owl, this can be one of the most reliable areas in East Sussex to encounter it, quartering over the fields at Rodmell in late afternoon. Barn Owls also vary from year to year but one can sometimes be seen at dusk. Merlin and Hen Harrier are both relatively frequent in the winter, again usually towards the Rodmell side, although the latter will also sometimes be spotted at Iford. Other species such as Marsh Harrier, Buzzard, Red Kite, Sparrowhawk and Kestrel may be seen at just about any time of year.

Little Egret and Grey Heron are both a common sight in the valley all year-round, owing to the heronry between Rodmell and Southease, while both Great White and Cattle Egrets are occurring here with increasing frequency, as both species continue to cement their presence across Sussex.

In keeping with other floodplains in Sussex, the Lewes Brooks area attracts a large congregation of Lapwings in the winter, with smaller numbers breeding in the summer. The wintering flock can sometimes draw in a Ruff or two, while other waders are always possible moving through in spring and autumn, including Dunlin, Greenshank, Black-tailed Godwit and Little Ringed Plover. Green Sandpipers often overwinter but can be hard to pin down.

The Ouse itself will often produce Kingfisher or Grey Wagtail, and Goosander in the winter, though the latter are most often slightly further south towards Piddinghoe. Common Sandpiper is possible along the river between Piddinghoe and Lewes at any time of year, with sometime three or more wintering here.

Passerine interest is varied here with Skylark, Cetti's Warbler and Reed Bunting among the species present all year, good numbers of Meadow Pipits in the winter (which sometimes attract a Water Pipit) and a host of warbler species breeding in the summer. Rodmell still hosts good numbers of Corn Buntings too, with flocks of several dozen possible in the winter. The likes of Wheatear, Whinchat and Yellow Wagtail will often drop in during spring or autumn. In winter the small sewage works at TQ 422065 attracts Chiffchaff and sometimes a Siberian Chiffchaff or two.

Although not able to boast a huge list of rarities, the area has nonetheless attracted some interesting species in recent years, including Green-winged Teal, Montagu's Harrier and Alpine Swift. Great Grey Shrike has also turned up occasionally, but not for a few years at the time of writing.

TIMING

This area is really at its best in the winter, with large congregations of wildfowl and the greatest potential for owls and scarce raptors.

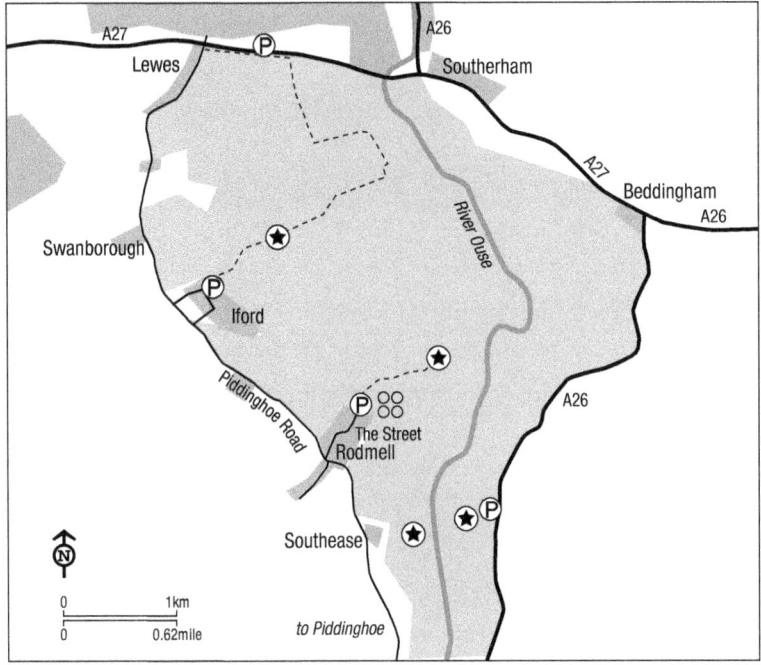

ACCESS

The nearest railway stations are at Lewes to the north and Southease to the south, both of which are easily reached from Brighton and Eastbourne, and only an hour or so from London Victoria. Bus services 123 (Compass Travel) and 132 (Community Transport for Lewes – Sundays only) stop at various points along the Kingston/Newhaven Road to the west of the Brooks. Various public rights of way criss-cross the area, including the Sussex Ouse Valley Way, though these are unsurfaced and can become very muddy in the winter. The best way to view Iford Brooks is to walk the footpath north-east from St Nicholas Church in Iford (TQ 408073) and scan north and south; parking is available at the church. This path loops round all the way to Lewes to enable viewing of Lewes Brooks. There is also parking available on Cockshut Road (TQ 412094) but this entails a longer walk to reach good habitat. For Rodmell Brooks, park either near Southease Bridge, 50m north-east of Monk's House (TQ 421064) or at Southease YHA (TQ 432055) and follow Itford Farm Lane west to reach the river.

Note that the paths at this site can become very muddy in winter and are not particularly suitable for wheelchair or pushchair users.

FACILITIES: A hide at Lewes Brooks is available by appointment only for members of the Sussex Ornithological Society. Otherwise, the nearest facilities are situated in Lewes town centre.

CALENDAR

All year: Egyptian Goose, Red-legged Partridge, Lapwing, Little Egret, Skylark, Cetti's Warbler, Reed Bunting.

November–February: White-fronted Goose, Shelduck, Shoveler, Wigeon, Pintail, Teal, Snipe, Marsh Harrier, Hen Harrier, Barn Owl, Short-eared Owl, Kingfisher, Grey Wagtail, Meadow Pipit.

March–May: Garganey, Cuckoo, Little Ringed Plover, Black-tailed Godwit, Wood Sandpiper, Redshank, Hobby, Sedge Warbler, Reed Warbler, Swallow, House Martin.

July–October: Green Sandpiper, Greenshank, chance of scarcer waders, hirundines and migrant passerines.

61 ARLINGTON RESERVOIR

OS Explorer OL25
OS grid ref: TQ 535073
Postcode: BN26 6UX

HABITAT

Built in 1971 by South East Water, which still owns and manages the site to this day, the 49ha Arlington Reservoir is a great example of how to balance the needs of both wildlife and people at a publicly accessible waterbody. The almost 3km 'Osprey Trail' around the dammed shore of the reservoir offers a pleasant walk for all and plenty of excellent views across the water, as well as of the surrounding scrub, woodland and farmland. Situated right next to the River Cuckmere, and just a few kilometres inland, it's not surprising that this site regularly hosts an attractive selection of migratory birds, as well as wintering wildfowl and gulls. The area has also been designated as both an SSSI and an LNR.

About halfway round from the car park, on the eastern side, a bird hide is tucked away down in the trees, offering good views across the whole reservoir. Another good area to view from is near the valve tower on the southern side, this spot being especially good in the winter when the sun will be behind you.

SPECIES

Although the likes of Tufted Duck, Pochard and Great Crested Grebe are present here all year, for the greatest abundance and diversity of waterbirds, a winter visit is a must as more than 2,000 ducks may be found; these are mostly Wigeon and Teal, but more than 50 each of Shoveler and Pintail can also assemble. The gull roost in the winter months is a big highlight, sometimes attracting upwards of 20,000 birds, including more than 10,000 Black-headed Gulls on occasion, and several thousand each of Herring, Common, Lesser Black-backed and Great Black-backed, plus smaller numbers of Mediterranean. The likes of Yellow-legged

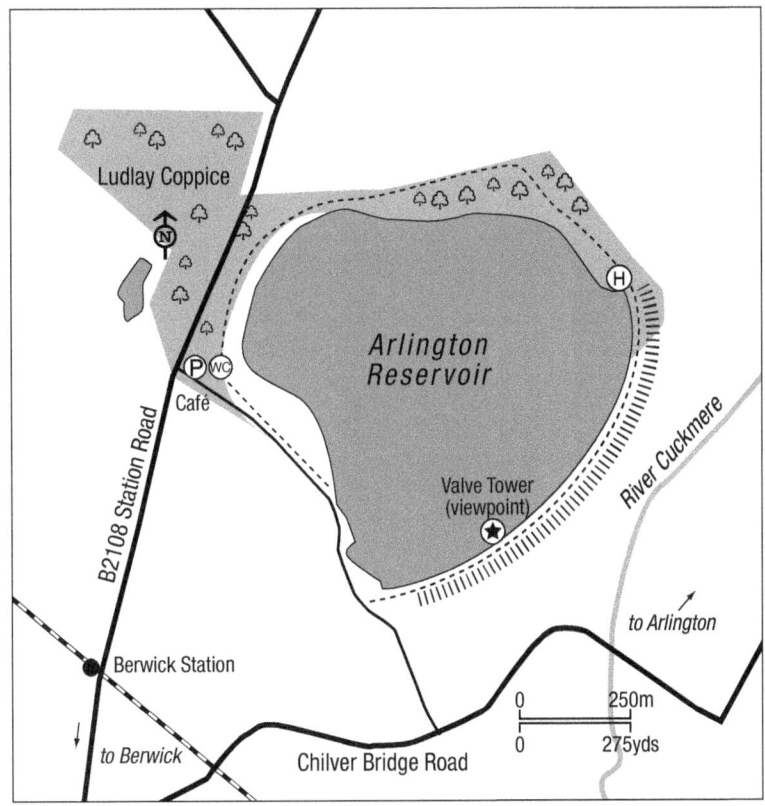

or Caspian Gulls are always a possibility too, with sometimes several of the latter occurring on a good day.

From mid- to late March onwards, any visiting birder's thoughts here should be turning to the possibility of scarcer migrant species dropping in, especially in grey, drizzly weather and north-easterly winds which will often produce Little Gull. Arlington Reservoir is also one of the most reliable sites away from the coast for Black-necked Grebe, the species being an annual occurrence here in early spring. Slavonian and Red-necked have both been recorded too, but less frequently.

After the arrival of the first Little Ringed Plover in March, wader passage really gets into gear, and just about all the regularly occurring species have appeared here. Common Sandpiper is perhaps the most frequently recorded, pottering around on the dam walls in the spring or autumn, but Green Sandpiper, Greenshank, Redshank, Whimbrel and Black-tailed Godwit are all annual too.

All three native hirundine species become a regular sight here, feeding over the water, from late March onwards, with numbers generally peaking in August and September, when many thousands will sometimes gather to feed up ahead of their southward migration. It's at this time of year when an Osprey is perhaps most likely, drifting overhead or even dropping in to catch a fish, and a Black Tern or two may stop off on migration.

Arlington also has a good selection of passerine species to offer throughout the year, with the year-round Grey Wagtails along the dam sometimes joined by Yellow or White Wagtails in spring or autumn. There's a chance of Kingfisher at any time of year. The areas of woodland on the western and northern flanks are home to a host of common resident passerines, joined by Blackcap, Chiffchaff and Garden Warbler in the summer. The more open, scrubbier landscape on the eastern side supports Yellowhammer, Whitethroat, Lesser Whitethroat and Nightingale, while Skylarks can often be heard singing over the surrounding farmland.

Rarer species found here over the years include Buff-breasted Sandpiper, Hoopoe, Bonaparte's Gull and Laughing Gull.

TIMING

The site offers an enjoyable circular walk at any time of year but really comes alive in the passage months, when almost anything is possible. From April there is also the added bonus of singing Nightingale and various migrant warblers.

ACCESS

Arlington Reservoir is ideally placed for access by train, with Berwick Station just 1km or so south down the B2108 (Station Road). By car, the reservoir is easy to find by turning off the A27 towards Berwick. Shortly after passing through the village, you will see the height barrier for the car park on your right. The car park is locked at night, so if you're planning to visit very early in spring, there is space for a few cars to park in the lay-by across the road. Various bus services also stop at Berwick Station, including 36, 40, 43 and 44 (all Cuckmere Buses) and 125 (Compass Travel). The Osprey Trail is relatively flat and wheelchair and pushchair friendly for the most part (around three quarters of it are hard surfaced), with two RADAR kissing gates along the dam wall (south-eastern side).

FACILITIES: A café and toilets are located by the car park.

CALENDAR

All year: Egyptian Goose, Pochard, Great Crested Grebe, Mediterranean Gull, Great Black-backed Gull, Little Egret, Kingfisher, Peregrine, Raven, Skylark, Grey Wagtail, Yellowhammer, Reed Bunting.

April–May: Garganey, Shoveler, Gadwall, Wigeon, Teal, Black-necked Grebe, Turtle Dove, Cuckoo, Swift, Little Ringed Plover, Whimbrel, Dunlin, Common Sandpiper, Green Sandpiper, Greenshank, rarer waders, Little Gull, Common Tern, Osprey, Hobby, Reed Warbler, Sand Martin, Swallow, House Martin, Willow Warbler, Cetti's Warbler, Garden Warbler, Lesser Whitethroat, Spotted Flycatcher, Nightingale, Whinchat, Wheatear.

July–September: Turtle Dove, Black-tailed Godwit, rarer waders, Yellow-legged Gull, Black Tern, hirundines and warblers as in April–May, returning migrant passerines.

October–March: White-fronted Goose, Brent Goose, Shoveler, Gadwall, Wigeon, Pintail, Teal, Goldeneye, Goosander, Jack Snipe, Snipe, Caspian Gull, Firecrest, Stonechat.

62 LULLINGTON HEATH AND FRISTON FOREST

OS Explorer OL25
OS grid refs: TQ 545016, TV 540596
Postcodes: BN26 5QJ, BN20 0AT

HABITAT

Situated roughly halfway between Seaford and Eastbourne lies the 62ha Lullington Heath National Nature Reserve, one of the largest areas of chalk heathland in Britain, which earned the site its NNR designation in 1954.

Although on the chalk bedrock of the South Downs, a fine layer of acidic soil on the surface is enough to support classic heathland plant species such as various heathers and Tormentil, alongside chalk specialists like Dropwort, Wild Thyme and Viper's-bugloss, as well as a small population of Burnt-tip Orchid. The reserve is grazed by Herdwick sheep, Exmoor ponies and occasionally goats, to keep the sward from being overrun by coarse vegetation and scrub. That said, there are some areas of scrub that have escaped the grazing efforts of the livestock and these provide habitat for various resident and migrant birds. Towards the north-western side of the reserve is as a reed-fringed dew pond known as Winchester's Pond, which is teeming with dragonflies in the summer. Moreover, 34 butterfly species have been recorded on the reserve including Small Blue, Adonis Blue, Dark Green Fritillary and Silver-spotted Skipper.

Immediately to the south of Lullington Heath lies Friston Forest, a 278ha Forestry Commission woodland planted in the 1930s and 1940s and dominated by chalk-loving Beech trees. Despite it being rather dark in places owing to the even age structure of the woodland, the sunnier rides can be good for butterflies such as

Silver-washed Fritillary and White Admiral, while Friston Gallops on the eastern side of the forest can produce spectacular emergences of Chalkhill Blue in good years.

SPECIES

The unusual combination of chalk downland and heathland at Lullington harbours an enticing mix of species. Although it can seem bleak in the winter, there is always the chance of Hen Harrier, Short-eared Owl or Merlin hunting across the open landscape. In spring and summer, the area bursts into life. Resident species such as Linnet, Skylark and Yellowhammer can be heard singing from as early as late January, joined by migrant warbler species in April. Wheatear and Whinchat may be found in varying numbers in spring and autumn, while the abundance of dragonflies from Winchester's Pond will often attract a Hobby to feed over the site. Given its elevation and combination of open downland and scrub, the reserve is also a good place to look for Ring Ouzel in the autumn.

The heavily wooded landscape of Friston Forest offers a rather different array of species, with the likes of Buzzard, Sparrowhawk and Tawny Owl breeding in the dense tree cover, while scrubbier areas along the woodland rides hold breeding Bullfinch and Garden Warbler. In contrast to Lullington Heath, Friston can be more productive in the winter than the summer, as the abundance of Beech trees here can attract flocks of Brambling, sometimes in large numbers during influx years. A good area to check for these is near the Butchershole car park. Similarly, Hawfinch can sometimes be encountered in the winter.

Other notable species recorded in the area include Pallas's Warbler, Wood Warbler, Montagu's Harrier, Hoopoe, Wryneck and Golden Oriole.

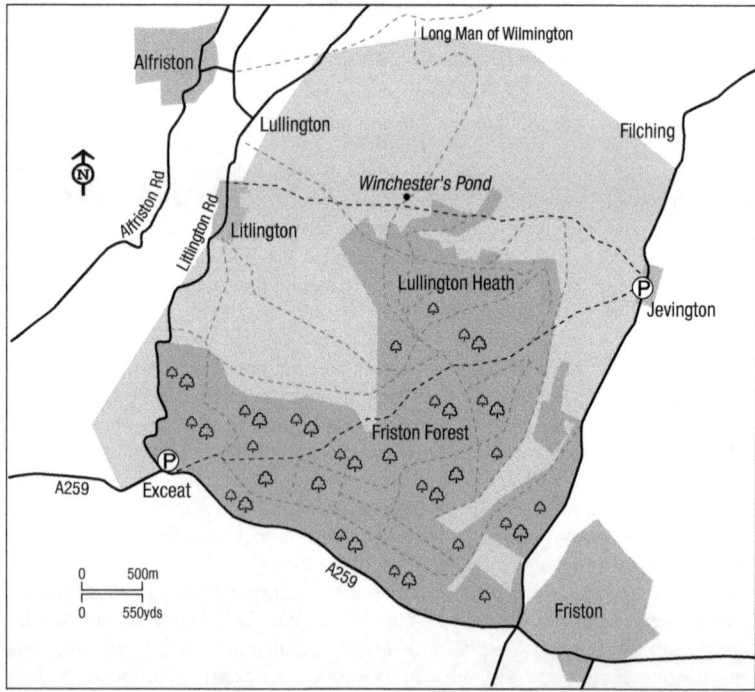

TIMING

Visit in spring and summer for the best selection of breeding species, and in late summer and autumn for a chance of scarce and rare migrants.

ACCESS

Bus service 12 from Brighton to Eastbourne (Brighton & Hove) stops at Exceat, from where you can take the public footpath east into Friston Forest, then north to Lullington Heath. The nearest train stations are at Seaford and Eastbourne. By car, either follow the A259 to Exceat and park at the visitor centre at TV 518995, or take the Jevington Road north off the A259 at Friston and park at the small car park in Jevington village at TQ 56200128. From here, walk up to Church Lane and follow the track (the South Downs Way) west, until the point where it veers to the right. Carry straight on for about 1km to reach Lullington Heath, although note that this is quite a steep climb! For access to the east side of Friston Forest, park at the Butchershole Bottom car park at TV 555993.

FACILITIES: A car park, visitor centre and toilets are located at Exceat, while other car parks can also be found in Jevington village and at Butchershole Bottom.

CALENDAR

All year: Tawny Owl, Raven, Skylark, Yellowhammer.

April–October: Hobby, Whinchat, Wheatear, Nightingale, Ring Ouzel.

November–March: Merlin, Brambling.

OTHER SITES IN SOUTH EAST SUSSEX

G1 DOLEHAM MARSHES

This area of wet grassland (TQ 842173) south of the River Brede floods in the winter and attracts good numbers of wildfowl. Cetti's, Sedge and Reed Warblers breed in the reeds. Little Egret and Great White Egret are also a common sight, particularly in the winter, while a Hobby will often be about in the summer. Doleham Station (plus car park) is just a short walk away (TN35 4LY).

G2 ALEXANDRA PARK

Situated in the middle of Hastings, this urban park (TQ 805104) has a few small reservoirs, woodland and grassland. Firecrest occasionally winters and other passerines move through on passage, including Pied Flycatcher, suggesting there is potential for scarcer finds. Indeed, a Night Heron visited in 2007.

G3 FORE WOOD RSPB

This small (20ha) area of ancient ghyll woodland (TQ 753128) near Crowhurst is designated as an SSSI and managed by the RSPB. It supports breeding Marsh Tit and Firecrest, and there are recent records of Lesser Spotted Woodpecker. Hawfinch also winters here from time to time thanks to a good amount of Hornbeam. The nearest station is at Crowhurst and parking is available at Crowhurst village hall.

G4 ASHBURNHAM FORGE

An enticing mix of ancient woodland, open fields and lakes (TQ 688142) make this a worthwhile place for a visit. The three lakes, Broad Water, Front Water and Reservoir Pond, support a host of wildfowl, while the woodlands hold breeding Marsh Tit, Firecrest and Spotted Flycatcher. Recent notable records include Smew, Quail and Yellow-browed Warbler.

G5 NORMANS BAY

This exposed stretch of shingle beach (TQ 686054) south of Pevensey Levels (site 49) has proven attractive to both Shore Lark and Snow Bunting in recent years, and even a Desert Wheatear. A few Sanderling may be among the wintering waders along the shore. At this time of year, check offshore Common Scoter flocks for Velvet, while also keeping scarcer grebes, divers and Fulmar in mind, and scan any structures or buildings for Black Redstart.

G6 PRINCES PARK

This 13ha town park in the heart of Eastbourne (TQ 627004) may seem like a strange choice for inclusion here but Crumbles Pond has form for attracting good birds. In particular, it can host lots of wildfowl and gulls and has turned up some impressive species in recent years including Slavonian Grebe, Black-throated Diver and Bonaparte's Gull, as well as regular Yellow-legged Gulls in late summer. Car parking is available on Royal Parade nearby, and Eastbourne station is around 2km away.

G7 HIGH AND OVER

This downland site offers great views across the Cuckmere Valley and the wider landscape of Sussex. All the expected farmland and downland species of the region – including Corn Bunting and Yellowhammer – are likely and it can also be good for grounded migrants and fly-overs in spring and autumn, especially Wheatear, Redstart and Ring Ouzel. The site attracted a lingering Great Grey Shrike a few years ago. It's also good for butterflies, including Wall Brown and Silver-spotted Skipper. Car parking is available at the tumulus (TQ 509010).

G8 TELSCOMBE TYE

This 72ha area of open downland and farmland between Brighton and Seaford (TQ 402029) is another great site for farmland birds, including Grey Partridge and Corn Bunting. It has also attracted Quail in recent years so is one site to check during influx years for this species. It's a good spot for passage migrants, including likely Ring Ouzel in the autumn, and Wryneck has been recorded here too.

G9 PORTOBELLO WATER WORKS

The small waterworks and adjacent beach between Saltdean and Telscombe Cliffs (TQ 393014) are most famous for hosting a Trumpeter Finch in 2008. This is a good site for Black Redstart and Rock Pipit in the winter, while there's an excellent chance of both Fulmar and Kittiwake offshore.

G10 ROTTINGDEAN

Another seaside village with some diverse birding potential. Seawatching from the beach (TQ 372020) can be rewarding and, even if it isn't, you can sometimes see waders and Wheatears along the beach in spring and enjoy views of the nesting Fulmars up on the cliff. The fields along Bazehill Road (TQ 372034) north of the village can produce Skylark, passage Wheatears and even held a Hoopoe a few years back. Ample parking is available along Marine Drive.

G11 OVINGDEAN

Just east of Brighton Marina, Ovingdean beach offers a good selection of waders, Fulmars and seawatching, while wintering Black Redstarts will often be found along the undercliff path (TQ 349029). If you have time, the farmland to the north of the village (TQ 354042) can produce Red-legged Partridge and some sizeable flocks of Linnet in the winter. Car parking is available at Rottingdean or Brighton Marina.

G12 MALLING DOWN AND SOUTHERHAM FARM

These two Sussex Wildlife Trust downland reserves on the outskirts of Lewes offer a mix of chalk grassland and scrub. Both can be good for common breeding warblers, but also passage migrants including Ring Ouzel, Spotted Flycatcher, Whinchat and Redstart. Hen Harrier has been recorded here on occasion. This is another great area for butterflies including Adonis Blue and Silver-spotted Skipper. Limited parking is available on Mill Road (TQ 422111) and off the A26 near Cliffe Industrial Estate (TQ 425092).

G13 ABBOT'S WOOD

This is a 360ha area of ancient woodland just a couple of kilometres east of Arlington Reservoir, with great Bluebell displays in the spring. Arguably the best place for Nightingale in East Sussex, the site also holds breeding Garden Warbler and Nightjar (the latter in clearfell areas), and is fairly reliable for Hawfinch in the winter. This is also a good site for butterflies, including Pearl-bordered Fritillary. A car park and picnic area are located at TQ 557072, while the nearest train station is in Berwick.

SOUTH WEST SUSSEX

South West Sussex, for the purpose of this book, is defined as everything south of a line running west from the East Sussex border near Burgess Hill to Pulborough, then roughly following the line of the South Downs where they reach Hampshire. In other words, all sites featured in this section are within or south of the South Downs National Park.

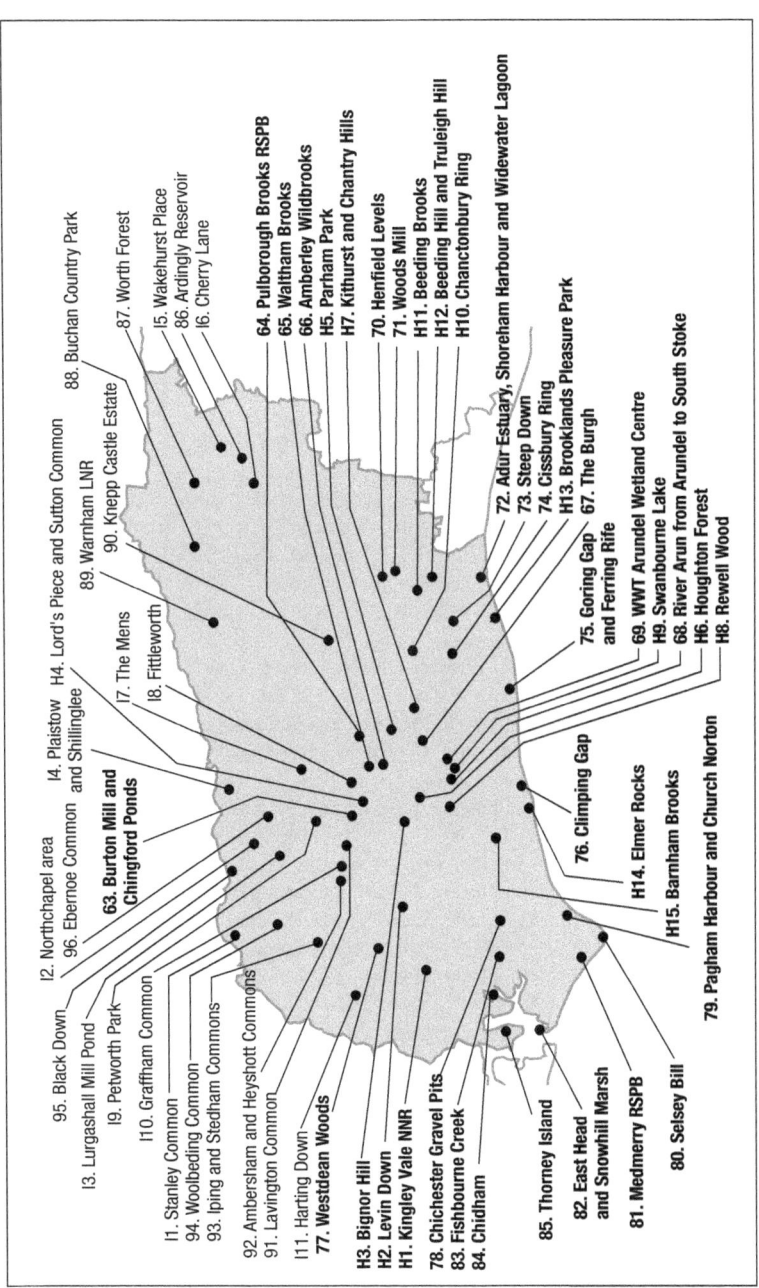

95. Black Down
I2. Northchapel area
I4. Plaistow
H4. Lord's Piece and Sutton Common
88. Buchan Country Park

13. Lurgashall Mill Pond
96. Ebernoe Common and Shillinglee
I9. Petworth Park
89. Warnham LNR
90. Knepp Castle Estate
I5. Wakehurst Place
86. Ardingly Reservoir
I6. Cherry Lane
87. Worth Forest

I1. Stanley Common
94. Woolbeding Common
93. Iping and Stedham Commons
I7. The Mens
I8. Fittleworth

92. Ambersham and Heyshott Commons
91. Lavington Common
I11. Harting Down
77. Westdean Woods
63. Burton Mill and Chingford Ponds

H3. Bignor Hill
H2. Levin Down
H1. Kingley Vale NNR
78. Chichester Gravel Pits
83. Fishbourne Creek
84. Chidham

85. Thorney Island

82. East Head and Snowhill Marsh
81. Medmerry RSPB
80. Selsey Bill

79. Pagham Harbour and Church Norton

64. Pulborough Brooks RSPB
65. Waltham Brooks
66. Amberley Wildbrooks
H5. Parham Park
H7. Kithurst and Chantry Hills
70. Henfield Levels
71. Woods Mill
H11. Beeding Brooks
H12. Beeding Hill and Truleigh Hill
H10. Chanctonbury Ring

72. Adur Estuary, Shoreham Harbour and Widewater Lagoon
73. Steep Down
74. Cissbury Ring
H13. Brooklands Pleasure Park
67. The Burgh

75. Goring Gap and Ferring Rife
69. WWT Arundel Wetland Centre
H9. Swanbourne Lake
68. River Arun from Arundel to South Stoke
H6. Houghton Forest
H8. Rewell Wood

76. Climbing Gap

H14. Elmer Rocks
H15. Barnham Brooks

The 50km stretch of coastline in West Sussex is naturally a big draw for birders and, unsurprisingly, accounts for around a third of the sites covered in this section. There are fewer coastal sites when compared to the South East Sussex region in this book, though this is largely because East Sussex has almost twice as much coastline as West Sussex. Despite this, West Sussex boasts some major coastal birding habitat, including sites that are nationally renowned. Probably the best known is the Pagham Harbour area – one of the most famous birding locales in England with a rich avian history. Along with Chichester Harbour to the west, Pagham is a Special Protection Area and both boast large concentrations of wintering wildfowl and waders, including internationally important numbers of some species. A fine day of birding can be had at the range of sites in these areas year-round. The Adur Estuary area at Shoreham-by-Sea is far smaller than Pagham or Chichester, but holds similarly interesting birdlife. Selsey Bill is the main seawatching spot in this region, though generally greater numbers of birds and a wider selection of watchpoints can be found in East Sussex. This is because less of the West Sussex coastline jut out into the English Channel, a factor that also reduces the number of 'headland' sites in West Sussex. Nevertheless, the West Sussex coast offers some truly excellent birding, with a diverse range of sites ranging from well-managed reserves to wilder, less-visited places.

The Arun and Adur rivers are dominant features both in terms of the geography and the birding in this region, being the only waterways in West Sussex to cut right through the South Downs to the English Channel. The Arun is the longest river entirely in Sussex, covering 37 miles from its source in the High Weald near Horsham down to its estuary at Littlehampton. Both rivers are tidal for several miles inland and readily flood large areas of their respective floodplains in this region in the winter months, especially during periods of spring tides combined with heavy rainfall. As a result, several wetland sites along the Arun have been awarded high levels of conservation status due to their national and international importance for breeding and wintering waders and wildfowl, such as Lapwing, Redshank, Wigeon, Teal and Pintail. Among such sites included here are RSPB Pulborough Brooks, Amberley Wildbrooks, Waltham Brooks, WWT Arundel, Henfield Levels and Beeding Brooks, many of which boast a great diversity of birdlife.

Away from the coast and the rivers, the other main features of South West Sussex are the chalk spine of the South Downs and the swathe of woodland tucked in along the southern slopes of the Downs. The downland birding on offer here can be particularly rewarding, with sites such as The Burgh, Cissbury Ring and Steep Down yielding many of the desired species typical of open country in this region. Among them are Corn Bunting, Yellowhammer, Grey and Red-legged Partridges and passage migrant passerines, as well as owls and scarce raptors in the winter and, sometimes, Quail in the summer.

The approximately 25km band of woodland that stretches right across this area of West Sussex, from South Harting to Houghton Forest, harbours some of the densest and most reliable populations of scarce breeding species, such as Goshawk and Hawfinch. Vast tracts are rather under-watched though, especially away from well-known sites such as Westdean Woods and Rewell Wood. Concealed among some of these more wooded areas are old mill ponds and other human-made lakes, many of which receive less attention from birders than nearby reserves. Even well inland, these wetland sites but can produce an interesting selection of wildfowl and other birds. Burton Mill and Chingford Ponds, and Swanbourne Lake are fine examples of this, both sites offering a potentially species-rich experience.

There is much in this region to entice, enchant and entertain any birder, be they visiting for a day or living here for decades, and it's no surprise that the coastal areas featured are some of the best-birded sites anywhere in Sussex. This section aims to cover the hotspots in good detail, while also highlighting some less-visited areas that may be of interest.

63 BURTON MILL AND CHINGFORD PONDS

OS Explorer OL10
OS grid ref: SU 979180
Postcode: GU28 0JR

HABITAT

Situated 5km south of Petworth, Burton Mill and Chingford Ponds offer some of the most dynamic birding in this part of inland West Sussex. Jointly managed by Sussex Wildlife Trust and Sussex County Council, together they form a single LNR. Burton Mill Pond is part of the wider Burton Park SSSI, designated as being of national importance for its wetland habitats, rare plants, bird and invertebrate populations.

Chingford Pond, meanwhile, has been designated as an SNCI. Both ponds are surrounded by mature woodland and some sizeable stretches of reedbed. Due to the rich diversity of the habitat here, a couple of hours can produce 60 or more bird species. Black Pond, just west of the trail when you join the minor road through Burton Park, is also worth a look as it can hold a reasonable number of wildfowl in the winter.

SPECIES

In winter good numbers of dabbling and diving duck gather here, especially Gadwall, Tufted Duck and Pochard. Numbers are high on Chingford Pond, which tends to be less disturbed. As herons have increased in the region, Cattle, Great White and Little Egrets are recorded more frequently and sometimes it's possible to encounter all three species in one visit. Great White Egrets in particular seem to enjoy the seclusion of the mature woodland edges and reeds to hide in and are often found in such areas in the winter. They are best looked for in late winter and early spring, particularly at dusk when they tend to be most active and may even fly past your watchpoint. Historically, this site was also reliable for wintering Bittern, though records have dropped off in recent years and this species is not to be expected. Cormorants roost in the trees to the southern end of Burton Mill Pond, sometimes in impressive numbers. In winter, Siskins frequent the Alders around the edges of the ponds, particularly on the northern side, and three-figure flocks can be encountered on occasion.

The woodland areas support common passerines, including all three woodpeckers, though Lesser Spotted Woodpecker is increasingly difficult to find. The most reliable spots have traditionally been the areas of woodland between Newpiece and Black Pond and the area known as 'The Black Hole' to the south-east of Burton

Mill Pond itself. Marsh Tits are relatively easy to locate around the wooded trail, usually being heard before they are seen.

The return path through Welch's Common with its mixed woodland and Holly understorey offers the best chance of encountering Firecrest year-round. Tawny Owl can often be heard in any of the wooded areas, while Burton Park Farm to the west hosts breeding Little Owl. Grey Wagtails are a reliable bet anywhere around the edges of both ponds, but especially on the outflow sluice at Chingford.

The areas of reedbed hold Water Rail and Cetti's Warbler, especially in winter, and come alive with Reed Warblers and Reed Buntings in the spring. In springtime, the courtship dance of Great Crested Grebes can be enjoyed, as the species breeds here along with Little Grebe, Coot and Moorhen. Burton Mill Pond is particularly good for Kingfisher at any time of year, even if views may be typically fleeting. Osprey passes through in spring and autumn. The heathland areas host breeding Woodlarks and Nightjars in the summer. All three hirundine species can be encountered here through the summer months, including sometimes good numbers of Sand Martin which breed at the nearby (private) Heath End Sand Pit.

While not necessarily a prime site for waders, Common and Green Sandpipers and Greenshank are possible in late summer, and a few Snipe will often return to spend the winter here.

Recent scarce species recorded at the ponds include Purple Heron, Red-crested Pochard and Yellow-browed Warbler.

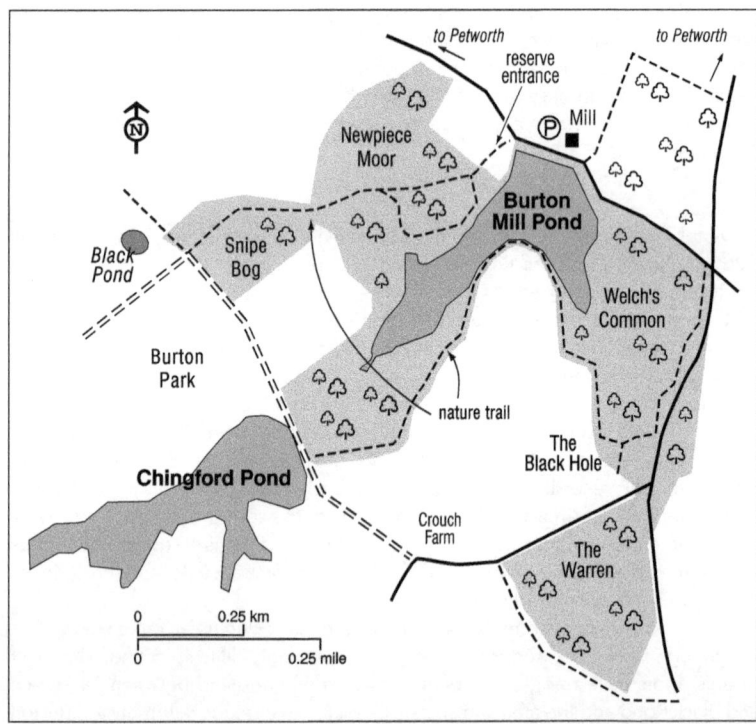

TIMING

Be aware that anglers sometimes launch boats out onto Burton Mill Pond, which can flush many of the birds, although often they will just relocate to Chingford Pond. Early morning is the most productive, especially if you can get there ahead of the anglers. Also, you may have a chance of encountering some of the egrets leaving roost or even a vocal Lesser Spotted Woodpecker in spring. An evening visit in summer should produce churring Nightjar.

ACCESS

There is a small car park just to the north of Burton Park Road at SU 978180. Bus route 99 from Chichester to Petworth has various stops along the A285, all within walking distance. Petworth is just 5km away, with cycling a straightforward jaunt down Station Road then onto Burton Park Road. From the car park a circular 5km trail takes in good views of both ponds plus some of the adjoining heathland, woodland such as Newpiece, Welch's Common and the rolling parkland landscape of Burton Park to the west. The northern edge of Burton Mill Pond offers the best viewing, particularly the large viewing platform towards the north-east corner. Chingford Pond is best viewed from the outfall causeway along its northern shore.

The terrain is mostly flat and the walking easy-going. There is just one stile around the whole trail, at Crouch Farm. Some sections can get muddy in the winter months but the very wettest areas have surfaced paths and a boardwalk at one point.

FACILITIES: There are no facilities other than the car park. The nearest toilets and restaurant facilities are in Petworth or Pulborough.

CALENDAR

All year: Egyptian Goose, Mandarin Duck, Gadwall, Little Grebe, Great Crested Grebe, Little Egret, Tawny Owl, Little Owl, Kingfisher, Lesser Spotted Woodpecker, Raven, Marsh Tit, Firecrest, Grey Wagtail, Bullfinch, Siskin, Reed Bunting.

April–September: Cuckoo, Nightjar, Common Sandpiper, Green Sandpiper, Osprey, Hobby, Reed Warbler, Sand Martin, Swallow, House Martin, Willow Warbler.

October–March: Shoveler, Wigeon, Pintail, Teal, Pochard, Water Rail, Woodcock, Snipe, Great Egret, Cattle Egret, Cetti's Warbler, Brambling, Lesser Redpoll.

64 PULBOROUGH BROOKS RSPB

OS Explorer OL10
OS grid ref: TQ 058164
Postcode: RH20 2EL

HABITAT

Pulborough Brooks is one of the RSPB's flagship reserves and its most visited in South-East England, attracting tens of thousands of visitors per year. The 171ha site was bought by the charity in 1989 as a unique wetland restoration project. The primary habitat is lowland wet grassland, home to breeding waders in the summer and huge numbers of wildfowl in the winter. There are pockets of woodland and grassland, and a modest area of heathland (Wiggonholt Common) on the south-eastern side of the main reserve.

The main wetland trail, starting from the visitor centre, is 3km long and incorporates four hides. West Mead and Winpenny both offer views across the south end of the reserve (South Brooks), while Little Hanger and Nettley's look out across the North Brooks. There are also two open viewpoints overlooking the North Brooks: The Hanger and Jupp's View. Other areas of note around the trail include Fattengates Courtyard, Sims Pond, Redstart Corner and Adder Alley – the latter is, unsurprisingly, a good place to look for Adders, of which there is a healthy population.

Heading south-west from the car park and skirting the edge of the heathland, you will come to Black Wood, turn off through the gate here and follow the trail to the end to reach Hail's View, which offers a vista out across the most southern part of the reserve.

SPECIES

Pulborough Brooks has more or less everything a birder could want, packed into a relatively small area. It is one of the most famous sites in South-East England for Nightingale, as well as hosting breeding Lapwing and Redshank (and Avocet in recent years).

The reserve is equally renowned for the thousands of dabbling ducks which spend the winter here and in the wider Arun Valley, largely Wigeon and Teal, but also somewhat smaller numbers of Pintail and Shoveler. Counts of over a hundred of either of the latter two species are not uncommon, while Wigeon and Teal numbers can often reach four figures. West Mead or Little Hanger hides offer the best chance of seeing these species up close; from the other hides and viewpoints they can be distant. The dwindling Arun Valley herd of Bewick's Swan uses the reserve as a roost site on rare occasions these days. Though it's hard to predict when, there's always a chance you might see them flying in to roost at dusk or flying off again in the early morning between December and February. It's also always worth scanning through the large flocks of Canada Geese and Greylag Geese for scarcer species such as White-fronted Goose or even Tundra Bean Goose. Brent Goose also turns up on occasion, usually in early or late winter when the species is moving to and from its wintering grounds on the Hampshire and Sussex coasts. A more recent phenomenon is the wintering flock of Black-tailed Godwits, which has been known to peak at over a thousand in some years, usually in December or January. Thousands of Lapwing also winter on the reserve, with a few Ruff and the odd Dunlin usually among them. Winter also sees dozens of Snipe out in the wet grassland, though they can sometimes be hard to spot.

The ducks and waders are often flushed into the air by a passing Peregrine or Merlin, which can make for a breathtaking sight and sound. Both are regular winter visitors – the latter exclusively so – and it's at this time of year that a Hen Harrier may frequent the reserve. Marsh Harrier is now a year-round presence, along with the common raptor species including Red Kite and Buzzard. Goshawk records are on the rise, though the species is still rare here. In recent years, Pulborough has become a reliable site to see White-tailed Eagle, as some of the birds from the Isle of Wight reintroduction project have taken a liking to the Arun Valley. Barn Owl and Tawny Owl regularly breed and are sometimes joined by Short-eared Owl in the winter.

Uppertons Field, immediately below the visitor centre, is worth a check at any time of year, often holding a flock of Woodlarks in the winter (some of which stay to breed) and Wheatears in spring and autumn. The 'ZigZag' path down from the visitor centre usually yields Bullfinch and Greenfinch, sometimes in good numbers, especially in winter when they'll often be joined by Redwings, Fieldfares and Lesser Redpolls. Fattengates at the bottom of the ZigZag has hosted a wintering Lesser Spotted Woodpecker on occasion in recent years, so it's worth having a scan through the trees here. This can also be a good spot for Firecrest and sometimes Marsh Tit.

As winter gives way to spring, late March and April sees migrant warblers including Blackcap, Chiffchaff, Garden Warbler, Whitethroat and Lesser Whitethroat and, of course, Nightingale returning to the reserve for the breeding season. Willow Warbler is always a possibility, though it has heavily declined as a breeding species. Sedge and Reed Warblers breed in good numbers and are best seen or heard along the path between Adder Alley and the Arun. Cetti's Warblers, too, are increasing in number and can sometimes be heard in the same area. Grasshopper Warbler is another possibility, particularly in late summer as passage birds move through and can turn up almost anywhere. Cuckoos breed and can usually be heard and occasionally seen from April to June, with juveniles sometimes seen later in the summer.

As well as the passerine arrivals in spring, it is also the time of year when wader passage really hots up at Pulborough. The breeding Redshank and Avocet are generally the first to arrive, from as early as late February, followed by the likes of Little Ringed Plover and Curlew from mid-March onwards. In years when the water levels are optimum, the area of the South Brooks nearest the river and the various pools on the North Brooks can be very attractive to a host of more unusual species. On a good day in late April or May, it's possible to see a satisfying selection, including some or all of Whimbrel, Ringed and Grey Plovers, Greenshank, Spotted Redshank and even Curlew Sandpiper, with many of these species looking resplendent in their breeding plumage. Wood Sandpipers are almost guaranteed at this time of year too – though note that many of these species won't stay very long in spring. Black-winged Stilt has been recorded here on several occasions in recent springs, so it's certainly one to have in mind at this time of year.

Other summer visitors returning to the reserve later in the spring (generally mid- to late April onwards) include Hobby, Nightjar and Spotted Flycatcher, with all three of these most reliably encountered on Wiggonholt Common, where an evening visit at this time of year may also produce roding Woodcock. If you time your visit right on a warm evening in May, you can enjoy a dusk stroll around the main trail enjoying the sounds of Nightingale, Redshank and Lapwing, then finish up on the heath to see and hear Nightjar and Woodcock. Please be advised that the central area of the heath is now permanently locked to prevent disturbance to the ground-nesting birds, so the path along the northern edge or the tumulus on the eastern side are the best spots from which to enjoy the Nightjars.

Osprey is always possible in the spring or autumn, though it's not especially regular at Pulborough. In late summer, persistence and luck may yield sightings of Honey Buzzards flying to and from the reserve, where they sometimes predate wasp nests.

From mid- to late August, return migration really gets into gear, as the likes of Green Sandpiper and the first returning wildfowl are seen on the wetland areas again. Return wader passage here is generally more diverse than in spring, with a better chance of scarcer species such as Little Stint and Pectoral Sandpiper. The flocks of common geese can also grow to impressive numbers at this time of year, and it's quite a spectacle to see them arriving and departing from roost on the North Brooks. Passage migrant passerines, too, may be found anywhere around the reserve in early autumn, especially Redstarts at Redstart Corner, possibly a Tree Pipit or two in the hedgerows and Yellow Wagtail among the cattle. A scan of the posts, gates and vegetation from any of the hides and viewpoints will often produce several Whinchat at this time of year, while any tit flock is worth combing through for a Pied Flycatcher or possibly even a Wood Warbler.

Given the quality of the habitat combined with the observer effort here, Pulborough unsurprisingly has a mouth-watering list of rare and scarce species on its site list. More unusual visitors in recent years have included Red-necked Phalarope, White-rumped Sandpiper, Common Crane, Richard's Pipit, American Wigeon, Black Stork and Great Grey Shrike.

TIMING

Pulborough Brooks offers something of interest at almost any time of day or year. For the best array of species, visit in April or May, especially to see and hear singing Nightingales and displaying Lapwings and Redshanks. A clear, crisp day in midwinter will offer the best opportunity to see the numbers of wintering ducks, and they will be looking their best ahead of their return to breeding grounds.

For Nightjars, Woodcocks, owls and Nightingales, choose a warm, still evening in mid- to late May or early June. In periods of prolonged wet weather and spring tides in winter, the River Arun can overtop and turn the whole reserve into one huge waterbody. This, although spectacular, can reduce the amount of wildfowl on-site, and those that remain can be even more distant from the hides and viewing areas, so it's worth checking before you travel.

ACCESS

Bus route 100 (Compass Travel) from Burgess Hill to Horsham stops at the main entrance to the reserve (though you will need to tell the driver of your intention to disembark there when you board), while Stagecoach route 1 has various stops on Mare Hill Road, in the village. Pulborough railway station is approximately 4km away to the north-west of the reserve, with most of the roughly 45-minute walk being on public footpaths. From the station entrance walk south onto Station Road, then follow this east onto Lower Street. After around half a mile, turn right onto Barn House Lane and follow this down to the footpath across the field to the footbridge that crosses the River Stor. Continue straight on to join the riverbank along the Arun then veer left at the fingerpost at the end. You are now on RSPB land. When you reach the crossroads with the nature trail, either turn left or right to join the trail or continue straight on across the middle of the reserve, then turn right at Wiggonholt Church to reach the visitor centre, café and shop.

Please be advised that the public footpaths, particularly those from the river to the reserve, can get very muddy in the winter and may even be totally impassable if the river overtops. By car, follow the A283 from Pulborough south for a couple of miles and the reserve entrance will be on your right.

With the exception of West Mead Hide, views of some birds (especially water-birds) can be distant at Pulborough, so a telescope is definitely recommended.

FACILITIES: Facilities include a visitor centre, shop, café, toilets, accessible toilets, baby-changing, picnic area, play area and binocular hire. At the time of writing, the parking charge of £3 all day includes access to the heathland, visitor centre, picnic and play area. An additional entry fee applies to the wetland nature trail and hides (£4 adults). Parking and entry fees are waived for RSPB members. The reserve also has two Tramper mobility vehicles available for hire, which need to be pre-booked.

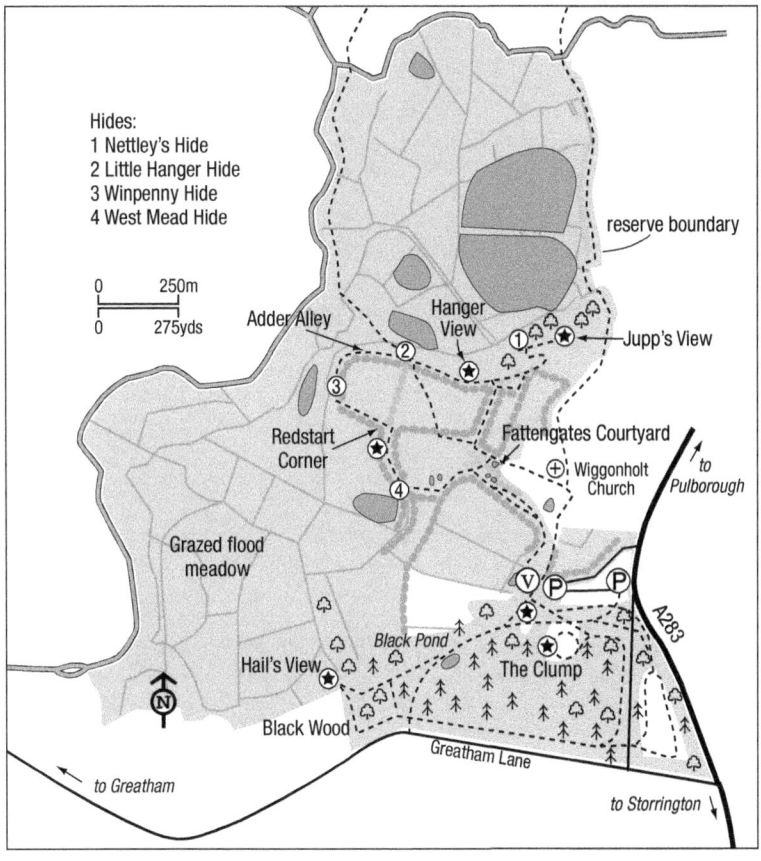

Hides:
1 Nettley's Hide
2 Little Hanger Hide
3 Winpenny Hide
4 West Mead Hide

reserve boundary

0   250m
0   275yds

Adder Alley
Hanger View
Jupp's View

Redstart Corner
Fattengates Courtyard
Wiggonholt Church
to Pulborough

Grazed flood meadow

Hail's View
Black Pond
The Clump
Black Wood
Greatham Lane

A283

to Greatham

to Storrington

CALENDAR

All year: Lapwing, Little Egret, Marsh Harrier, Barn Owl, Kingfisher, Peregrine, Woodlark, Cetti's Warbler, Stonechat.

April–May: Garganey, Little Ringed Plover, Whimbrel, Greenshank, Redshank, Hobby, Sand Martin, Lesser Whitethroat, Sedge Warbler, Reed Warbler, Nightingale, Spotted Flycatcher, Yellow Wagtail.

June-September: Nightjar, Wood Sandpiper, Osprey, Honey Buzzard, Pied Flycatcher, Redstart, Whinchat, Wheatear, Tree Pipit.

November–March: White-fronted Goose, Bewick's Swan, Shoveler, Wigeon, Pintail, Teal, the largest gatherings of Black-tailed Godwit, Ruff, Hen Harrier, Short-eared Owl, Merlin.

65 WALTHAM BROOKS

OS Explorer OL10
OS grid ref: TQ 026159
Postcode: RH20 1LS

HABITAT

This delightful 42ha pocket of wetland beside the River Arun is nestled between Pulborough Brooks and Amberley Wildbrooks. It greatly increases the ecological value of the wider SPA by providing a wetland corridor between those two much larger sites. In winter the main lake can swell to flood much of the reserve, offering a handy overspill for wildfowl displaced from other nearby sites. The lake here is surrounded by a fairly substantial margin of reeds, making it ideal for reduced disturbance to the birds as well as offering plenty of cover for various species to breed and roost. There are areas of willow carr around the edges, with more scrubby woodland areas towards the western and northern boundaries. The small sewage works along the western boundary has silt pools and further scrub with patches of rough grassland and is often teeming with flying insects, even in the winter.

SPECIES

For a relatively small site, the habitat diversity here offers an attractive mix of species. In winter the main lake hosts a variety of wildfowl including Shelduck, Shoveler, Pintail, Wigeon, Teal and Gadwall, while the spring and summer months can sometimes produce a Garganey. Unlike nearby Pulborough Brooks or Amberley Wildbrooks, Waltham Brooks offers much closer views of these ducks, which is beneficial for photography. Water Rail can often be heard squealing in the reeds and willow carr. In the winter months, many thousands of Starlings congregate to roost in the area and can be seen murmurating overhead at dusk. Barn Owl is a regular sight hunting over the reedbeds and rough grassland in the evenings, while Short-eared Owl occasionally occurs in the autumn and winter.

Marsh Harrier is an increasingly common sight throughout the year in this area, and at least one Hen Harrier often winters in the valley and can sometimes be seen flying to roost on the reserve. Another species to have on your radar here is Water Pipit, as the reserve has form for hosting at least one wintering bird. They can be elusive unless accidentally flushed, so it pays to familiarise yourself with the flight call. The sewage treatment works towards the western boundary of the site is a real honeypot for overwintering Chiffchaffs and attracts at least one Siberian Chiffchaff in most years. The areas of thorny scrub and gorse near the railway bridge occasionally draw in a wintering Dartford Warbler and also once regularly hosted a Great Grey Shrike, though this is sadly a rare event these days.

Breeding warblers include Blackcap, Whitethroat and Lesser Whitethroat, Garden and Willow Warblers, and sometimes Grasshopper Warbler. Cetti's Warblers are an ever-present soundtrack to a visit to the reserve. In some years Nightingale breeds, favouring the scrubbier areas near the railway line. Various wader species are possible, particularly in the spring and summer when the water level of the main lake recedes. Lapwing and Redshank both breed locally, and Avocet is increasingly likely as the species now breeds at Pulborough Brooks. Black-tailed Godwit, Greenshank and Green Sandpiper sometimes drop in outside the breeding season and Snipe are a constant presence in the winter, usually flushed up from the edges of the lake. It's worth being vigilant for Jack Snipe too, as they are often skulking out in the marshy areas but tend to remain characteristically elusive unless almost trodden on.

Given its relatively small size, the site has played host to an impressive selection of scarce species over the years, including Scaup, Red-rumped Swallow, Night Heron, Melodious Warbler and Red-backed Shrike.

TIMING

Waltham Brooks is diverse enough to offer a rewarding visit at almost any time of the year, with the relatively small size of the site making it an attractive choice. Evening visits, particularly in the winter months, can be rewarding in terms of raptors, wildfowl and owls. Early morning in spring will produce plenty of singing warblers and possibly a Garganey or waders on the main lake. For the most impressive numbers of wildfowl on the lake, visit any time between late November and February. One thing worth bearing in mind in the winter is the angle of the sun; bright sunny mornings can be problematic when viewing the lake from the western side, so an afternoon visit will provide more helpful light.

ACCESS

The car park near Greatham Bridge is closed at the time of writing, but there are various spots for a car or two to safely park along Brook Lane, including just east of the bridge itself, and to the west of the railway bridge by the allotments. Bus routes 69 (Alford–Worthing) and 71 (Storrington–Chichester) both stop on the London Road near the junction with Brook Lane, just a few minutes' walk to the north-west of the reserve. The nearest railway station is Pulborough, 3.5km away and this is easily accessible by bicycle or bus or, if you are feeling a little more adventurous, around an hour's walk following the A29 and then the Wey-South Path. The paths here are mostly level but can get very muddy in the winter so may not be suitable for wheelchair and pushchair users.

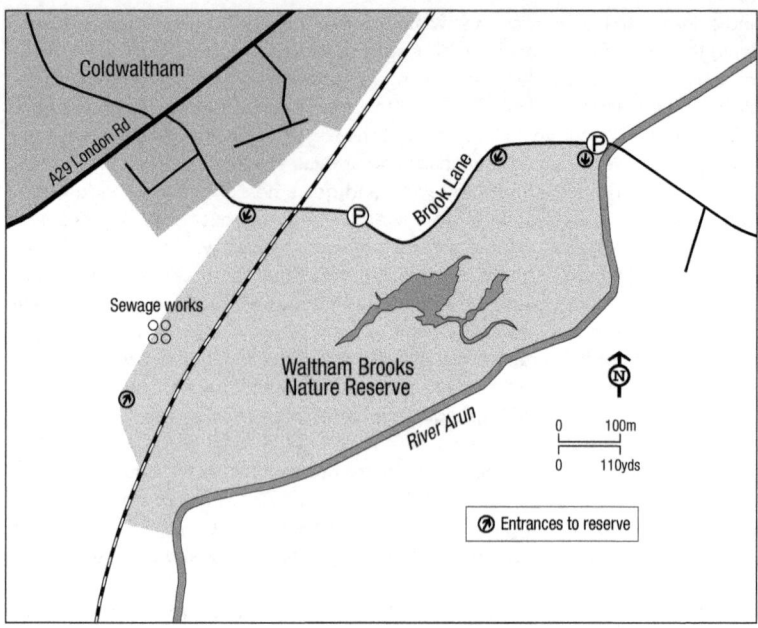

FACILITIES: The nearest facilities are in Pulborough.

CALENDAR

All year: Shelduck (largest numbers in spring), Shoveler, Gadwall, Teal, Water Rail, Lapwing, Snipe, Little Egret, Marsh Harrier, Red Kite, Barn Owl, Chiffchaff, Cetti's Warbler, Stonechat, Grey Wagtail, Reed Bunting.

April–June: Garganey, Cuckoo, Little Ringed Plover, Redshank, Sedge Warbler, Reed Warbler, Grasshopper Warbler, Willow Warbler, Garden Warbler, Lesser Whitethroat, Nightingale.

July–September: Greenshank, Green Sandpiper, Black-tailed Godwit, Common Sandpiper, Spotted Flycatcher, Redstart, Whinchat, Yellow Wagtail.

October–March: Wigeon, Pintail, Pochard, Hen Harrier, Short-eared Owl, Merlin, Great Grey Shrike, Siberian Chiffchaff, Firecrest, Water Pipit.

66 AMBERLEY WILDBROOKS

OS Explorer 121
OS grid ref: TQ 038145
Postcode: RH20 2ES

HABITAT

At just over 322ha, Amberley Wildbrooks is the largest of the three main reserves that form the wider Arun Valley Ramsar site. It was one of the first sites in the country to be designated as an SSSI and has been described as 'the jewel in the crown' of the Arun Valley. Jointly managed by the RSPB and Sussex Wildlife Trust, the reserve is one of the best examples of grazing marsh and wet grassland in West Sussex, if not the whole of the south-east of England. It is criss-crossed with ditches, home to the very rare Little Whirlpool Ramshorn Snail (one of only three sites left in the UK that can claim to have this species) and various rare aquatic plants.

In contrast to nearby Pulborough Brooks, Amberley lacks any visitor facilities or marked trails but, if anything, this only adds to the wild beauty of the place. From anywhere you can enjoy wonderful, panoramic views across the marsh, with the incredible backdrop of the South Downs to the south. From various elevated view-points around the edges of the reserve, most notably Rackham Viewpoint on the east side, you can look out across the whole site and the wider landscape. In keeping with much of the rest of the Arun Valley, Amberley Wildbrooks floods readily in the winter, which can lead to an impressive 'inland sea' look to the site, attracting thousands of wintering wildfowl in the process. Another species here in abundance is Fallow Deer, hundreds of which can be seen grazing across the site, although numbers are kept in check as part of the ongoing habitat-restoration work.

SPECIES

The size of the reserve combined with its open landscape and seasonal flooding offers a diverse range of species throughout the year. As with all similar sites across the Arun Valley, for sheer numbers of birds and the spectacle of great flocks of wildfowl, then winter visits will always be the most rewarding, when hundreds or thousands of Wigeon and Teal and smaller numbers of Shoveler and Pintail can be seen on the flood water. In spring, a Garganey is always possible among the ducks that remain. Huge numbers of geese also congregate here outside the breeding season, mostly Canada and Greylag Geese, but there is a chance of White-fronted, Brent or Tundra Bean Geese in winter. The remnant Arun Valley Bewick's Swan herd still frequents the reserve in the winter months, usually as a roost site when the water levels are high, before moving off to feed in the fields at Burpham during the day. Equally, an evening visit may offer the opportunity to see them returning to roost.

Amberley is very popular with Lapwings which breed in the summer but also congregate in huge flocks in the winter. It's always worth scanning through these large gatherings as there will often be Ruff among them in the winter, and perhaps a Golden Plover or Dunlin. All three egret species are now a regular sight here among the local Grey Herons from the nearby Rackham heronry. The ducks and waders will often be flushed into the air by a passing Peregrine. Indeed, Amberley is an excellent site for raptors, and many individuals of various species can be seen patrolling the marsh, especially in the winter. Marsh Harrier and Red Kite are the most numerous, but Buzzard, Kestrel, Sparrowhawk and Peregrine are also frequently seen, and Hobby is a regular sight in the summer. Less common, but not unusual in the winter months, are Hen Harrier and Merlin. There is also a growing chance of seeing a Goshawk outside the breeding season, as the species continues to increase in West Sussex. Barn Owl and Short-eared Owl are always a possibility quartering over the grassland at dusk, the former year-round and the latter in the winter. A more recent feature of the site has been the arrival of a pair of White-tailed Eagles from the Isle of Wight reintroduction project, which are often seen perched in one of the lone trees out in the marsh or tucking into carrion on the ground.

In the more wooded and scrubby areas one can find all the expected species, including Firecrest and, in the summer months, many species of warbler. This is perhaps the best West Sussex site for Grasshopper Warbler, though it is erratic in its appearance each year. The area of scrub along the Wey-South Path near Middle Gutter and Amberley Swamp is particularly good for warblers and other migrant songbirds, with an early morning visit in May often producing various *Sylvia* warblers, Cuckoo and Nightingale singing in close proximity.

Given the amount of floodwater here in wet winters almost anything can turn up, including coastal species like Little Gull and Grey Phalarope.

TIMING

The spectacle of thousands of ducks and waders taking to the air when a bird of prey passes through is really something to behold, and makes a winter visit very worthwhile, but the reserve offers interest year-round. Spring and early summer visits are best for singing warblers and displaying Lapwings, while an evening visit at any time of year can produce a Barn Owl, if you're lucky.

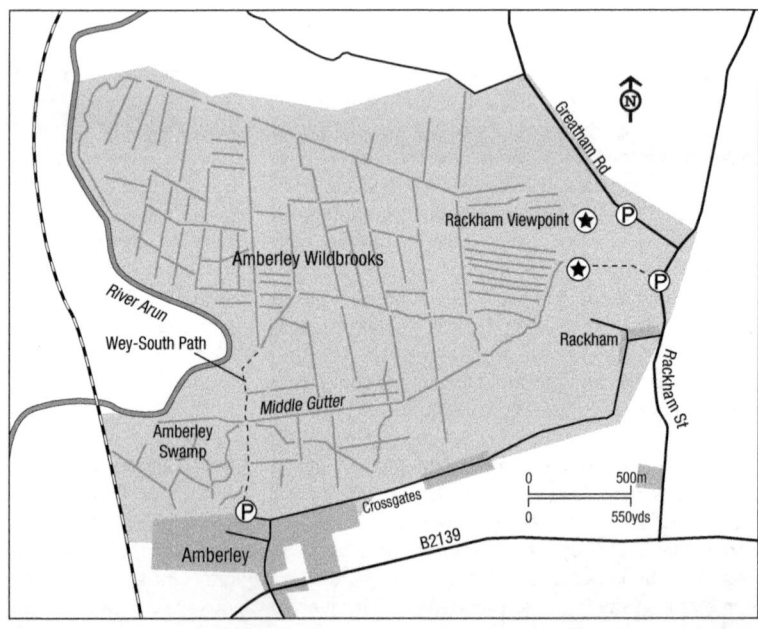

ACCESS

The nearest train station is in Amberley village, about a mile to the south. Bus route 74A (Compass Travel) from Horsham stops in Amberley, but this is subject to change, so it's worth checking before you travel. If travelling by car, there is room for 3–4 cars at the car park along Greatham Road (at TQ 048146) from where it's a short walk up the hill to Rackham Viewpoint. Otherwise, limited parking is available on Rackham Street and also on East Street and Hog Lane in Amberley Village. Please respect residents' access and do not block driveways, etc. From Amberley village, you can join the Wey-South Path from an entrance on Hog Lane. Follow this to the north and you will eventually reach the River Arun and the edge of Waltham Brooks.

Note that many of these paths can become very muddy or, indeed, totally impassable in the winter months. Wellies are always advisable from October until the spring. Amberley is sadly not very accessible for visitors with limited mobility, though views across the site can be enjoyed from Rackham Viewpoint or the Sportsman pub in Amberley.

There are no hides or trails close to where many of the birds are located, so a telescope really is essential for being able to pick out and identify the various species on show.

FACILITIES: The nearest facilities are in Pulborough or Amberley village.

CALENDAR

All year: Water Rail, Lapwing, Little Egret, Marsh Harrier, Red Kite, Barn Owl, Skylark, Cetti's Warbler, Stonechat, Reed Bunting.

April–June: Shelduck, Garganey, Cuckoo, Swift, Little Ringed Plover, Redshank, Great White Egret, Hobby, Sedge Warbler, Reed Warbler, Grasshopper Warbler, Sand Martin, Swallow, House Martin, Willow Warbler, Lesser Whitethroat, Nightingale.

July–September: Willow Warbler, Lesser Whitethroat, Spotted Flycatcher, Redstart, Whinchat, Wheatear, Yellow Wagtail.

October–March: Shelduck, Shoveler, Wigeon, Pintail, Black-tailed Godwit, Ruff, Great White Egret, Hen Harrier, Short-eared Owl, Merlin, Firecrest.

December–February: Bewick's Swan.

67 THE BURGH

OS Explorer OL10
OS grid ref: TQ 045109
Postcode: BN18 9RN

HABITAT

Nestled on the southern slopes of the South Downs between Amberley and Arundel, The Burgh is one of the best examples of well-managed downland farm-land anywhere in Sussex and, indeed, the south-east of England. Once the site of a Saxon hillfort, the area is now part of the wider Norfolk Estate, where the current Duke of Norfolk and his team place a strong emphasis on wildlife-friendly land management techniques.

Hundreds of metres of dense hedgerows, interspersed by small areas of scrub, woodland and rough grassland surround the arable fields which are bordered by wide margins and packed with wildflowers in the summer. It offers a very rewarding visit at any time of year, and not just for birds, as the area also supports hundreds of Brown Hares – making it one of the best places to see this enigmatic mammal in West Sussex. The sensitive farming methods here have also seen a marked increase in Short-tailed Field Voles and Harvest Mice, both of which of course provide plentiful food for birds of prey.

SPECIES

Skylark, Yellowhammer, Linnet and Corn Bunting all breed and can form large flocks in the fields in the winter, along with Chaffinch, Reed Bunting and Brambling. These species, as well as Goldfinch, Greenfinch and tits, also benefit from the wild-seed mixes and game and woodland bird feeders that have been dotted about the landscape as part of the management. The small patch of woodland at TQ 043109, in particular, is a good place to check for small birds, while the areas of scrub and hedgerows nearby may well yield a Ring Ouzel in the autumn. Indeed, anywhere in this part of the South Downs can be good for encountering passage migrants in spring and especially autumn, when overhead movements of hirundines and Meadow Pipits, for example, can be impressive.

The various feeders are often a good place to check for Grey and Red-legged Partridges, both of which do very well here, especially Grey Partridge which was reintroduced to the area in 2003. Even if you don't see any, you are almost certain to hear their strange ratcheting calls. Lapwing are another year-round sight, with many pairs breeding and sometimes big winter flocks using the fields for feeding. Another species that can be encountered here in large numbers in winter is Common Gull.

With its expanse of open fields, patches of woodland and abundance of food, The Burgh is hugely attractive to raptors with as many as 10 species possible here during a winter visit. Buzzard, Peregrine, Kestrel, Red Kite and Marsh Harrier can be seen all year, with a particularly impressive Red Kite roost in the winter sometimes numbering 50 or more birds. Winter also sees the arrival of scarcer, non-breeding species such as Merlin and Hen Harrier, taking full advantage of the abundance of small birds and mammals on offer. Patience and a bit of luck is often needed to score these two, though you stand a good chance in most winters. Unsurprisingly, the area has attracted rarer birds of prey, such as Pallid Harrier and Rough-legged

Buzzard. One or more of the introduced Isle of Wight White-tailed Eagles can often be seen too. Barn Owl can be encountered at any time of year but is perhaps most obvious in the winter months when they will hunt at almost any time of day. In influx years, these will also be joined by one or more Short-eared Owls quartering over the field margins and patches of rough grassland, particularly in the area near the Dew Pond (TQ 053108) where the birds can be easily watched from the main track.

In high summer, this can be an excellent place to visit to listen out for the 'wet my lips' song of Quail, with sometimes multiple individuals present in influx years.

TIMING
The Burgh has something to offer year-round but for the best selection of raptors and owls, visit in winter. Note that the limited parking spaces can fill up quickly on weekends and bank holidays.

ACCESS
The nearest railway stations are at Amberley (approx. 2km away) and Arundel (approx. 4km). The immediate area is not well served by bus routes, though the 74A (Compass Travel, weekdays only) from Horsham stops near the junction of Stoke Road and the B2139 in Amberley. There are two relatively easy options for accessing The Burgh by car, either by following the rather bumpy single-track road east from North Stoke to park at 'Canada Barn' (TQ 038111) or at 'The Triangle' on Peppering Lane just north of Burpham (TQ 041094), where there is space for more cars.

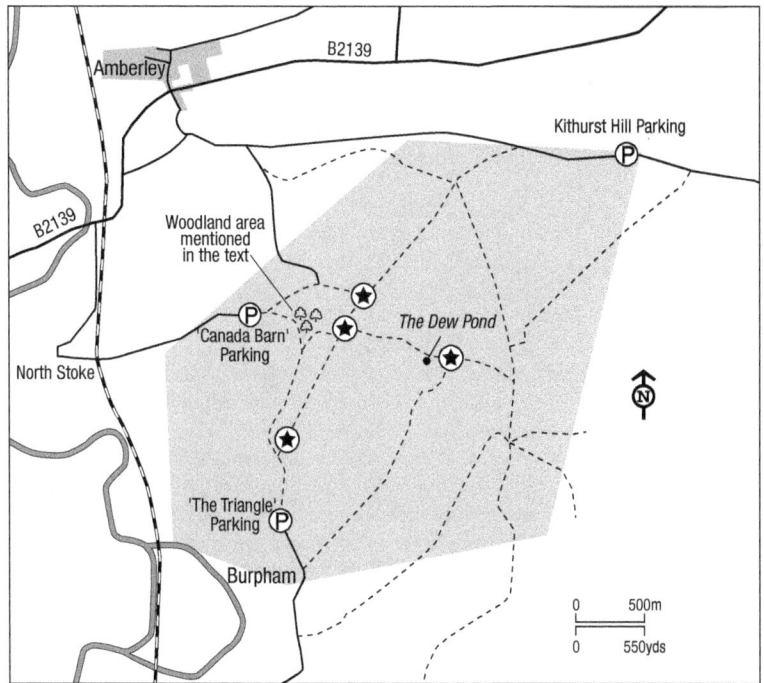

To access The Burgh from The Triangle simply follow Peppering Lane north, turning right at Peppering High Barn, then right again after about 1km to reach the Dew Pond. From the parking at Canada Barn follow the main path east, taking either the left or right fork, both of which bring you out on the main north–south footpath to the north of Peppering High Barn. Another option, for the more energetic birder, is to park at Kithurst Hill (TQ 069124) and follow the footpath west then south-west from here but note this is a good 2–3km walk each way.

Due to the undulating and sometimes steep paths here – which can become very slippery in the winter – the site is unfortunately not very accessible for visitors with limited mobility, although views across some of the fields can be enjoyed from the gate right by the car parking area at The Triangle.

FACILITIES: The nearest facilities are in Arundel.

CALENDAR

All year: Grey and Red-legged Partridges, Lapwing, Marsh Harrier, Sparrowhawk, Red Kite, Buzzard, Barn Owl, Peregrine, Raven, Skylark, Stonechat, Corn Bunting, Yellowhammer.

April–September: Cuckoo, Quail (possible in influx years), Hobby, Swallow, House Martin, Willow Warbler, Lesser Whitethroat, Redstart, Wheatear.

October–March: Golden Plover, Common Gull, Hen Harrier, Short-eared Owl, Merlin, Ring Ouzel (October), Brambling.

68 RIVER ARUN FROM ARUNDEL TO SOUTH STOKE

OS Explorer OL10
OS grid refs: TQ 019069 to TQ 028100
Postcodes: BN18 9PA to BN18 9PF

HABITAT

The Arun is the longest river in West Sussex and offers great birding along its length, with several of the main sites covered in more detail elsewhere in this book. The 3km section from Arundel to South Stoke can be particularly productive, especially in the winter, and the primary destinations along the way are described here.

Stretching up the western side of the valley north of Arundel Castle all the way past South Stoke is Arundel Park, a 134ha old deer park and an SSSI in its own right. A chalk stream, known as the Mill Stream, runs off the Downs through the Park, feeding Swanbourne Lake (site H9) before continuing east, immediately south of WWT Arundel (site 69) and into the Arun. To the south of this are the Mill Road Water Meadows, which is part of the wider floodplain in this area. Indeed, the whole area of open land from here across to Warningcamp and up towards the Downs can be significantly flooded in the winter.

For the best views across the valley one can either choose to head just west of the village of Offham, from where a roadside pull-in offers expansive views to the east; or on the opposite side of the valley, where the original course of the river deviates from the canalised section past Offham, the little village of Burpham close to The Burgh (site 67) offers views to the west along with sheep fields and a small sewage treatment works.

SPECIES

This area offers the best chance of encountering the diminishing herd of Bewick's Swans which resides in the area from December to late February. The most reliable spot for them is the section of the water meadows below the village of Burpham. These can often be seen from the road near the church or the cricket pitch, otherwise you will need to head up to Offham or get on the footpath down by the river. Sometimes they can also be seen from the lay-by along the road in Warningcamp (TQ 038080). There will be tens of Mute Swans grazing here too, but the more diminutive Bewick's should stand out. These days the herd is normally only single figures, and it seems a sad reality that this captivating species may one day no longer winter in Sussex.

All three egret species are possible in one visit here, with Cattle Egret now perhaps the most likely of the three at any time of year. Several will often be out in the Mill Road Water Meadows, among the various wildfowl – Greylag Geese, Egyptian Geese, Gadwall, Wigeon and Teal in particular – or in the sheep fields at Offham or Burpham. Little Egret is more likely to be encountered along the river itself or in the Mill Stream. The Mill Stream also often holds a Water Rail, which can be quite showy here, as they are used to lots of people walking by. Similarly, the Moorhens and Coots here can almost be hand-fed. Mandarin Duck is invariably present, again along the Mill Stream or on the edges of the water meadows or the various ditches. The ditches are also worth checking for Snipe, sometimes even right by Mill Road. Lapwings are present in good numbers in the winter, but can be seen at just about any time of year, as some nest on the Downs north of here. Other waders are possible too, especially Common Sandpiper which quite often winters along the river and Oystercatcher which breeds at the WWT reserve.

Cetti's Warblers are possible anywhere along the river where there are reedy sections, and will be joined by Sedge and Reed Warblers in the summer. Grey Wagtail and Kingfisher will also often be encountered, the former especially along the Mill Stream, which also attracts Firecrest and Chiffchaff in the winter, though both these species may be found in the area year-round. Nightingale still breeds in a few spots, near Warningcamp village and just east of South Stoke. The little sewage works at Burpham is a reliable wintering site for Chiffchaff, with a *tristis* occasionally present.

The whole area is also good for raptors and owls, with Marsh Harrier a year-round sight here now, along with Buzzard, Red Kite, Peregrine, among others, as well as Barn Owl and sometimes Short-eared Owl, Hen Harrier and Merlin in the winter. A Hobby will often be seen in the summer months, hawking over the river catching dragonflies.

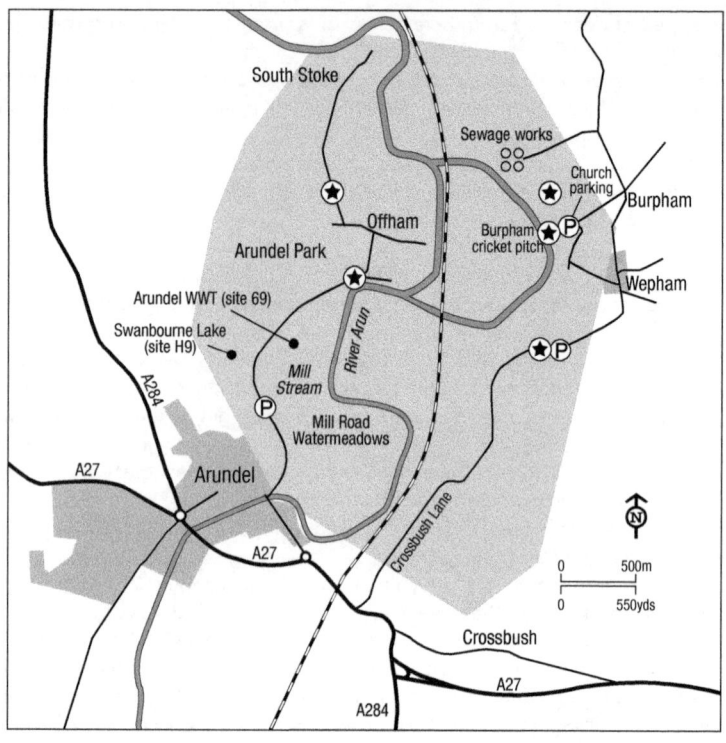

TIMING

Winter is best for the full selection of wildfowl. The area can become busy during the summer, especially at weekends.

ACCESS

The area is well covered by public footpaths, with rights of way following both sides of the river most of the way here, along with others cutting off away from the Arun east to Warningcamp, Wepham and Burpham. It's possible to spend an entire day exploring the various areas described earlier in this section on foot or, for a more whistlestop tour, there are parking spots at each of them. For Burpham either park near the church (TQ 038090) or the George & Dragon pub (TQ 039088) and follow the footpath south from the village to the river. The sewage works can be accessed via the little track off the road at TQ 037091. For Offham Viewpoint, follow the South Stoke Road north past WWT Arundel and the Black Rabbit pub, and park at one of the pull-ins along the road at roughly TQ 023089. For the Mill Stream and water meadows there is ample parking along Mill Road or at the car park at TQ 020071. From the latter you can get straight onto the various public footpaths following the river. Bus service 9 (Stagecoach South) from Shoreham to Arundel stops in the town centre (Mon–Sat) as does the number 85 (Compass Travel) from Chichester, while Arundel railway station is just a short walk from the river.

FACILITIES: The nearest facilities are in Arundel.

CALENDAR

All year: Egyptian Goose, Mandarin Duck, Gadwall, Lapwing, Common Sandpiper, Little Egret, Cattle Egret, Marsh Harrier, Kingfisher, Barn Owl, Chiffchaff, Cetti's Warbler, Firecrest, Grey Wagtail.

April–September: Oystercatcher, Hobby, Sedge Warbler, Reed Warbler, Nightingale.

October–March: Bewick's Swan (Dec–Feb), Wigeon, Teal, Snipe, Hen Harrier.

69 WWT ARUNDEL WETLAND CENTRE

OS Explorer OL10
OS grid ref: TQ 022080
Postcode: BN18 9PB

HABITAT

One of 10 reserves owned by the Wildfowl and Wetlands Trust around the UK, the Arundel Wetland Centre is situated immediately to the north-east of the town of Arundel, tucked away between the River Arun and the steep and rugged slopes of Offham Hanger.

Although small compared to some wetland reserves, Arundel packs a lot into its 25ha, with a mosaic of reedbed, wader scrapes, wetland pools and little pockets of woodland – plus, of course, the River Arun which runs right alongside it. There are six hides and a viewing screen offering views across the various scrapes and pools, while the collection of captive birds offers the novelty of seeing up close a range of species from around the world.

SPECIES

Perhaps the best way to describe the birding at Arundel is as a 'tasting menu' of what's on offer at other Arun Valley wetland sites such as Pulborough Brooks and Amberley Wildbrooks, in lesser numbers and condensed into a smaller site, though often affording closer views.

In the winter months, good numbers of wildfowl gather here including dabbling duck such as Wigeon, Teal and Pintail, and diving species such as Tufted Duck and Pochard. Waders, too, can accumulate in the many hundreds, especially Lapwing but also smaller numbers of Snipe and sometimes a Jack Snipe. The latter is sometimes seen particularly well from the Lapwing hide. Kingfisher can be sighted at any time of year as it breeds on-site. Both Cattle Egret and Marsh Harrier roost here in impressive numbers, making for an entertaining winter's evening spectacle. The former species is now a year-round sight and has attempted to breed recently. When the reserve is closed, a good place to view both egrets and harriers flying to roost is the car park of the Black Rabbit pub at TQ 025084. It's an evocative sight watching the harriers circling in and down into the reedbed, with the backdrop of Arundel Castle behind. The Black Rabbit car park is an excellent place to see Barn Owl at dawn or dusk as well. Pied Wagtail, Reed Bunting, Yellowhammer and Linnet often join the reedbed roost in the winter too.

Water Rails are present all year and offer perhaps the best views of this species anywhere in Sussex, especially under the feeders in the woodland lodge area.

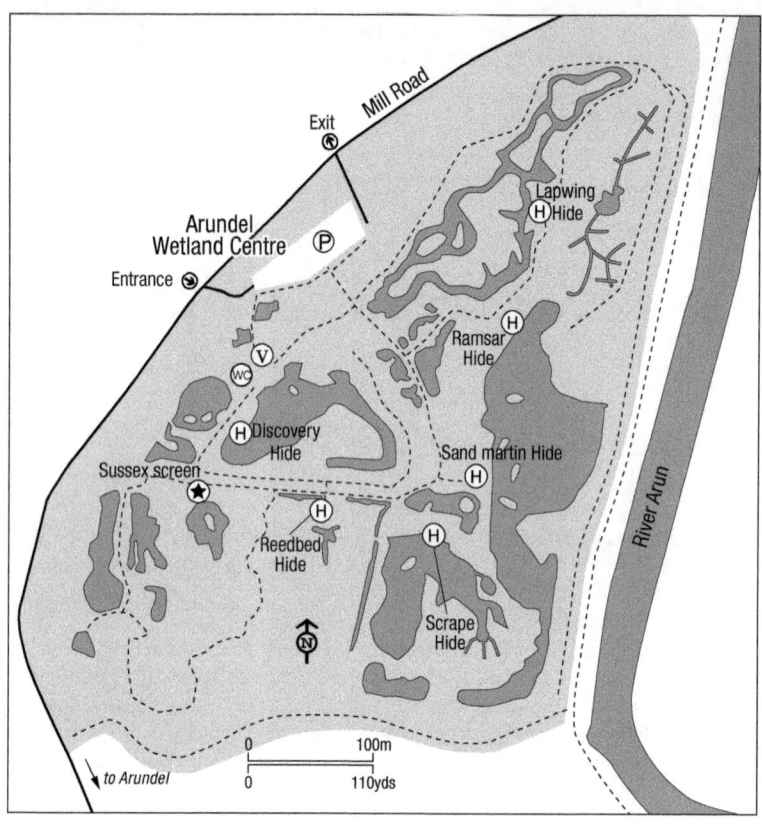

Coots, Moorhens and even Mandarin Ducks are often seen beneath the feeders too, with close views from the nearby viewing screen. Unsurprisingly, the seed attracts passerine species, including sometimes Brambling, Siskin and Lesser Redpoll in the winter.

The more wooded and scrubby areas of the reserve can produce a Firecrest or Marsh Tit at any time of year, while all the common warblers of the region return to breed in the spring, including Garden Warbler and Whitethroat. Sedge Warbler and Reed Warbler nest in the reedbeds, where Cetti's Warblers are present year-round.

Spring sees the arrival of passage waders and wildfowl such as Little Ringed Plover, Black-tailed Godwit, Whimbrel, Common and Green Sandpipers and sometimes a Garganey a two, as well as those species which return here to breed, including Oystercatcher and Sand Martin, which nest in the artificial nest bank.

In addition to the aforementioned Marsh Harriers, all the common raptors of the region can be seen, with Peregrine likely at any time of year and Hobby a frequent sight in the summer. The reintroduced White-tailed Eagles from the Isle of Wight occasionally put in an appearance, while a Hen Harrier will occasionally be seen coming to roost among the Marsh Harriers in the winter. Arundel is also a good spot for passage Osprey, especially in the autumn when juveniles may linger for a few days.

Scarce and rare species to have been seen here in recent years include Little

Crake, Spotted Crake, Common Crane, Ortolan Bunting, Great Grey Shrike, Black Kite and Night Heron.

TIMING
The reserve is open 10:00 am–4:30 pm every day except Christmas Day when it is closed and Christmas Eve when it shuts at 2:00 pm.

ACCESS
Bus service 9 (Stagecoach South) from Shoreham to Arundel stops in the town centre (Mon–Sat) as does the number 85 (Compass Travel) from Chichester. The wetland centre is around a 25-minute walk up Mill Road from the town centre and approximately 35 minutes from the train station just south-east of the town.

By car, follow the brown duck signs from Arundel town centre onto Mill Road, past the castle and you will see the entrance to the reserve just past Swanbourne Lake on your right. The car park is free at the time of writing and there is also a bicycle park near the entrance. Admittance is free to WWT members, otherwise admission fees apply (please check for the latest prices before travelling).

The main trails are all level and well surfaced, with easy, step-free access to the hides; assistance dogs are welcome but must be kept on a lead.

FACILITIES: There is a café, accessible toilets, baby changing, picnic area, car park, bicycle racks, first aid station and children's play area.

CALENDAR
All year: Egyptian Goose, Shelduck, Mandarin Duck, Shoveler, Gadwall, Pochard, Little Grebe, Water Rail, Lapwing, Little Egret, Cattle Egret, Marsh Harrier, Red Kite, Kingfisher, Raven, Marsh Tit, Cetti's Warbler, Firecrest, Grey Wagtail.

April–September: Oystercatcher, Little Ringed Plover, Common Sandpiper, Green Sandpiper, Redshank, Mediterranean Gull, Common Tern, Hobby, Sedge Warbler, Reed Warbler, Sand Martin.

October–March: Wigeon, Pintail, Teal, Great Crested Grebe, Jack Snipe, Snipe, Stonechat, Brambling.

---

70 HENFIELD LEVELS

OS Explorer OL11
OS grid ref: TQ 199150
Postcode: BN5 9QY

HABITAT
The River Adur rises from both Slinfold in West Sussex and near Ditchling Common in East Sussex, with these western and eastern arms converging just northwest of Henfield, before flowing south through the low-lying agricultural land at Henfield Levels. Just like the River Arun in Pulborough, the Adur is tidal as

far upstream as Henfield. A combination of spring tides and excessive rainfall during the winter will often see the river overtopping its banks onto the levels, flooding much of the land from Henfield down to Upper Beeding. There are essentially two main sections to Henfield Levels: one larger area to the south and west of the village extending as far as where the old railway line – now called the Downs Link – crosses the Adur at Stretham Manor and Farm, and another to the west/northwest of the village, extending north from Bineham Bridge up as far as Betley Bridge.

SPECIES

In winter, when much of the area will be very wet (if not entirely underwater), most of the interest will come from the impressive gatherings of wildfowl, especially Teal, with many hundreds sometimes present. There will also be good numbers of Wigeon among them along with smaller counts of Pintail and Shoveler and Gadwall in some areas (particularly the floods just north of Rye Farm). Very occasionally, there have been records of more unusual wildfowl, such as White-fronted and Tundra Bean Geese. Bewick's Swan was once regular here in the winter but is now a very occasional visitor. When the area is very flooded, diving ducks such as Tufted Duck and Pochard will occur. If there is still enough standing water around in March and April, then a passing Garganey or two may drop in, occasionally lingering for several days. The pools and bushes beside the track to Rye Farm (which is a private track, but a public footpath) are worth checking in the winter for the chance of Water Rail, and sometimes a few Chiffchaff. Water Pipit has been found on occasion in the winter and early spring, with almost any ditch or patch of wet grassland worth scanning for one.

Lapwing are present all year with many hundreds wintering and a few pairs still breeding in the area, particularly in the fields either side of the Downs Link between Rye Farm and Stretham Farm. In the winter these will be joined by Snipe (often dozens though they can be difficult to spot except in icy weather) and, very occasionally, a Golden Plover or two. In spring and autumn, various wader species are possible, with the likes of Whimbrel, Black-tailed Godwit, Dunlin, Little Ringed Plover, Ruff, Redshank and Green and Common Sandpipers occasionally dropping in.

There is a small heronry in the trees west of the Downs Link near Stretham Bridge, so Grey Herons are a familiar sight, as are Little Egrets, although the latter species doesn't breed here. Great White Egret is beginning to become a somewhat regular visitor too, as the species spreads across the South-East.

A pleasing selection of passerines can be found in the area with resident Skylark, Yellowhammer and Cetti's Warbler joined in spring and summer by Nightingale, Cuckoo, common warblers and sometimes Turtle Dove. Along the river you have a reasonable chance of finding a Kingfisher. The Mill Stream, which runs from the Downs Link across to Woods Mill on the eastern side of the Levels, is a particularly good area to hear Nightingale and Turtle Dove in spring (May is best), though the latter have become less reliable in recent years.

As Henfield Levels has increased in popularity and more birders are now visiting the area, an impressive selection of notable species have been found in recent years, including Grey Phalarope, Wryneck, Glossy Ibis and Scaup.

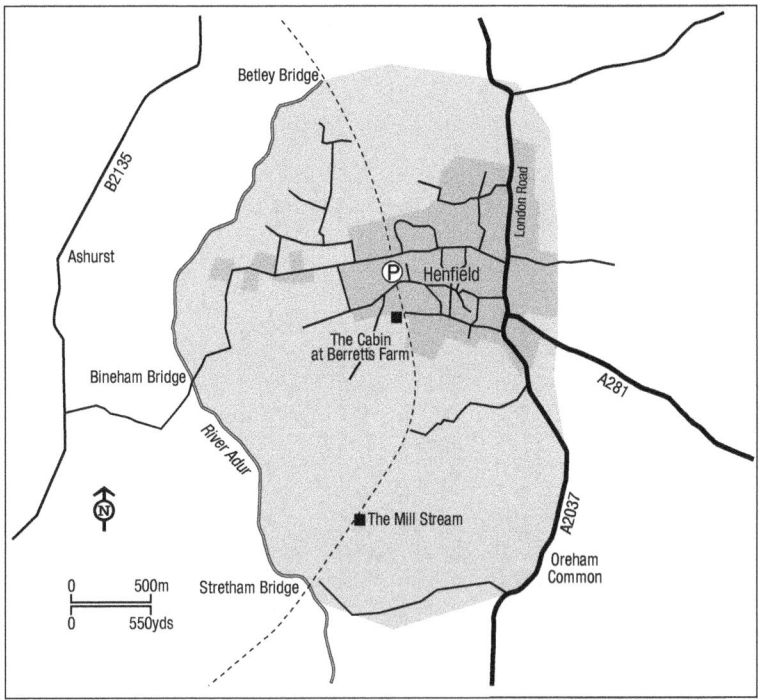

TIMING

Go in winter for the best selection of wildfowl, and summer for breeding migrants such as Turtle Dove and Nightingale. The main paths, especially the Downs Link, can get busy with dog walkers and cyclists, so early mornings are recommended.

ACCESS

The most convenient railway stations in terms of transport links are Brighton and Horsham. Bus service 17 (Stagecoach South East), which travels between the two towns, and route 100 (Compass Travel) from Burgess Hill to Pulborough both stop on Henfield High Street, from where it is a relatively short walk to join the Downs Link path and the many other public footpaths which cross the levels. Note that neither of these services run on Sundays.

If travelling by car, the best spot to park is the small car park by the Old Railway pub at TQ 205161 or, if this is full, there are usually a few spaces on West End Lane. The whole area can become very flooded in the winter, sometimes making some footpaths very muddy or, in wet winters, entirely impassable. Wellies are essential.

FACILITIES: There is a car park on West End Lane, and The Cabin at Berretts Farm serves hot drinks and snacks, and has portaloos. There are public toilets by the bus shelter on the High Street.

71 WOODS MILL

OS Explorer OL11
OS grid ref: TQ 217136
Postcode: BN5 9SD

HABITAT

This delightful little reserve packs a lot of habitat into its 19ha, as well as being the headquarters of the Sussex Wildlife Trust. Traditionally managed pockets of grassland and pasture are criss-crossed with hedgerows and ditches, reed-fringed ponds and areas of ancient woodland. In the spring and summer the meadows burst into colour with an array of wildflowers and spectacular displays of Bluebells carpet the woodland floor.

In addition to an impressive selection of birdlife for its size, the reserve also offers much in the way of insects and other species of interest. The streams and ponds host Water Shrews, while many species of dragonfly can be found here, including Scarce Chaser and Downy Emerald. Dogs are not permitted anywhere on the reserve.

SPECIES

As you arrive from the car park, Woods Mill Lake is right in front of you as you cross the bridge, and this is always worth a check. It can often yield Moorhen, Little Grebe and Grey Heron, and Teal in the winter. Be vigilant for the loud call of a Grey Wagtail as you follow the tracks along the ditches past the main lake, especially in the winter. Similarly vocal is Water Rail, which will usually be heard rather than seen, but you may catch a glimpse of one from the bridge by the car park.

The hedgerows and areas of scrub are home to a wealth of passerines, particularly in the summer months when the resident Bullfinches and thrushes are joined by the likes of Willow Warbler, Chiffchaff, Blackcap, Whitethroat and Lesser Whitethroat. Listen out near the reedbeds for the songs of Cetti's Warbler and, in summer, Reed Warbler. The real stars of the show though are the Nightingales, and this reserve is one of the best sites in West Sussex to hear this enigmatic species through late April and May. A little less obvious, but still very much worth listening out for, is the Turtle Dove, which can be seen and heard here in the summer months – there are normally two or three purring males.

In the winter much of the reserve becomes very wet indeed and can sometimes host Snipe and Green Sandpiper. Barn Owl and Tawny Owl are resident, while

Short-eared Owl sometimes occurs in the winter, quartering over the meadows towards the southern end of the reserve.

TIMING

Woods Mill offers a pleasurable walk at any time of year and is ideal for an unchallenging walk for the whole family. For the best soundtrack of Nightingales and summer warblers, visit between late April and early June.

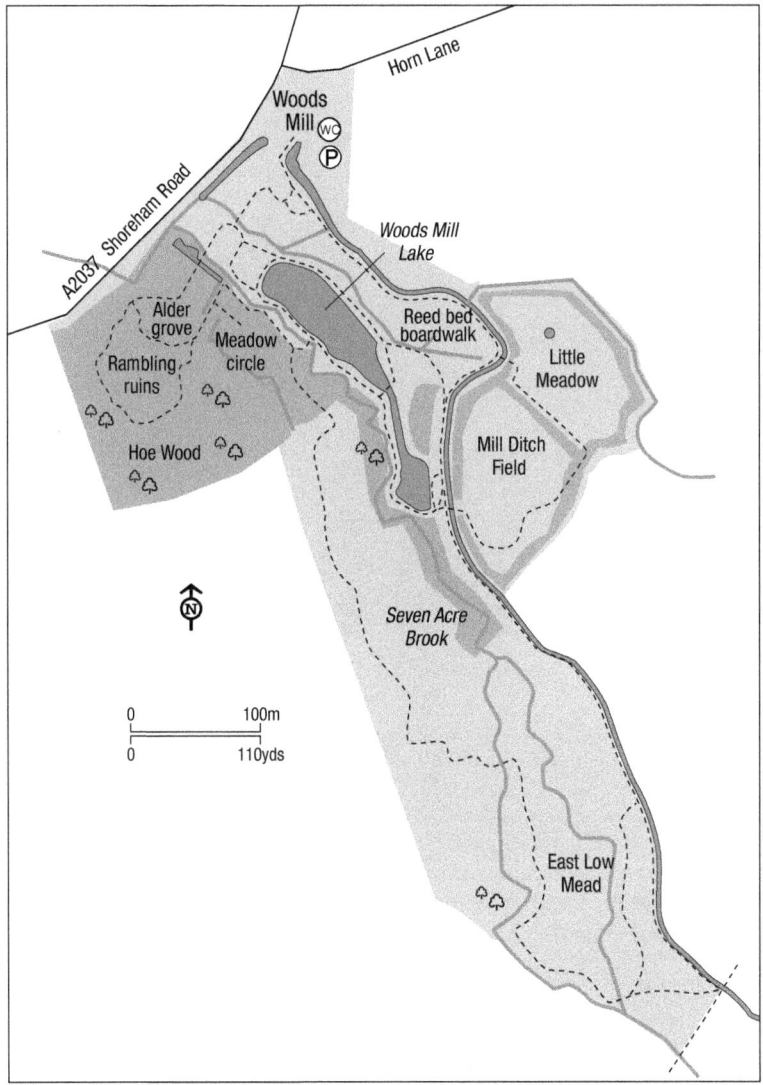

ACCESS

The main entrance to the reserve and car park lies on Horn Lane, just east of the A2037 between Henfield and Small Dole. The nearest railway station is Hassocks, approximately 13km to the east. Bus route 100 (Compass Travel) stops near the junction of Horn Lane and the A2037. There are bicycle racks near the mill buildings, just across the bridge from the car park. The car park has space for around 30 cars and much of the reserve trails are hard surfaced. The meadows towards the southern end of the reserve can get quite muddy in the winter, however, so are less suitable for pushchairs or wheelchair users.

FACILITIES: There is a free car park and toilets.

CALENDAR

All year: Little Grebe, Water Rail, Little Egret, Barn Owl, Tawny Owl, Kingfisher, Kestrel, Skylark, Yellowhammer, Reed Bunting.

April–September: Turtle Dove, Cuckoo, Swift, Lapwing, Willow, *Sylvia*, Sedge and Reed Warblers, Swallow, House Martin, Nightingale.

October–March: Snipe, Green Sandpiper, Short-eared Owl.

72 ADUR ESTUARY, SHOREHAM HARBOUR AND WIDEWATER LAGOON

OS Explorer 122
OS grid refs: TQ 212048, TQ 237048, TQ 200042
Postcodes: BN43 5NG, BN41 4WD, BN15 8JT

HABITAT

The River Adur reaches the sea at Shoreham-by-Sea and, despite being so close to busy roads, a railway line and other human activity, provides some productive estuary birding. Part of the lower stretch of the river – known as the Adur Saltings – has been managed by the RSPB since 1987, when the whole area of mudflats here were also designated as an SSSI. This status is due to the site's value to roosting and feeding waders, as well as the assemblage of saltmarsh specialist plants such as Sea Purslane, Glasswort and Sea Aster. The embankment near the car park just west of Norfolk Bridge is also home to a large colony of Common Lizards.

Further downstream, the river reaches the Shoreham Harbour complex. The high-sided harbour walls here – especially in the Southwick Ship Canal to the east of harbour entrance – provide a sheltered refuge for a host of species, including sometimes wind-blown seabirds and divers. On the western side of the harbour entrance is the old Shoreham Fort, home to a colony of Wall Lizards and popular with Black Redstart in the winter.

Widewater Lagoon is a naturally created tidal lagoon, which has now been reinforced thanks to artificial supplementation of the shingle bank between it and the sea. It is 1.2km long but just 50m across at its widest point. The lagoon boasts 140 plant species and is an important refuge for various resident birds, as well as vagrants and seabirds taking shelter during or after bad weather.

SPECIES

Although small in comparison to other similar sites, the mudflats of the Adur Estuary host a variety of waders throughout the year, but especially in the winter months. Oystercatcher, Redshank, Dunlin, Turnstone and Ringed Plover are generally the most frequently encountered, but Avocet, Grey Plover, Bar-tailed and Black-tailed Godwits, Whimbrel, Curlew and Greenshank are all semi-regular too, especially in the spring and autumn. Good numbers of Snipe can sometimes be present on the saltmarsh in the winter and are most easily seen when they are flushed by a high spring tide. Little Egrets maintain a year-round presence now, in keeping with any estuarine site in the south of England, and it's always worth scanning through the assembled gulls for Mediterranean Gull, which is now rather common on the Sussex coast. Yellow-legged or Caspian Gulls occur here from time to time. Kingfisher is a fairly regular sight outside of the breeding season. The open fields of the airport to the west attract Lapwings and a few Skylarks and Meadow Pipits.

Widewater offers standard fare in terms of wildfowl for much of the year but has been known to turn up some more unusual species. Red-breasted Merganser and sometimes Goosander can be found here in the winter, as well as Brent Goose, Teal and Little Grebe. The shingle margins are always worth checking for waders, especially in the passage months, with Turnstone, Redshank, Dunlin, Ringed Plover and Oystercatcher fairly regular, along with more infrequent occurrences of Little Ringed Plover, Whimbrel, Bar-tailed and Black-tailed Godwits, Knot, Sanderling, Little Stint, Curlew Sandpiper and Common Sandpiper. Look out for Grey Wagtail around the edges too, especially in the winter. Sandwich Terns are sometimes seen here in the spring and summer, on the shingle or perched up on the posts in the lagoon. Rarities and scarcities have included Red-necked and Grey Phalaropes, Sabine's Gull, Gull-billed Tern and Baird's Sandpiper.

The Shoreham Harbour complex is the most likely area to encounter seaduck, grebes and divers in the winter, with the likes of Great Northern Diver and Long-tailed Duck having lingered in Southwick Ship Canal for several weeks, or even months, in recent winters. Species such as Red-breasted Merganser occur more frequently, along with the occasional Red-throated Diver. Both Glaucous and Iceland Gulls have been known to stay in the area for several weeks in recent winters. The harbour mouth along with the occasional hold the likes of Shag, Kittiwake, Guillemot and Razorbill, especially after storms, and Gannets can often be seen fishing offshore.

A check of Turnstone flocks in the winter can occasionally produce Purple Sandpiper. Indeed, Shoreham Harbour is one of the few sites in Sussex for this species, which is unfortunately becoming less regular here. The most favoured spots are the wooden pier inside the west harbour arm and on the concrete revetments on the seaward side, halfway along the east arm. The harbour is also an excellent site for Black Redstart, although this species can be found almost anywhere along this stretch of coast. Indeed, the whole harbour and estuary area can be productive for migrant passerines such as Wheatear and Yellow Wagtail, especially in late summer and autumn. Stonechats and Rock Pipits routinely winter in the area too.

Peregrine can be encountered year-round, while more unusual raptor species such as Osprey or Hobby may be seen, especially in spring and autumn.

The beach from the harbour along to Widewater can be good for roosting waders such as Dunlin, Ringed Plover and Sanderling, although they are prone to dog disturbance here. This is an excellent place to find recently arrived Wheatears in the spring, sometimes in good numbers. In March, it's also worth checking this area for Black Redstart.

A scan of the sea from Widewater or the harbour entrance can be rewarding, with species such as Pomarine Skua, Arctic, Black and Common Terns and Little Gull passing on good days, although it is never quite as spectacular as other, more famous seawatching sites in the county. Birds also tend to be more distant from this stretch of coast, so a telescope is essential.

TIMING

As Shoreham is a popular seaside destination, there is inevitably a fair bit of disturbance, so early morning is best if possible. It's also worth bearing in mind that the eastern side of the harbour offers better viewing conditions in the morning with the sun behind you, and likewise the western side is more favourable in the afternoon. Widewater is a very popular site for dog walkers, with off-lead dogs an ever-present problem, especially on weekends and bank holidays.

ACCESS

The mudflats of the main estuary reserve can be viewed from either the road or pedestrian bridges which cross the river at TQ 212050 and TQ 216048, or from the path that runs south down the west bank of the river from Norfolk Bridge. The path can also be followed north up towards the A27, just south of which you can cross the Old Shoreham footbridge to reach the east side of the river, which can then be followed back towards Norfolk Bridge for a pleasing circular route. To get to Shoreham Fort and the west side of the harbour, follow Beach Green east/south-east off the roundabout on the A259 all the way to the pay and display car park at the end of Forthaven. There is space for around 40 cars here, and it offers immediate access onto the fort and the beach. Parking is also available on Harbour Way at TQ 231048 and at the small car park at TQ 217047. For the east side of the harbour mouth, follow Kingsway east all the way along to Hove

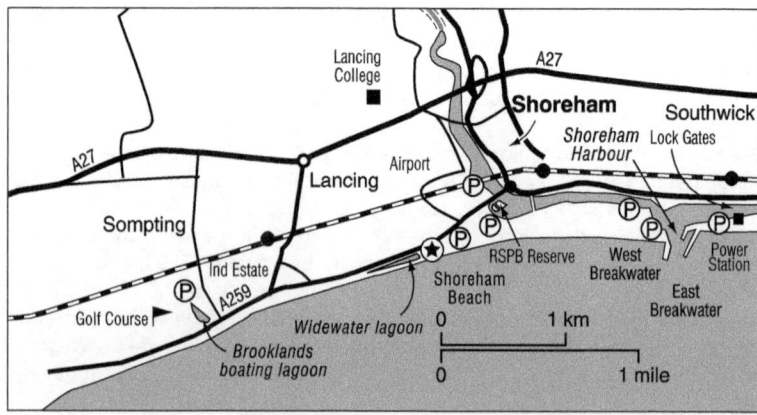

Lagoon, then turn south onto Wharf Road at the traffic lights and follow this down to Basin Road South, which eventually winds its way down to a car park near Carat's Café at TQ 244047.

On foot, you can also cross the lock gates across the East Arm of the harbour from the A259. The whole harbour and estuary complex lends itself to exploration by bicycle, especially owing to the rather challenging and circuitous road routes required to reach both sides of the harbour (it's worth noting that from Widewater round to the east side of the harbour mouth is a 20-minute drive!). Widewater Lagoon can be accessed from the car park at TQ 204042, where one can join the public footpath which runs right along the southern side of the lagoon towards South Lancing. The area is also well served by public transport, with railway stations at both Shoreham-by-Sea and Southwick and regular buses – 700 (Stagecoach South), 2 (Brighton & Hove) and others – stopping at various points in the area.

FACILITIES: There are public toilets and a food and drinks kiosk by the car park at Widewater (£1 for two hours or £4 all day at the time of writing). Public toilets can also be found at the Adur Recreation Ground car park (free at the time of writing), just to the west of Norfolk Bridge. Carat's Café and public toilets can be found on the south side of Southwick Ship Canal at TQ 244047, and The Port Kitchen restaurant and bar is on the north side of the lock gates at Southwick (TQ 242049).

CALENDAR

All year: Oystercatcher, Turnstone, Little Egret, Peregrine.

April–September: Whimbrel, Bar-tailed Godwit, Common Sandpiper, Mediterranean Gull, Sandwich Tern, Sedge Warbler, Reed Warbler, Wheatear, Yellow Wagtail.

October–March: Teal, Red-breasted Merganser, Little Grebe, Grey Plover, Lapwing, Ringed Plover, Dunlin, Purple Sandpiper, Greenshank, Redshank, Kingfisher, Cetti's Warbler, Black Redstart, Grey Wagtail, Rock Pipit, Reed Bunting.

---

73 STEEP DOWN

OS Explorer 122
OS grid ref: TQ 168075
Postcode: BN15 0AY

HABITAT

Steep Down is a fabulous example of Sussex downland farmland. The rolling chalk hills make for a dramatic landscape, particularly from the peak of Steep Down itself (149m high), which offers very impressive panoramic views, taking in the likes of Brighton, Shoreham and Worthing, as well the Sussex coast to Beachy Head in the east. In summer, the fields are adorned with wheat and barley, edged with rows of Chicory, poppy, Greater Knapweed and a host of other wildflowers. The northern slope of the hill, meanwhile, hosts Pyramidal Orchid and an abundance of other chalk grassland species.

SPECIES

Steep Down is perhaps the most reliable location in Sussex to encounter Quail, with the species recorded in most years. Skylark and Corn Bunting breed in high numbers, but Grey Partridge is more elusive. It's also an excellent spot for both visible migration and encountering grounded passage migrants, with the likes of Wheatear, Whinchat and Redstart all commonly occurring every year. More unusual have been regular sightings of Hen Harrier in recent years, as well as records of Crane and Dotterel.

TIMING

High summer is best for Quail, as well as passerines such as Corn Bunting, while winter can produce Grey Partridge and possibly Hen Harrier. Spring and autumn, meanwhile, offer the chance of encountering migrant passerines.

ACCESS

This is not a particularly easy site to access via public transport. Bus route 7 from Lancing to High Salvington (Stagecoach South) stops on Howard Road, which is about a 25-minute walk from the south of Steep Down.

There is a free car park on Titch Hill at TQ 161079. From there head east across the road and through the gate onto the public footpath. A roughly 5km loop around, up and over the main prominence takes in much of the habitat on offer, as well as giving excellent views across the wider landscape.

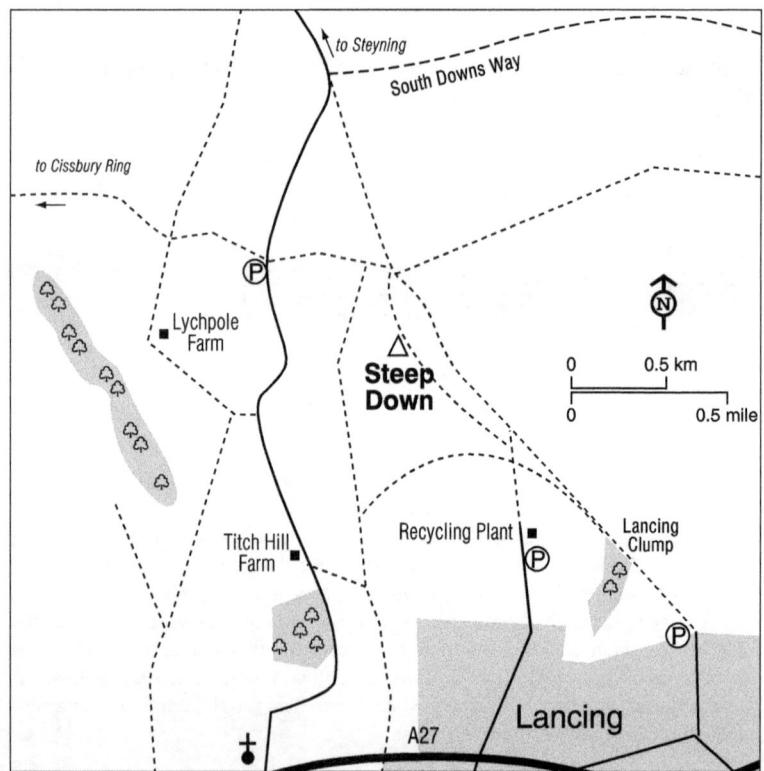

FACILITIES: The nearest facilities are in Lancing.

CALENDAR

All year: Grey and Red-legged Partridges, Raven, Skylark, Yellowhammer, Corn Bunting.

April–September: Hobby, Lesser Whitethroat, Wheatear, Redstart, Whinchat.

June–August: Quail.

October–March: Hen Harrier.

---

74 CISSBURY RING

OS Explorer OL10
OS grid ref: TQ 140080
Postcode: BN14 0HT

HABITAT

Situated just a few kilometres north of Worthing, Cissbury Ring is the largest hillfort in Sussex, with a history dating back to around 4000 BC. The site covers around 84ha in total, encompassing areas of open grassland, mature oak and Yew trees and some enticing patches of mixed scrub, most notably gorse. Rising to a height of 184m above sea level, the top of the hill offers commanding views across the surrounding landscape, as far as the Isle of Wight to the south-west, the Seven Sisters to the south-east and the Surrey Hills to the north.

The grassland here is a precious habitat, home to calcicole plants such as Round-headed Rampion, Carline Thistle and Yellow-wort, and many butterflies including Adonis Blue, Green Hairstreak, Dark Green Fritillary and Wall Brown.

SPECIES

An array of classic downland species can be encountered on and around the hillfort all year-round, such as Corn Bunting, Yellowhammer, Skylark and, some-times, Grey and Red-legged Partridges. In the summer, there is a chance of a singing Quail in the arable fields to the north. The scrub on the hill holds breed-ing warblers including Lesser and Common Whitethroat, Garden Warbler and Blackcap.

From a birding point of view, Cissbury really comes into its own in spring and especially autumn, when the areas of scrub can fill with all manner of migrant passerines. The South Downs are one of the last places to stop and feed up before reaching the south coast, and the high and open landscape at Cissbury helps to gather birds on the move. Impressive movements of hirundines are often observed in the autumn, with the likes of Tree and Meadow Pipits and Yellow Wagtail also passing overhead in numbers. The grassland and scrub are a popular stop-off point for migrating Wheatear and Whinchat, most frequently in August and September. These months also offer the chance of Pied Flycatcher and Nightingale in the bushes, while it is possible to encounter double-figures of Redstarts on a circuit of the Ring. Other passerine migrants often recorded here in the spring and/or

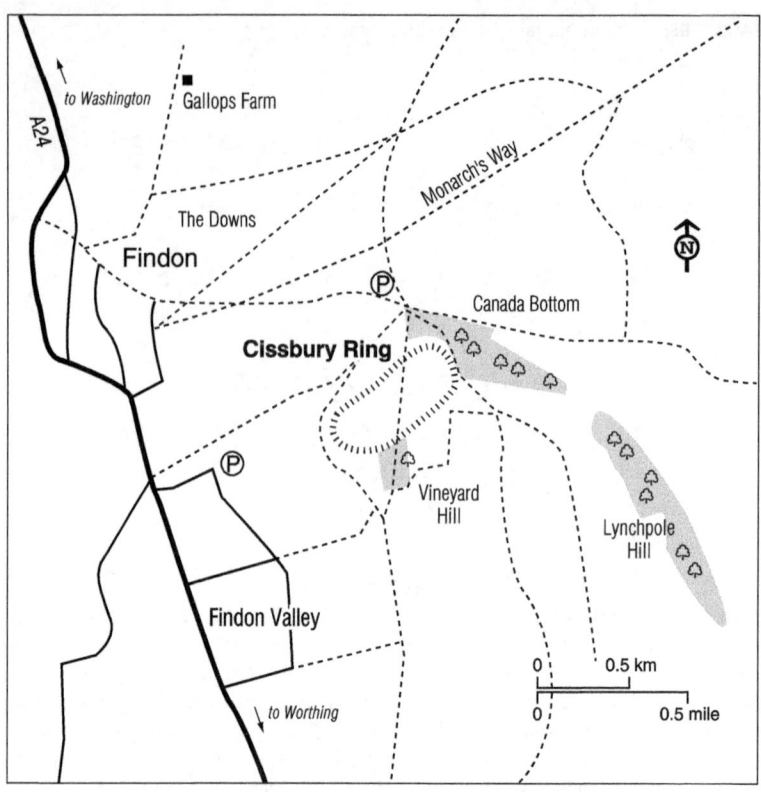

autumn include Willow Warbler, Grasshopper Warbler, Firecrest (which also breeds on the lower slopes) and Spotted Flycatcher. In September, there is the outside chance of a Honey Buzzard migrating overhead on a fine day.

This is arguably the best site in West Sussex for Ring Ouzel, with a visit in mid- to late October very likely to produce one of these charismatic thrushes feasting on berries in one of the two isolated Yew trees. They are most often found in the old Yew with the metal fence around it at TQ 13670796. It is at this time in the autumn that Brambling, Lesser Redpoll, Siskin and other finches move through. Hen Harrier can sometimes be encountered in the vicinity of the site, occasionally wintering, while Short-eared Owl and Marsh Harrier are possibilities flying through in the later autumn.

Increasingly, Dartford Warblers can be found in the gorse near the trig point in autumn and winter, when Marsh Tits sometimes appear on the edges of the site. Early starts in winter can produce views of Woodcock flying over the wooded slopes, while Tawny Owls hoot and Fieldfares and Redwings leave their roosts.

More uncommon species to have been recorded here in recent years include Cattle Egret, Golden Oriole, Red-backed Shrike, Great Grey Shrike, Olive-backed Pipit, Yellow-browed Warbler and Wryneck.

TIMING

Spring and autumn are best, especially the latter if you are hoping to find grounded migrants. Get there early though, especially on weekends, before the dog walkers.

ACCESS

Worthing station is 5km away. Stagecoach South bus route 1 from Worthing to Midhurst stops at May Tree Avenue in Findon Village, which is about a 25-minute walk from the edge of Cissbury Ring. Bus route 23 (Metrobus) from Crawley to Worthing stops at Bost Hill, a similar distance from the Ring. There are two car parks nearby: one on the north side along the Monarch's Way, east of Nepcote, at TQ 139085. The other is on the west side, just off Storrington Rise, at TQ 129076. The northern car park is closer but smaller (with room for half a dozen or so cars), while the western car park is a good 15-minute or so walk from the summit but offers a lot more space. Both car parks are free at the time of writing.

FACILITIES: Other than the car parks mentioned above, all other facilities are in Findon and Findon Valley.

CALENDAR

All year: Grey and Red-legged Partridges, Red Kite, Raven, Skylark, Firecrest, Stonechat, Corn Bunting, Yellowhammer.

April–May: Lesser Whitethroat, Garden Warbler and other common breeding migrants of the region.

June–July: Quail.

August–November: Redstart, Ring Ouzel, Dartford Warbler, Marsh Tit, Pied Flycatcher, Nightingale, Tree Pipit.

---

75 GORING GAP AND FERRING RIFE

OS Explorer 121
OS grid refs: TQ 102021, TQ 089019
Postcodes: BN12 4QW, BN12 5QU

HABITAT

As the name suggests, Goring Gap is a 59ha stretch of undeveloped coastline and adjoining farmland between Goring-by-Sea and Ferring, split down the middle between the districts of Arun and Worthing, and one of several such 'gaps' along the West Sussex coast. Only at Goring and Climping does the arable land extend right up to the sea, however, and the flat, open landscape, flanked by small areas of woodland and hedgerows is almost reminiscent of North Norfolk to the birder's eye. The elevated slopes of Highdown Hill to the north serve as a reminder that one is not far from the South Downs, though. The northern boundary is marked by the Ilex Avenue, a near mile-long avenue of over 400 Holm Oaks, planted in the mid-1800s. Along the eastern boundary is situated a plantation of largely Sycamore, with a mix of other deciduous species.

The Ferring Rife stream is one of five watercourses flowing to the sea from the South Downs. The quiet, vegetated banks offer secluded breeding habitat for many bird species as well as being a refuge for migrants in the spring and autumn. In keeping with similar sites along the south coast, both Goring and Ferring serve as both a refuge and a funnel for passerines either recently arrived on or soon to be departing from this stretch of coast.

It's easy to walk between the two sites along the coastal path, and a raised concrete platform with benches at TQ 099015 offers a convenient viewpoint for seawatching, which can be rewarding here.

SPECIES

The open arable fields abutting the beach at Goring Gap offer ideal habitat for foraging and roosting waders and gulls. Indeed, the south-western field nearest Ferring hosts a daily wader roost, including species such as Ringed Plover, Grey Plover and Dunlin. It's always worth scanning through this flock for something rarer, as the likes of Little Stint do occur in some years. Likewise, from late summer into the winter months a scan through assembled flocks of gulls in the fields can produce Yellow-legged or Caspian Gulls on occasion; Goring Gap is probable the best site in West Sussex for the latter. There are recent records of Iceland and Glaucous Gull here too. The Ilex Avenue and the plantation on the eastern side are both good areas to search for passerines, as is the pumping station at TQ 101018.

The section of rough grassland that separates the southeasternmost fields from Marine Drive often attracts Stonechats outside of the breeding season and Whinchats in spring and autumn. Other migrant passerines such as Wheatear and Yellow Wagtail can also regularly be found in the fields, particularly in the autumn, as well as in the larger country centre paddocks at Ferring.

Ferring Rife holds breeding Reed Warbler in the summer months and regular Water Rail, Jack Snipe and Snipe in the winter. Various warblers and other

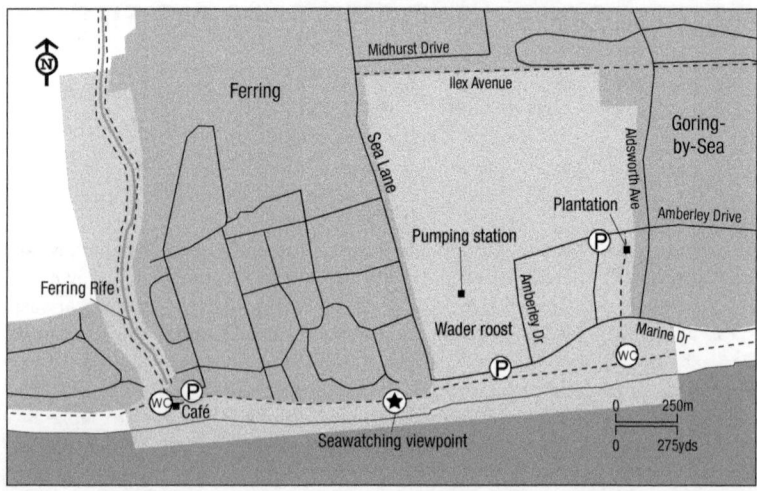

passerines can be found in the passage months including Spotted and Pied Flycatcher, Redstart and any of the standard set of *Sylvia* and *Phylloscopus* warblers.

The area has enjoyed its fair share of rarities before, including Desert Wheatear, Snow Bunting, Hoopoe, Serin, Common Rosefinch and Lapland Bunting.

TIMING

Both sites are popular with dog walkers and beach visitors, so an early morning visit is always advisable if possible, particularly when it comes to catching roosting waders before they're disturbed by dogs. Of course, the wader roost is dictated by the timing of high tide, so it's always worth checking the tide times before setting off. Between 10:30 am and 2:00 pm is the best time for checking loafing gulls.

ACCESS

The area is well served by public transport. Goring train station is only a 10- to 15-minute walk from Ilex Avenue at the northern end of Goring Gap, and just over a 20-minute walk from Rife Way, at the north end of Ferring Rife. Bus route 8 (Compass Travel) stops at various points along Sea Lane in Goring and West Drive in Ferring. If travelling by car, there is ample free parking anywhere along Marine Drive or Amberley Road at Goring Gap, and a decent-sized car park at the south end of West Drive in Ferring.

FACILITIES: A free car park can be found at Ferring Beach as well as public toilets, both of which are open only during the opening hours of the Bluebird Café. The public toilets at the south end of the plantation at Goring Gap are open from 1st April to 30th September, 9:00 am–7:00 pm.

CALENDAR

All year: Common Scoter, Oystercatcher, Ringed Plover, Curlew, Turnstone, Sanderling, Dunlin, Redshank, Mediterranean Gull, Gannet, Little Egret, Skylark, Reed Bunting.

April–June: Skuas, waders and wildfowl moving offshore, Reed Warbler, Swallow, Willow Warbler, Blackcap, Whitethroat, Wheatear, Whinchat, Yellow Wagtail, Meadow Pipit.

July–September: Yellow-legged Gull, Lesser Whitethroat, Spotted Flycatcher and passage migrants as listed for spring.

October–March: Brent Goose, Shoveler, Wigeon, Red-breasted Merganser, Great Crested Grebe, Water Rail, Grey Plover, Jack Snipe, Snipe, Guillemot, Razorbill, Kittiwake, Caspian Gull, Red-throated Diver, Great Northern Diver, Stonechat, Meadow Pipit.

76 CLIMPING GAP

OS Explorer OL10
OS grid ref: TQ 005007
Postcode: BN17 5RN

HABITAT

Climping Gap is the westernmost and largest of the remaining undeveloped 'gaps' dotted along the Sussex coastline. This stretch of wild coast, where arable fields and woodland meet the beach, stretches from Poole Place just east of Elmer all the way – just over 3km – to West Beach and the mouth of the River Arun at Littlehampton. The eastern end offers the most open expanse of beach, flanked by large sand dunes, behind which is a golf course. Various ruined walls and other sea defences all along the shore here have become over-grown with brambles, ivy and scrub, providing fantastic habitat for breeding and migrant birds.

At Climping Beach (an SSSI in its own right), a series of storms in recent years have taken a serious toll on the old sea defences, which the Environment Agency has announced may now not be repaired. As a result of the encroachment of these storm surges, various interesting little channels and pools have formed, offering sheltered high-tide feeding opportunities for wading birds. The strewn boulders, crumbled sea wall and arable fields here also offer great resting places for waders, gulls and arriving and departing migrant passerines.

SPECIES

In winter, the shoreline is home to flocks of waders, especially Sanderling, Grey Plover, Turnstone, Ringed Plover and Oystercatcher, although all of these species may be seen here in small numbers at just about any time of year. Scarcer waders occur from time to time, although inevitably they run the gauntlet of dog walkers and other recreational beach users. Where they are able to find sheltered spots, some species can show remarkably well, particularly on some of the newly created pools between the arable fields and the beach. Little Stint have been known to show down to a few metres here, while the ditches alongside the arable fields can hold both Snipe and Jack Snipe in the winter. The odd Common Sandpiper may often be found overwintering along the river too.

The patches of woodland at Poole Place, near the Atherington car park and Long Wood (accessed from the footpath behind the golf course), hold the typical resident passerines. However, it is in spring and especially autumn that these little pockets of habitat can really come into their own, when they will often attract the likes of Spotted and Pied Flycatcher and sometimes a Yellow-browed Warbler.

The scrubby fields between Atherington and West Beach often hold Stonechat, and sometimes Whinchat in the passage months, while the banks of gorse and other scrub behind the dunes at West Beach have form for turning up a Dartford Warbler in the winter, and so are always worth checking. From the West Beach car park, a scan of the sea wall or the ruined fort behind the café may sometimes produce a Black Redstart in the winter too. Wheatears will often be found hopping about anywhere along the beach, in the arable fields or on the ruined sea defences in spring and autumn. Indeed, in the autumn in particular numbers of the likes of Wheatear and Yellow Wagtail can build quite impressively here, particularly the latter which may be seen moving along the coast in the tens or even hundreds some days; likewise, hirundines, which track the Sussex coast in their many hundreds or thousands in September.

The arable fields hold Skylarks all year-round, often forming big flocks in the winter, along with Linnet, Yellowhammer and sometimes a few Corn Buntings. Both Grey and Red-legged Partridges may be found skulking about along the farm tracks and hedgerows – Climping is one of the last remaining coastal sites for the former species. The farm hedges are also worth checking for passerine migrants such as Redstart or even Wryneck in the autumn.

Seawatching at Climping can be rewarding, especially in late winter and spring, when large movements of Brent Geese, Common Scoter and Red-throated Diver are possible, along with wildfowl, terns and skuas later into the spring. In the winter, a scan of the sea may produce a rarer grebe among the many Great Crested Grebes and Red-breasted Mergansers offshore. Mediterranean Gulls are present here all year-round, sometimes in large numbers. Yellow-legged Gull may be found occasionally as well.

The site is well watched and the list of scarcities and rarities found here in recent years, including Booted and Barred Warblers, Short-toed Lark, Kumlien's Gull, Golden Oriole, Red-throated Pipit, Ortolan Bunting and Hoopoe, is testament to the hours put in by local birders.

TIMING

Preferably early morning to avoid the crowds on sunny weekends and bank holidays, but otherwise the birding here can be good at any time of year. For the greatest number of waders and Brent Geese, visit in winter. For the best chance of scarce passerine migrants, visit in September–October.

ACCESS

The nearest railway station is at Littlehampton, from where it's around a 20-minute walk across the harbour bridge and down Rope Walk to reach West Beach. At TQ 022019, take the right-hand fork to follow the footpath south-west across the arable fields, which brings you out west of the dunes, further along the beach towards Atherington. Once you reach Atherington there are several options open to you: either continuing along the beach towards Poole Place or taking one of the footpaths back inland across the fields towards the campsite and Kent's Farm, with a return route towards Littlehampton along Ferry Road. Stagecoach South bus routes 69, 700 and 665 (Monday–Saturday only) all stop on the A259 near the junction with Ferry Road. There are two main car parks providing easy access to the area, one at the end of Climping Street at Atherington (TQ 004007), as well as some on-street parking on the lane, and another at West Beach, Littlehampton (TQ 028011).

FACILITIES: Cafés and toilets can be found at both the West Beach and Atherington car parks, though note these are only open during spring and summer.

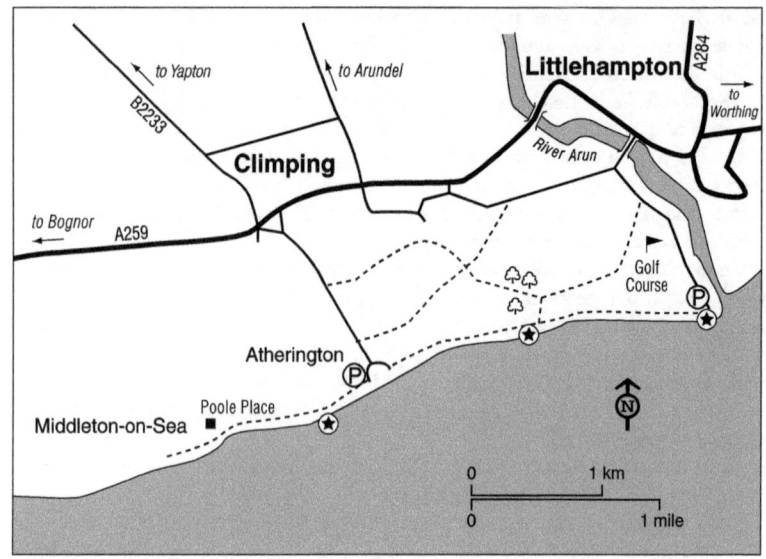

CALENDAR

All year: Common Scoter (at sea – especially Feb to May), Grey Partridge, Great Crested Grebe, Oystercatcher, Grey Plover, Ringed Plover, Turnstone, Sanderling, Dunlin, Common Sandpiper, Redshank, Mediterranean Gull, Gannet, Little Egret, Skylark, Stonechat, Meadow Pipit, Yellowhammer, Reed Bunting.

April–September: Whimbrel, Curlew, Bar-tailed Godwit, Common Tern, Sandwich Tern, Sedge Warbler, Reed Warbler, Pied Flycatcher, Redstart, Whinchat, Wheatear, Yellow Wagtail, scarcer passage migrants (especially in autumn).

October–March: Brent Goose, Shelduck (sometimes into May), Red-breasted Merganser, Golden Plover, Jack Snipe, Snipe, Red-throated Diver, Great Northern Diver, Peregrine, Black Redstart, Rock Pipit.

77 WESTDEAN WOODS

OS Explorer OL08
OS grid ref: SU 850160
Postcode: PO18 0RU

HABITAT

Halfway between Chichester and Midhurst, and stretching some 15km east to west along the South Downs, lies an extensive area of mixed woodland. At the western end of this is Westdean Woods, which is the most popular and well known in terms of its birding potential and access. A 17ha section of it is managed by Sussex Wildlife Trust, with the rest owned and managed by the West Dean

Estate, largely for commercial forestry, but with Pheasant and partridge shooting as well as stalking to control deer numbers. There are records of traditional woodland management in this area dating back to the 1600s and the woods are a fantastic example of the ecological benefits of such work being maintained over the centuries.

The habitat is mixed deciduous woodland, much of it ancient, interspersed with blocks of dense Beech and conifer plantations. In areas of regular coppicing such as on the SWT reserve, there is a rich array of ground flora, including Fly Orchid and Greater Butterfly-orchid and White Helleborine.

SPECIES

Marsh Tit, Firecrest and Siskin all breed and will likely be encountered during a visit here in any month. Cuckoo, Tree Pipit, Spotted Flycatcher and various warbler species including Garden Warbler, arrive in April and May, and Nightjar sometimes breeds in areas of clearfell. In influx years, Crossbill may be present and sometimes breeds. Finches in general can be numerous here, particularly during good winters when there is a good beechmast crop for the likes of Brambling and Lesser Redpoll, with flocks of the former in particular sometimes into three figures among Chaffinch flocks that can number over a thousand.

Westdean is one of the best sites in Sussex to see Hawfinch, with birds present throughout the year. A particularly good place to watch from is about 1.3km along the footpath leading north from Yewtree Cottage to Monkton Farm. Stand above the small flint barn at roughly SU 829164 and scan the treetops and sky, preferably for up to an hour or so from sunrise. There is a late-summer gathering of 30–50 Hawfinches, sometimes more, from July to September, but high counts can be seen through the winter months until March. A few can also be observed from the same footpath, viewing from a few hundred metres north of Yewtree Cottage. Two to three hours before sunset is likely to give the best chance of seeing Hawfinches from this spot. This area can also be good for Chaffinches and Bramblings and offers good skywatching, with an expansive view to the south-east.

All the regularly occurring raptors may be seen here too, including Buzzard and Red Kite all year and Hobby in the summer. Goshawks are frequently seen and in summer there is a chance of Honey Buzzard too. There are good viewpoints along the Monkton Track northwest of Yewtree Cottage, looking north/north-east, and from the South Downs Way looking south. Another option is the public footpath above Colworth Barn. A telescope is preferable when watching for both Hawfinch and Goshawk here, as they can sometimes be rather distant.

The farmland around the woods holds many Red-legged Partridge, and occasionally Grey Partridge too, as well as breeding Skylark and Yellowhammer. Woodlark sometimes breeds in the areas of clearfell and may be heard singing in these areas or over the surrounding farmland. Passage migrants such as Ring Ouzel and Pied Flycatcher may be found, especially in the autumn, and there has even been a recent record of a Snow Bunting on the Monkton Track.

TIMING

Westdean can offer a rewarding visit all year but is perhaps best in spring and summer, especially late summer for the largest Hawfinch gatherings. Afternoons in winter can also be good for seeing the Hawfinches come in to roost.

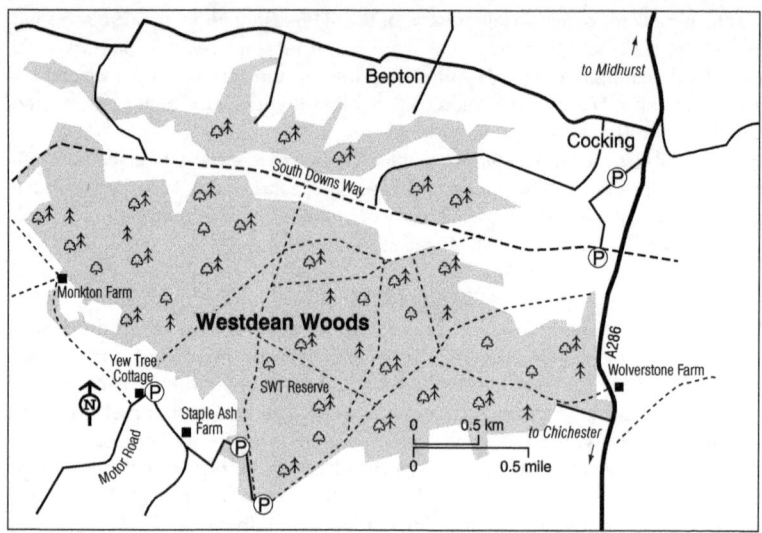

ACCESS

Bus service 60 from Chichester to Midhurst stops at Wolverstone Farm on the A286 between Singleton and Cocking, from where you can get straight onto the public footpath that leads west to Westdean Woods.

By car, there are several options for parking and accessing the area. There is limited parking at the 'Oil Well Road' (SU 875173) and a small car park at SU 875166 where the A286 is intersected by the South Downs Way, Middlefield Lane and Hillbarn Lane. From the former, cross the road and take the footpath heading west; from the latter, follow Middlefield Lane west for about 1km, then take the footpath southwest and then south to reach the woods. There are also small car parks at Hylters Lane (SU 847147) and SU 844151, from where you can follow footpaths heading into the woods. Another parking spot is by the sharp bend on Motor Road at SU 837155, which offers easy access to the Monkton Track for Hawfinch watching. From there you can also follow a permitted access track leading north into the woods, joining the footpath at SU 840161.

Please note that the SWT reserve itself and much of the surrounding woodland is private, so do respect rights of way and keep to the main tracks.

FACILITIES: The nearest facilities are in Midhurst.

CALENDAR

All year: Red-legged Partridge, Goshawk, Little Owl, Raven, Marsh Tit, Woodlark, Skylark, Firecrest, Hawfinch, Crossbill, Siskin, Yellowhammer.

April–September: Cuckoo, Hobby, Honey Buzzard, Garden Warbler, Willow Warbler, Lesser Whitethroat, Spotted Flycatcher, Tree Pipit.

October–March: Brambling, Lesser Redpoll.

78 CHICHESTER GRAVEL PITS

OS Explorer OL08
OS grid ref: SU 872033
Postcode: PO20 1NP

HABITAT

Situated immediately to the east and south-east of the city of Chichester, this group of 20 or so flooded former gravel pits forms the largest collection of inland waterbodies in West Sussex.

The pits can be quite heavily disturbed by watersports and some have no public access, but a visit is worthwhile, with good numbers of wildfowl and other water-birds present year-round. The westernmost cluster of pits, dominated by Ivy Lake, are the easiest to access and attract most of the birds seen on the other lakes. Ivy Lake also benefits from more marginal vegetation than the other pits, offering rather more interest than just open water. Drayton Pits further east are somewhat trickier to access but again offer more diverse habitat for waders and reedbed species.

Leythorne Meadow SNCI nearby is celebrated for its rich botanical and dragon-fly life in the summer months, although access here is difficult and requires prior arrangement (see the Access section).

SPECIES

As is often the case with former gravel pits, diving duck and Coots are plentiful, with counts of over a thousand of the latter possible in the winter. Tufted Duck and Pochard are typically numerous, with counts of over 300 and 150 typical. Dabbling ducks occur too, mostly Mallard, Shoveler (100+) and Gadwall (60+), though Wigeon, Teal and Pintail are occasionally present in small numbers. Given the amount of open water on offer here and the site's proximity to the coast, it also attracts scarcer ducks, grebes and divers on occasion, with records of Scaup, Goldeneye, Red-throated Diver and Red-necked Grebe in recent years. Long-tailed Duck is a possibility in winter too. Historically, the site was quite reliable for Smew, but this species is barely annual as a winter visitor in West Sussex these days; the more likely sawbill is Goosander, with one or two present in most winters.

Another occasional winter visitor is Bittern, with a few records from Drayton Pits in recent years. Perhaps the best chance of seeing one is to wait and watch towards dusk to see if one drops into the reedbeds. The reedy areas at Drayton and Ivy Lake are also good to check for Water Rail, more often heard than seen, but you may be lucky enough to see one skulking around in the marginal vegetation. Although the site isn't best known for passerines, Firecrest and Siberian Chiffchaff are sometimes found, with the small copse at Peckhams Copse Lane (SU 875028) a good place to look for such species.

In spring, the resident ducks on any of the lakes may be joined by a Garganey or Red-crested Pochard, while the likes of Sedge Warbler and Reed Warbler return to breed in the reeds, joining the ever-present Cetti's Warblers and Reed Buntings. The main passage migrants to consider here on drizzly days in spring are gulls and terns, particularly Little Gull and Arctic Tern in March and April, followed by Black Tern in late April and May. Kittiwake is possible too, with a few spring records of

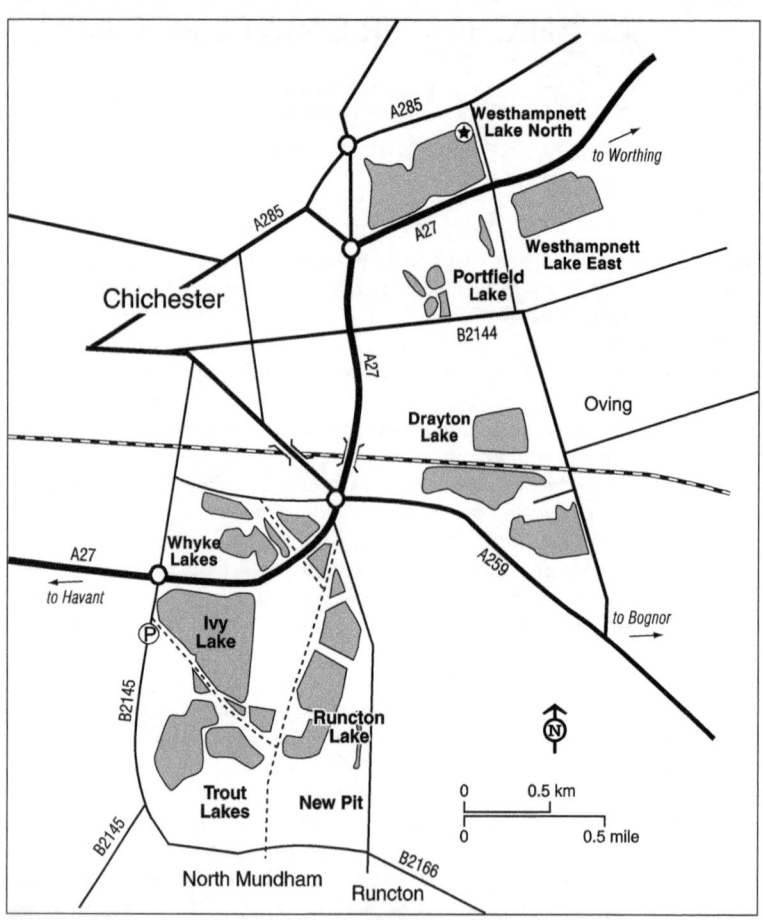

birds dropping in or moving through. Common Tern also returns to nest on the artificial rafts, along with many pairs of Black-headed Gull. Great Crested and Little Grebes also breed.

Wader passage is rather unpredictable here. Common Sandpiper is the most likely to be seen in spring around the margins of any of the lakes, but other species largely depend on the water levels, especially in the autumn. Towards the end of a dry summer, if a good amount of mud is exposed, then Green Sandpiper, Greenshank, Ruff, Black-tailed Godwit and almost any scarcer wader species are possible, but the disturbance on the pits means their visits are often fleeting.

Rarities and scarcities found here in recent years have included Alpine Swift, Bonaparte's Gull, Red-footed Falcon and Dusky Warbler.

TIMING

Winter is best for the highest numbers of wildfowl. Early mornings in summer and migration periods are recommended – especially on weekends – to beat the dog walkers and watersports activities, which can flush any birds that have dropped in.

ACCESS

Bus routes 51, 600 and 651 (all Stagecoach South) all stop on the B2145 near Chichester Free School, offering easy access to the southernmost group of lakes. For Drayton and Westhampnett Pits, buses 85 and 85A (both Compass Travel) both stop on Drayton Lane and on the A27 at Westhampnett. The nearest railway station is in Chichester and all the aforementioned bus services stop in the city centre.

By car, there are a number of options for where to park, depending on which pits you wish to visit. For Ivy Lake and the other pits in this area, there is a pull-in on the B2145 at SU 868034, with room for several cars to park. From here follow the public footpath south-east which take you alongside Ivy Lake, Copse Lake and East and West Trout Lake. After about 500m, you will reach a T-junction at New Lake. Take the left turn here to walk alongside Runcton Lake, Vinnetrow Lake and Peckham Lake. There is also space for two or three cars to park in the lay-by on Vinnetrow Road, allowing access to the eastern pits in this area and also Leythorne Meadow nearby (to arrange access to the latter, please phone the Land Management team on 01273 041819). For Whyke Lake, Quarry Lake and Long Lake, the best place to park is on Quarry Lane at SU 873040. For Drayton Lakes, there are pull-ins on the Bognor Road and Drayton Lane at SU 887039 and SU 891042. Westhampnett Pits are even trickier, with only a couple of available parking/viewing spots. For Westhampnett Lake East, park in the pull-in off the A27 at SU 884057 and follow the footpath south through the trees, then take the left track to view the lake from its north-western corner. For the North Lake, follow Coach Road north off the A27, then park by the junction of the access road at SU 882060 and view the lake from the northeastern corner.

The main tracks are generally level and well surfaced so are reasonably accessible for wheelchair or pushchair users. Views of Ivy Lake in particular can be enjoyed without travelling too far from the car.

FACILITIES: The nearest facilities are in Chichester city centre.

CALENDAR

All year: Egyptian Goose, Gadwall, Pochard, Little Grebe, Great Crested Grebe, Kingfisher, Cetti's Warbler.

April–September: Shelduck, Garganey, Swift, Mediterranean Gull, Little Gull, Black Tern, Common Tern, Hobby, Sedge Warbler, Reed Warbler, Sand Martin, Swallow, House Martin.

October–March: Shoveler, Goldeneye, Great White Egret, Cattle Egret.

79 PAGHAM HARBOUR (INCLUDING CHURCH NORTON)

OS Explorer 120
OS grid refs: SZ 878976 and SZ 856966
Postcodes: PO20 7NE and PO20 9DT

HABITAT

One of Britain's best-known birding locations, Pagham Harbour is a truly wonderful site with a wide range of habitats and species. One of the few undeveloped stretches of the Sussex coast, combined with easy access, it is rightly one of the most popular destinations for birders in southern England. A major wintering area for wildfowl and waders, this 323ha expanse of habitat also holds rare breeding species and is famous for turning up rare and scarce birds.

The area is made up of saltmarsh and tidal mudflats with shingle, open water, lagoons, reedbeds, swamp and wet permanent grassland habitats, as well as farmland. The sheltered harbour and wider area is designated as a nature reserve and is internationally recognised as a site for breeding shorebirds, especially the famous tern colony. Cattle and sheep graze the grasslands and the water levels are managed to suit the vast numbers of wintering birds as well as breeding waders.

The Pagham Harbour area encompasses various hotspots that collectively make up an impressive landscape between Selsey and Pagham. At the southern end, closest to Selsey, is Church Norton. This part of the site includes: Church Norton Spit, a shingle bank sheltering the harbour; The Severals, two reed-fringed pools adjacent to the pebbly beach; Glebe Meadow, a productive area for migrant passerines between St Wilfrid's Church and a hide overlooking the extensive mudflats; and Tern Island. St Wilfrid's Church itself has form for migrant passerines too.

Footpaths along the West Side, where one walks between the harbour and arable fields, eventually lead to the Sidlesham Ferry area, passing the reedy Long Pool and Ferry Channel. Ferry Pool is the main centre of attention here, though, and the new hide can offer excellent and comfortable views of waders. The RSPB visitor centre is situated a little further to the north.

Continuing north, footpaths pass through picturesque Sidlesham Quay before reaching the Halsey's Farm area. Another incredibly productive part of the complex begins here, running to the North Wall, which offers views over the harbour to the south and wet grassy fields to the north. Breech Pool sits at the eastern end of the North Wall and is another spot for close-range wader views, although water levels can vary (the pool is not managed by the RSPB).

Further east is Pagham Lagoon and Pagham Spit. These areas are less visited but the lagoon, a former entrance to the harbour, can be productive. The spit can turn up surprises as well.

SPECIES

The volume of birds present in the winter can be spectacular. Dark-bellied Brent Goose is the most obvious, with a few thousand wintering. At low tide they forage in the harbour, but at high tide huge feeding flocks will use the plentiful arable fields in the area. Groups are worth scanning through for Pale-bellied Brent or

Black Brant, as well as Barnacle Goose. Rarer grey geese are not to be expected at Pagham, though cold continental conditions can produce a flock of White-fronted Geese on occasion. Thousands of dabbling ducks are conspicuous at this time of year, usually including triple-figure counts of Pintail. Diving ducks tend to be present in the harbour at high tide and are best looked for off Church Norton Spit. Indeed, a winter watch off the spit can be productive (albeit rather cold), with Red-breasted Merganser and Goldeneye likely, along with Gannet, auks and Red-throated Diver further out. However, the star species here is Slavonian Grebe. Wintering concentrations of this species are not what they once were, but you can still expect at least a few birds most winters if you 'scope patiently. Occasionally, a Black-necked or Red-necked Grebe may be present too. Long-tailed Duck is also a possibility, though uncommon; there is a better chance of Eider.

Another prominent winter feature are waders. As many as 1,500 Dunlin may be present, along with large numbers of Oystercatcher, Grey Plover, Curlew, Ringed Plover, Redshank and Turnstone. Avocet – a breeding species here – is a regular presence though in smaller numbers. Counts of Knot have dropped off but it's still a species to be expected on a winter's day. Bar-tailed Godwit is usually seen as well, though never in big numbers. Church Norton is a reliable place for wintering Whimbrel, with one or two present in most years. Greenshank can be found in the winter months, along with Spotted Redshank, with the latter species having a fond-ness for Ferry Channel. Black-tailed Godwit has increased as a winter visitor and nearly 2,000 have been recorded, mainly using Breech Pool and the adjacent flooded fields, which they usually share with thousands of Lapwing and a few hundred Golden Plover.

Cattle Egret colonised Sussex in 2020, when five pairs bred at Pagham in the Owl Copse heronry. This species is readily encountered here and indeed anywhere along the North Wall and Halsey's Farm in spring and summer, but is more localised in the winter, when usually the birds flock up and move to one of the nearby farms. Glossy Ibis has become a more frequent winter visitor, too, but is somewhat erratic regarding where it turns up. Spoonbill is a winter possibility as well.

Such gatherings of birds inevitably attract raptors, with Peregrine and Sparrowhawk roving around the harbour. Harriers are rarer, even though Marsh Harrier has recently begun breeding in the area. This species is best looked for along the North Wall to Halsey's Farm, which is also the optimum stretch for Short-eared Owl, a somewhat irregular winter visitor and passage migrant. A Merlin may be seen whizzing low over a field, but is uncommon. Wandering White-tailed Eagles from the Isle of Wight programme may be encountered too.

Passerine action is somewhat limited in the winter, though regular bursts of Cetti's Warbler song can be expected. Despite their status on the other side of Chichester Harbour, Bearded Tit remain uncommon and less than annual. A Kingfisher can sometimes be seen zipping along a reedy dyke or channel.

The first big departures of wildfowl commence in February and, from March, the first Wheatears of the year can be anticipated. Of the breeders, Sandwich Tern is among the earliest returnees, along with Little Ringed Plover, which nests at Breech Pool and North Fields most years. Before long, the reedbeds are full of *Acrocephalus* warblers, Whitethroats and Lesser Whitethroats scratch away in the hedgerows, egrets are sporting their full breeding regalia and hirundines are flitting overhead. Sitting on the south coast, Pagham is well placed to receive migrants, and a spring day in the field here can be wonderfully rewarding. Redstart and Pied Flycatcher may be found in the bushes, while the quieter pools and lagoons could hold Garganey or Wood Sandpiper. Scarce overshoots are a possibility too – Black-winged Stilt, Hoopoe and Golden Oriole are the kinds of species to have in mind.

Late spring and midsummer may seem like quiet times, but they can be very productive. Tern Island is alive with activity and watching from one of the Church Norton benches (ideally with a telescope) will produce good views of the three breeding species (Common, Little and Sandwich). Little Tern was absent for a decade by the start of the millennium but has bounced back, and is a highlight of any spring or summer visit to Pagham. Hundreds of Mediterranean Gulls noisily go about their business too. In recent years, the presence of non-breeding Roseate Terns has been notable here, with as many as three individuals counted. A mega rare isn't out of the question either – look no further than the 2017 Elegant Tern or 2018 Royal Tern, both of which were drawn to Tern Island. Indeed, this time of year has produced several major rarities down the years, including Hudsonian Whimbrel, Oriental Pratincole, Terek Sandpiper and Collared Flycatcher.

Summer breeding is still well underway by the time return wader passage commences in early July. From now until mid-October there is a regular turnover of birds, with Church Norton, Ferry Pool and Breech Pool the best places to target. The water levels of the two pools are important at this time – if either is full of water, then wader passage is negatively impacted. Green and Common Sandpipers are often among the first to move through, while numbers of Black-tailed Godwit swell.

Into August the chance of Wood Sandpiper increases, along with Spotted Redshank and Whimbrel – double-figure counts of the former species are possible. Later in the month and into September is an excellent time for Curlew Sandpiper and Little Stint, with rarer possibilities very much in play, including Pectoral Sandpiper and Temminck's Stint. Pagham has historic form for North American rarities, including Baird's, Least and White-rumped Sandpipers and Wilson's Phalarope. Late summer is a decent time for Yellow-legged Gull in the harbour, too. Continental birds – often juveniles – can be sought out from mid-July onwards, and the mudflats at White's Creek and the east side of the harbour are a reliable roost spot.

Pagham is not known for raptor migration, but there's a decent chance of an Osprey going through. Honey Buzzard is far rarer. Post-breeding flocks of egrets and Grey Herons build and rarer species can turn up. Squacco Heron has been found in the Halsey's Farm area in recent years, while Purple Heron may frequent a reedy pool or ditch. These unusual species are not to be expected, of course.

Autumn passerine migration is already well underway by mid-August. In suitable conditions, it can be a lively time, with Wheatears, Whinchats and Yellow Wagtails conspicuous, Tree Pipits occasionally passing overhead and the bushes often teeming with warblers. Hedgerows and areas of trees can be a happy hunting ground for species such as Pied and Spotted Flycatchers and Redstart; the area behind the hide at Church Norton and the churchyard itself are particularly reliable area for flycatchers and produces Pied almost annually. Pagham does well for Wryneck, too, with the area of brambles by The Severals a traditionally productive spot for this desirable species.

The makeup of species shifts as September progresses and eventually moves into autumn proper. Red-backed Shrike occasionally puts in an appearance in the first half of the month. Chiffchaff becomes the most obvious migrant passerine, and searching through roving mixed-species flocks might reveal a Yellow-browed Warbler. October can be a good month for Black Redstart and Firecrest as well. Finches and thrushes pass overhead but are never numerous on the deck. The nearby farmland can sometimes hold decent flocks of seed-eaters, including Yellowhammers, while both Grey and Red-legged Partridges are possible.

Soon, the harbour will be filling up with wintering wildfowl and waders again as the year comes full circle.

TIMING

Pagham is a site than can be enjoyed at any time of day, at any time of year. However, for the sheer spectacle of masses of birds, midwinter is a great time to visit. Passage seasons are also excellent, and at these times the entire site is worthy of attention. Midsummer can be quiet, but there is still plenty to see. Early mornings are best. Church Norton can get busy, as can the North Wall, though rarely are they overcrowded. The car park at Church Norton can fill up very quickly, especially on Sundays or if a rare bird is present. The North Wall parking area along Church Lane can also be extremely busy at times with general visitors from mid-morning.

Keeping an eye on tide times is important as well. A rising tide can be productive for Church Norton and from the North Wall, with waders moving closer. At low tide they can be distant and a telescope is essential. At high tide there is no mud to be found, and many waders will move to Breech and Ferry Pools, or flooded fields north of the harbour boundary.

On bright mornings, viewing the harbour from Church Norton and the sea from Church Norton or Pagham Spits can be difficult. Conversely, on bright afternoons the Ferry Pool can be similarly difficult, with species largely in silhouette.

ACCESS

There are lots of footpaths around Pagham Harbour, though covering the entire site is some undertaking. Focusing on certain areas is best, with some of these outlined here. On the Pagham side, park at the western end of Church Lane (SZ 879975). From here, take the footpath north then west along North Wall. Upon entering onto the main North Wall footpath (by the small Salt House building) you are on top of the sluice gates. From there, White's Creek will be immediately on your left/straight ahead, the creek running south towards the harbour mouth. Proceeding away from the sluice and creek towards Halsey's Farm, the Breech Pool is then on your right. A suggested route is to go along as far as level with Halsey's Farm (SZ 86949764) where you can cross back north-west through farmland towards Honer Farm (SZ 87439832). You can then head back south towards the North Wall and back to Church Lane.

In the middle section of the site, the RSPB visitor centre is a sensible place to park (SZ 85659658). You can walk north or south, though the latter is best as it takes you to Ferry Pool and then along Ferry Channel and Long Pool, following the West Side as far as Church Norton, and you can double-back to return to the car park.

The car park at Church Norton (SZ 87169567) offers quick access to this part of the site and is advised if you're wanting to stake out Tern Island and the mudflats, work The Severals, churchyard and Glebe Meadow for passerines or scan the sea off Church Norton Point.

Chichester is the nearest station, some 8km away. To cycle from the station, turn south over the level crossing down Stockbridge Road for approximately 100m, then turn left into Canal Basin. Immediately to the right is the canal path which is the start of Route 88, currently ending at the visitor centre. Alternatively, the 51 Link service from Chichester Bus Station to Selsey stops outside the visitor centre. The

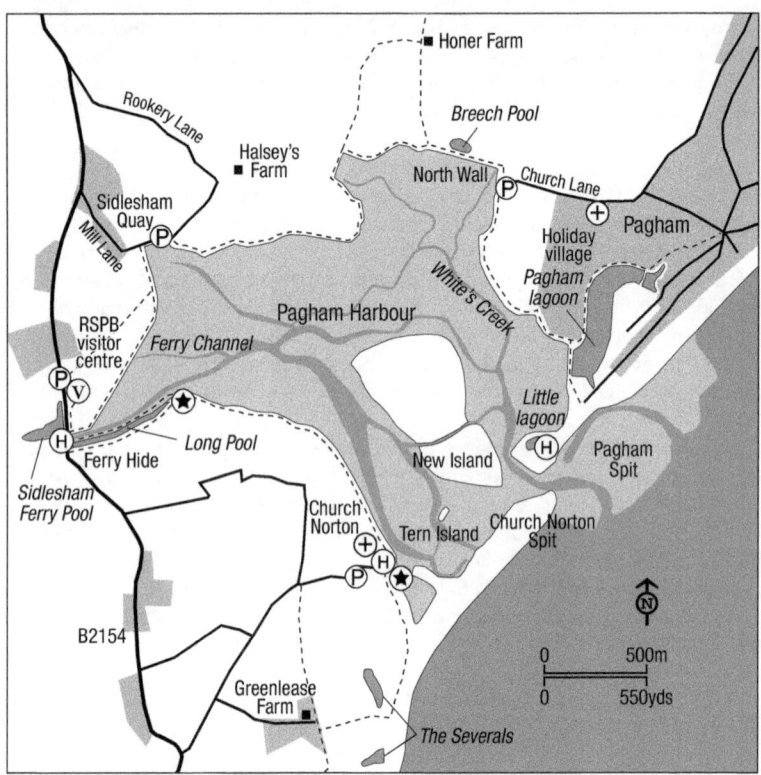

bus station is 2 minutes' walk from the train station and the bus journey takes 20 minutes. A taxi rank is situated outside the station as well.

If travelling by car, from the A27 at Chichester, take the B2145 south towards Selsey. After about 1km at the first roundabout, turn right continuing on the B2145 to Selsey. Remain on the B2145 for about 10km. The turning into the visitor centre car park is half a mile on from Sidlesham on the left-hand side. To reach Church Lane by car, follow Pagham Road until it merges into Church Lane. Church Norton is accessed by taking Rectory Lane off the Chichester Road between Selsey and Chichester. Alternative parking can be found in Sidlesham Quay (SZ 86139729) and Rookery Lane (SZ 86139729).

Before a visit to Pagham, it's worth checking the excellent Selsey Birder blog. This website publishes recent sightings daily for the entire Manhood Peninsula and can be most useful ahead of a trip to the area.

FACILITIES: The aforementioned parking at Church Lane, Sidlesham Quay and Rookery Lane is free. At the time of writing, the parking at the visitor centre is free to RSPB members, blue badge holders and Sussex Wildlife Trust members (place your membership card on your dashboard). Otherwise, it's £2 for up to four hours and £3 for more than four hours. Coaches, motorhomes and minibuses are £12. It's open from dawn until dusk, has room for 18 cars and the overflow closes at 4:00 pm.

Church Norton car park is free to church visitors, RSPB members and blue badge holders. Otherwise, it's £2 for up to four hours and £3 for more than four hours. Coaches are asked to park at the visitor centre.

The RSPB visitor centre is open daily. Seasonal nest and wildlife cameras are viewable and you can hire binoculars. Refreshments can be purchased daily between 10:00 am and 4:00 pm, including coffee. The feeders outside the centre can be worth a check too, with Brambling and Turtle Dove having visited before. There are toilets and baby changing facilities here too, open daily from 9:00 am until 5:00 pm.

Ferry Hide (SZ 85649635), overlooking Ferry Pool, has indoor seating and decking with a screen and has access for wheelchair and pushchair users. Note that you can't go inside beyond the hours of 10:00 am and 4:00 pm, though the adjacent viewing platform is always open.

The hide at Church Norton SZ 87279578 is open 24 hours. A ramp allows access to wheelchair and pushchair users. The footpath near this hide can get very muddy, although the embedded railway sleepers makes access possible even in the depths of winter; there is ramp access to the boardwalk as well. Note that there is no access to Church Norton Spit during the summer to safeguard nesting waders such as Ringed Plover.

CALENDAR

All year: Shelduck, Grey and Red-legged Partridges, Cattle and Little Egrets, Marsh Harrier, Peregrine, Avocet, Oystercatcher, Ringed Plover, Lapwing, Redshank, Mediterranean Gull, Cetti's Warbler, Yellowhammer.

November–February: Dark-bellied Brent Goose, rarer geese, Pintail, Goldeneye, Red-breasted Merganser, Common Eider, Long-tailed Duck, Red-throated Diver, Slavonian Grebe, Barn and Short-eared Owls, Avocet, Grey and Golden Plovers, Knot, Bar-tailed and Black-tailed Godwits, Spotted Redshank, Whimbrel, Rock Pipit.

March–June: Garganey, Osprey, Hobby, Little Ringed Plover, passage waders, Common, Little and Sandwich Terns, migrant passerines, Reed and Sedge Warblers, Lesser Whitethroat.

July–October: peak time for passage waders including Spotted Redshank, Wood Sandpiper, Curlew Sandpiper and Little Stint, Ruff, Yellow-legged Gull, passerine migrants including Redstart, Pied Flycatcher and Wryneck.

80 SELsey BILL

OS Explorer 120
OS grid ref: SZ 856922
Postcode: PO20 0LF

HABITAT

Selsey Bill is well known as one of the premier seawatching locations in South-East England. Sitting at the southernmost tip of the Selsey Peninsula, the Bill juts out some 8km into the English Channel and as a result one can enjoy exciting seabird movements, especially in spring. Visible migration of passerines is also a feature of the Bill, though the development of Selsey village means that nowadays there is little habitat for grounded migrants. That said, there are a number of long, vegetated gardens backing onto the beach and some other interesting ones just inland from those, as well as a couple of playing fields (including Oval Field).

SPECIES

Scanning the sea off Hillfield Road can be productive during the winter. Gulls often congregate here with Mediterranean Gull present regularly; Little Gull can appear after storms. Great Northern Diver is reliable these days, with a few usually detectable – by March, pre-departure numbers can reach double figures. Red-throated Diver is likely to be seen passing, but Black-throated Diver is uncommon. Great Crested is the most likely grebe species to be encountered, though never in significant numbers. Slavonian Grebe is a possibility too, but numbers tend to fluctuate. There's also an outside chance of Red-necked Grebe. Wildfowl action is usually rather limited, with Red-breasted Merganser a regular presence. Common Scoter flocks may be located offshore, and Long-tailed Duck or Eider could be found as well. Auk numbers are unpredictable but winter counts of thousands moving have been logged. Razorbill is the most regular species identified. A small and tame flock of Turnstones frequents the car park and beach and, on occasion, a Purple Sandpiper may be around. Rock Pipit aren't too hard to find in the winter either.

Early March heralds the first migrants, and Wheatear in particular can be numerous here as birds make landfall. Meadow Pipit passage can be strong in late March and early April, sometimes with a trailing Merlin. Early April is also when the first arrival of hirundines commences, though big counts are generally achieved later in the season. The built-up nature of Selsey means larger falls of migrants are mostly a distant memory, but it is still well worth checking the seafront gardens for species such as Whinchat, common warblers and Redstart. Scarcer species such as Black Redstart and White Wagtail are possibilities as well. Selsey is one of the most regular sites in Britain for Serin, with this species a classic spring overshoot here. Most records are of birds in-off the sea flying over, so be tuned in to their distinctive soft, trilling call. All sorts of unusual species have been seen arriving at Selsey down the years, including Bee-eater, Nightjar, Stone-curlew, Alpine Swift and Golden Oriole, so it pays to keep an open mind.

The main draw of Selsey Bill is the spring eastbound passage of birds up the English Channel. Passage can commence as early as February, with Brent Geese, Red-throated Diver, Common Scoter (occasionally with Velvet Scoter in tow), Kittiwake, Sandwich Tern, Guillemot and Razorbill featuring until early April. From then until mid-May, seawatching cranks up a gear and, in the right conditions, can be spectacular. Various duck species, waders, divers, grebes, skuas, gulls and terns can be logged, though you want to hope the wind is in the east or, ideally, south-east. A light north-easterly with rising pressure can also do the trick.

Arctic and Great Skua are regular from mid-April. Towards the end of the month, Pomarine Skua becomes a possibility – Selsey is renowned as an excellent site for this desirable species. That said, 'Poms' can be more unpredictable in their appearances at Selsey than at other seawatching sites further east (such as Splash Point). One reason for this is the 'Isle of Wight effect', which means birds may skirt around the southern tip of the island as opposed to cutting through the Solent, thus being too far out to view from land when they pass Selsey. When they do move past the Bill, however, views can be excellent and often involve small groups with resplendent 'spoon' tails. Long-tailed Skua is very rare.

Late April and early May is the best time to connect with Black-throated Diver. It's also the peak time for 'commic' terns, which can move in their thousands. Such days are often good for far smaller movements of Black Tern and Little Gull. Roseate Tern is rare. A small Bar-tailed Godwit window at the end of April and start of May can sometimes be exciting, with species such as Whimbrel, Grey Plover and Sanderling among the other wader species on the move at this time.

Large raptors arriving in-off the sea can be a spectacular sight, with Marsh Harrier and Osprey the most likely, although Hen Harrier and Honey Buzzard are possible. Hobby is quite regular in-off or along the shore at this time, with double-figure counts not exceptional.

Shearwaters and petrels are not typical seawatching fare in the South-East but a few Manx Shearwaters will pass by. Late spring, midsummer and early autumn can sometimes be better for such species, and Selsey Bill does relatively well for Balearic Shearwater. Numbers fluctuate year-on-year and generally it's an unpredictable species best looked for between July and September. European Storm Petrel is barely annual but mid-June can be a good time. The summer is typically quiet, although a scan offshore will usually produce Little and Sandwich Terns commuting to and from feeding grounds. The beach and Hillfield Road car park can occasionally be productive for Yellow-legged Gull at this time of year – a few loaves of bread can often entice birds in closer!

As is the case anywhere on the south coast, autumn seawatching is rarely much to write home about. Rough weather or strong southerlies can sometimes produce some skua passage, shearwaters (including Sooty) or Grey Phalarope, which may linger offshore. Species that one may hope for during an east coast seawatch at this time of year, such as Sabine's Gull, Leach's Storm Petrel and Little Auk are very rare indeed but can occur. Finch passage can sometimes be notable during October and November. The biggest movements are when most birds are heading east, with a bit of east/north-east in the wind. At such times, Goldfinch passage can involve thousands of birds.

Away from the Bill tip there are other sites that can offer good birding. The small bay at West Beach, at the bottom of West Street, sometimes holds Black Redstart and can be a good place from which to scan the sea. On the north side of Selsey village, an area of fields and hedges around Northcommon Farm can be productive for passerine migrants, especially in the autumn. The area has good form for Pied Flycatcher, with Black Redstart another possibility. Further south-west, the fields towards Chainbridge Lane can occasionally hold wintering Short-eared Owl.

TIMING

Spring passage is the best time to visit Selsey, with the peak period between mid-April and mid-May. Wind direction is a very important factor to consider. Ideally, a moderate to strong south-easterly will be blowing; this pushes birds closer to land while also acting as a headwind, slowing them down. A light north-easterly can also be good. You are unlikely to see much on the sea during northerly winds (especially north-westerlies), although it can be more productive for passerine migrants (both overhead and on the deck) that have arrived in-off, especially if the sea is fogbound. In such conditions, it's worth being patient and checking the beachfront and gardens as the fog clears.

There is less science behind the best summer and autumn conditions, although an offshore breeze of some sort is best. An unseasonal storm can also produce something unexpected.

Before a visit to Selsey, it's worth checking the excellent 'Selsey Birder' blog. This website publishes recent sightings for the entire Manhood Peninsula and can be most useful ahead of a trip to the area.

ACCESS

Seawatching is best undertaken from the end of Grafton Road or the car park at Hillfield Road. Most people sit on the seaward side of the Bill House garden (SZ 856922) with its landmark tower. Once an old coastguard lookout, this locally famous building is now a care home and its south-facing wall offers shelter. The tip of the Bill can be a better option in rough weather. On spring days with promising conditions, there can be as many as 50 birders around and joining them is recommended.

There is a regular bus service (51) from Chichester Bus Station to Selsey. If driving, follow the B2145 through the village and turn left into Seal Road, then right at

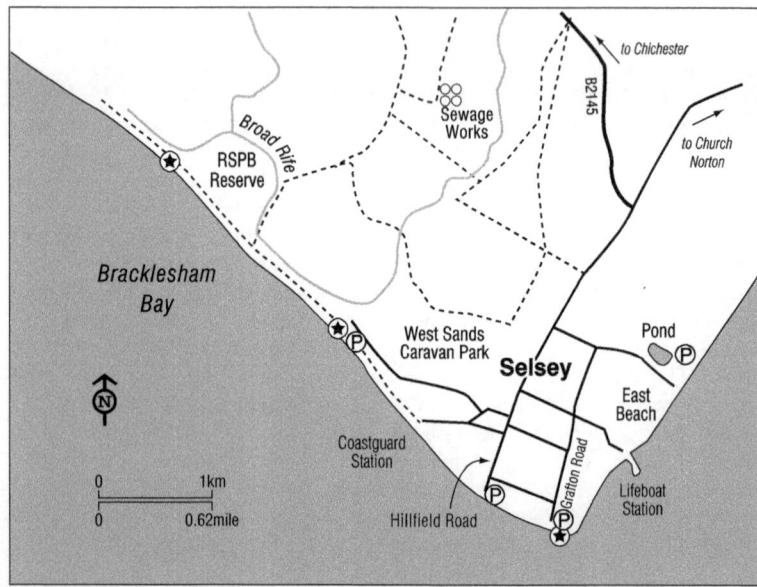

the T-junction into Grafton Road, where there should be parking (SZ 85599216). Alternatively, drive all the way down the B2145 until you reach the large Hillfield Road car park (SZ 85119233).

Northcommon Farm is best accessed along the footpath running north-west from the B2145 at SZ 85879420, or from Golf Links Lane or Paddocks Lane.

FACILITIES: Hillfield Road car park is free from December to February, but otherwise a small fee is levied. There is a height barrier and no overnight parking. Free-to-use public toilets are located a short distance north along the B2145 (SZ 85059246). There is no car park at Grafton Road, though typically there is room for several cars. A variety of amenities, including shops, can be found in Selsey village, with a fairly new supermarket and coffee shop (with toilets) on the roundabout as you enter the village from the north.

CALENDAR

All year: Gannet, Mediterranean Gull.

December–February: Dark-bellied Brent Goose, Red-breasted Merganser, Common Scoter, Great Northern and Red-throated Divers, Turnstone, Sanderling, Kittiwake, Razorbill, Guillemot, Rock Pipit.

March–May: Dark-bellied Brent Goose, Common and Velvet Scoter, Eider, Black-Throated, Red-throated and Great Northern Divers, Fulmar, Marsh Harrier, Osprey, Hobby, Sanderling, Whimbrel, Bar-tailed Godwit, Grey Plover, Arctic, Great and Pomarine Skuas, Little Gull, Kittiwake, Arctic, Black, Common, Little and Sandwich Terns, Yellow Wagtail, Meadow Pipit passage, hirundines, Swift, Wheatear, Black Redstart, overshoots including near-annual Serin.

June–August: Common Scoter, shearwaters (usually Manx; occasionally Balearic), European Storm Petrel in some years, occasional Yellow-legged Gull, Little and Sandwich Terns.

September–November: Dark-bellied Brent Goose, occasional shearwaters, skuas and rarer seabirds/phalaropes, flycatchers (Northcommon Farm).

81 MEDMERRY RSPB

OS Explorer 120
OS grid ref: SZ 832966
Postcode: PO20 7NE

HABITAT

Created following an Environment Agency decision in 2011 to proceed with one of the largest managed realignment schemes of its kind in Europe, Medmerry is a wonderfully wild coastal reserve spanning some 440ha. The lengthy shingle bank that previously protected this low-lying area required expensive annual maintenance, and thus the creation of a large bay on the inland farmland took place, with clay banks built surrounding it and the shingle bank breached to allow the sea in – roughly 184ha of the now thriving reserve are fully intertidal.

Medmerry's mudflats and saltmarsh continue to develop naturally and are home to a wide range of wildlife that can offer a dynamic day in the field year-round. Saline lagoons provide a haven for waders, with several species breeding on the islands. A vast network of freshwater ponds and ditches weave their way around the site – they are home to Water Voles and various breeding passerines. Much of the farmland is managed with birds in mind, with winter stubble, uncropped margins and sown bird and bee mixes providing winter food. Extensive rough grassland is used by wintering wildfowl, breeding waders and birds of prey. Further habitats include scrub and shingle beach.

Medmerry is not exactly the easiest of reserves to access and get around, with long walks required to reach all of the best spots. Furthermore, it is not possible to cross the breach area, meaning there are two distinct 'sides': east and west. The west side offers most interest and is where the official car park lies, just beyond Earnley Concourse. From there, a 2.5km walk beside a bank to the sea takes you through the variety of habitats, including the famous freshwater Stilt Pool (named after a breeding pair of Black-winged Stilts in 2014). Two viewpoints – Earnley and Easton – are situated along the bank and offer panoramic views of the site. The east side is trickier to reach and is generally less productive than the west side, but Breach and Ham Viewpoints can be good for views of raptors.

SPECIES

Medmerry can be a rather bleak and unforgiving place in the winter, but the birding can be great. Large flocks of wildfowl use the grasslands and Brent Goose flocks are worth sifting through for Pale-bellied Brent. Occasionally, White-fronted Geese use the area. Lapwing and Golden Plover are present in big numbers as well, with groups routinely spooked by raptors, which are a theme here at this time of year. Peregrine and Kestrel are the most likely falcons to be encountered, although Merlin is regularly seen. Marsh Harrier is increasingly reliable here year-round, and in winter Hen Harrier is a possibility. In good winters multiple Short-eared Owls will be present, though Barn is more likely; Long-eared Owl is an unlikely but remote possibility. It can be worth keeping an eye offshore at this time of year as well, with the potential for Great Northern Diver and seaduck. The rocky sea-defence groynes have played host occasionally to Purple Sandpiper but they can be easily overlooked. Dartford Warbler is a pleasingly increasing presence in areas of gorse across the site as well. The caravans near Breach Viewpoint are relatively reliable for Black Redstart in the winter.

Early spring migrants include Wheatear, hirundines and Little Ringed Plover, with the latter species breeding on Stilt Pools. Eventually the main arrival of summer visitors happens and, before long, the soundtrack of the resident Skylarks, Yellowhammers and Corn Buntings is boosted by Whitethroats and Cetti's, Reed and Sedge Warblers. Ospreys are a regular passage migrant, though autumn is better, when birds are more inclined to linger.

As well as Little Ringed Plover, Avocet also breeds on Stilt Pools, while Lapwing and Redshank do so in the grassland. Unfortunately, despite their 2014 breeding success, Black-winged Stilts have not returned to the pools named after them. Midsummer is typically quiet, although Medmerry has decent recent form for Quail and should be on your radar, especially in a 'quail year'.

Wader passage can be exciting in autumn when it might pay to give Stilt Pool

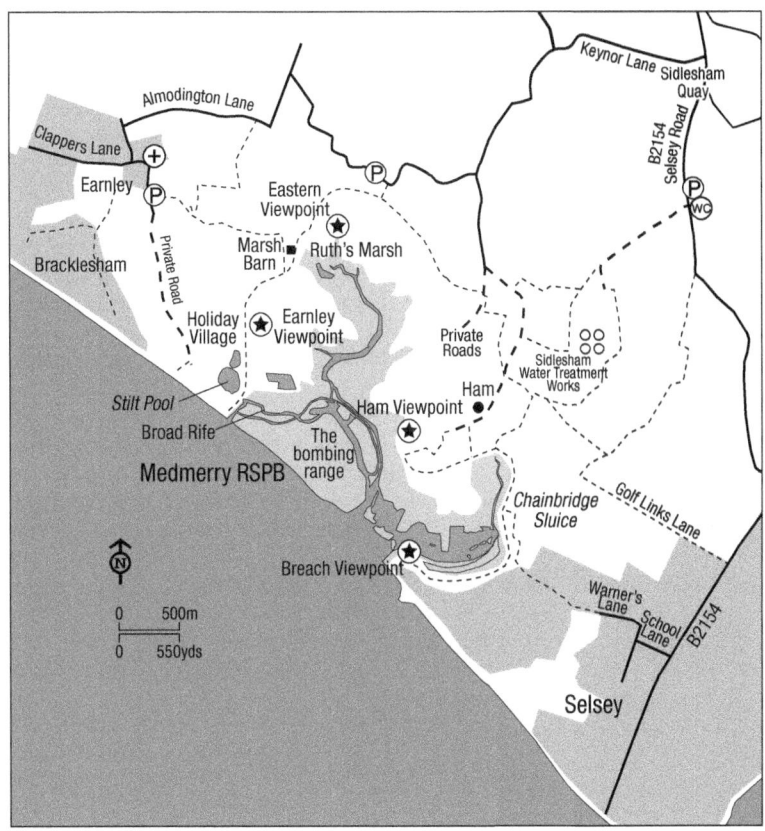

attention. Little Stint and Curlew Sandpiper are virtually annual, with species like Ruff, Wood Sandpiper and Spotted Redshank even more of a possibility. Medmerry is also developing something of a reputation for rarer species, too, with Buff-breasted and White-rumped Sandpipers having occurred.

Passerine migration can be entertaining, especially in August and September when Wheatears can, on some days, appear to be dotted about everywhere. Whinchats also enjoy the multitude of fence lines from which to forage, while Yellow Wagtails are commonly encountered. Hirundine numbers can be spectacular at this time of year as well, with four-figure counts possible on busy days. Unsurprisingly, Medmerry thus offers the chance of unearthing a scarcity at this time, with Red-backed Shrike and Wryneck both outside possibilities.

Indeed, Medmerry can feel rather 'rare' year-round, with some of the other unusual species recorded here since the reserve was created including Purple Heron, White-winged Tern, Montagu's Harrier, European Bee-eater and an 'eastern stonechat' that unfortunately wasn't assigned to a species.

TIMING

It's important to check the weather when visiting Medmerry. The lengthy walks, coupled with the fact none of the viewpoints offer shelter, mean that you want to avoid a day of heavy rain. Similarly, the low-lying and open nature of the site can make it uncomfortably cold in windy conditions. Early mornings are best, as Medmerry can get surprisingly busy with dog walkers and cyclists (the latter of which are welcome on the designated tracks). Note that, on bright days, you will be looking into the sun from the Easton and Earnley Viewpoints in the mornings.

ACCESS

The nearest railway station is Chichester, which is 11km away. The bus station is 2 minutes' walk from the train station; the 52 Witterings service stops at the end of Clappers Lane, from which you can walk along Clappers Lane to Earnley Church where the reserve is signposted.

If coming by car, the main Earnley car park is situated south-east of Earnley village, at SZ 816966. If approaching Earnley from the north along the B2198, turn left into Clappers Lane at the T-junction. Continue past the church on your left and towards Earnley Concourse and holiday village – the car park is just beyond the Concourse on your right. The track opposite the car park leads across fields to the reserve. A smaller car park can be found at Easton, on a bend along Easton Lane next to some sewage treatment beds (SZ 834967). Note that this is a small car park with room for only 10 cars.

Access to the east side is more complicated. You can either walk 5km along public footpaths from the Pagham Harbour Visitor Centre, or from Selsey West Beach, passing Medmerry windmill and the Seal Bay caravan site, which is shorter but not particularly scenic.

FACILITIES: There are no toilets or hides at Medmerry. The aforementioned view-points – Breach, Earnley, Easton, Ham – have wooden benches but are uncovered.

CALENDAR

All year: Avocet, Lapwing, Redshank, Grey Partridge, Marsh Harrier, Mediterranean Gull, Barn Owl, Peregrine, Cetti's Warbler, Corn Bunting, Yellowhammer.

November–February: Dark-bellied Brent Goose, Pintail, Great Northern Diver, raptors including Hen Harrier and Merlin, Golden Plover, Black-tailed Godwit, Black Redstart.

March–May: Osprey, Hobby, Little Ringed Plover, Whimbrel, Greenshank, terns, hirundines, Swift, Wheatear, *Acrocephalus* warblers.

June–July: Chance of Common Quail in June and July, chance of scarce overshoots.

August–October: Wader passage including Wood Sandpiper, Ruff, Curlew Sandpiper and Little Stint, Osprey, Wheatear, Whinchat, Yellow Wagtail, scarce passerines.

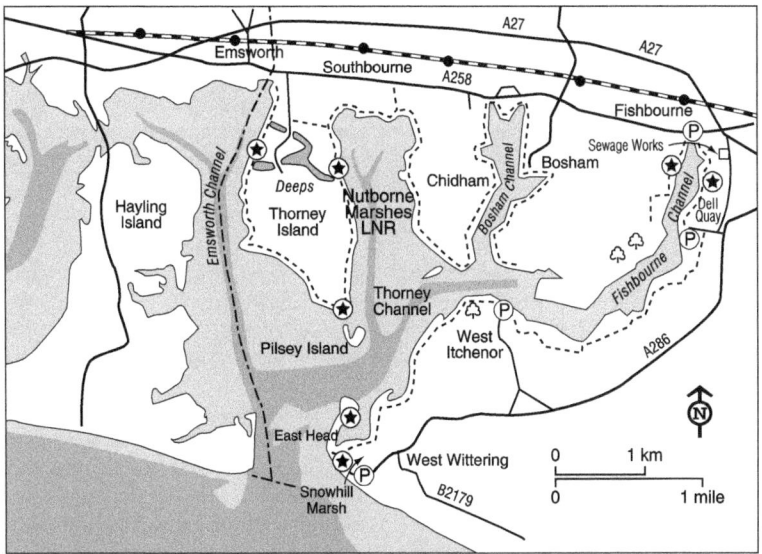

Chichester Harbour complex, sites 82, 83 and 84

82 EAST HEAD AND SNOWHILL MARSH

OS Explorer 120
OS grid ref: SZ 765985
Postcode: PO20 8AJ

HABITAT

Sitting on the Sussex side of Chichester Harbour entrance are the popular sand dunes of East Head. On the seaward side of this National Trust site lies an extensive area of sand, mud and shingle that extends into the harbour channel. A saltmarsh is on the landward side. East Head is 'joined' to the mainland by an area known as The Hinge – unfortunately rising sea levels threaten this narrow strip. The tidal Snowhill Marsh is further north, behind the sea wall, and is well known for its high-tide wader roost. As well as these habitats, there are large areas of arable farmland and pastures.

SPECIES

Winter is the most rewarding time of year and the birding often begins in the car park or even on the entry road from West Wittering village. The fields here hold a flock of up to 3,000 Brent Geese, and during a rising tide they move from the saltmarsh to feed here. It can make for a seriously impressive spectacle and you can often enjoy superb views from the car park. With so many geese around, it's no surprise that rarer species have been found among them down the years, including Black Brant and Red-breasted Goose; Pale-bellied Brent is virtually annual. The fields usually have Golden Plover (sometimes hundreds) in among the Lapwing, and Ruff is not impossible. The pines around the car park will occasionally hold wintering Firecrest.

The beach is likely to yield groups of Sanderling and little else, but a scan offshore can be productive, with divers, scarcer grebes and seaduck, including Common Eider and Long-tailed Duck, possible. The dunes at East Head can be attractive to Snow Bunting – although, if there are any birds on site, they can be fairly mobile due to the considerable human disturbance. The saltmarsh on the landward side of East Head, Snowhill Marsh, supports a large wader roost in the winter months, including Grey and Ringed Plovers, Knot, Redshank, Oystercatcher and Dunlin, as well as a few Snipe (sometimes Jack Snipe) and Black-tailed Godwit. Snowhill is a reliable winter site for Greenshank and Water Rail, too, with Spotted Redshank and Avocet also fairly regular. Such wader numbers mean visits from Peregrines are frequent.

East Head is not known for its seawatching but some species, especially terns, can be seen from spring onwards. Winter can be rewarding – all three species of diver have occurred in recent years. Eider and Slavonian Grebe are possible and even the odd Velvet Scoter may be prised out.

Summer is quiet for birds, but incredibly busy with people – West Wittering beach is one of the most popular on the South Coast. It is not really worth visiting at this time of year. While the area is not known for producing particularly exciting birding during the passage seasons, it's probably under-watched in this regard.

TIMING

It's important to note that this area is extremely popular with visitors throughout the year. In summer it is usually heaving with beachgoers, while an incredible number of dog walkers visit in the winter. As a result, it pays to be savvy with your visits – early starts are best, and ideally not on days when the forecast is fine. On busy summer days there are often traffic queues from West Wittering village.

ACCESS

The nearest station is Chichester, some 11km away. The bus station is 2 minutes' walk from the train station – from there, take the Stagecoach South 52/53/652/653 circular bus, alighting at the Old House at Home stop, before walking down the unclassified road on your left, off Pound Road (with obvious 'Welcome to West Wittering Estate' signage).

If driving, the vast East Head car park is at SZ 769981. From West Wittering village, take the unclassified road off Pound Road until you reach the car park. It's important to note that car park spaces are only available to buy via the Just Park app or website: select West Wittering Beach from the choice of locations, pick the day you wish to travel and follow the instructions from there. You will need to show your parking confirmation to the staff on the entrance gates through your car window upon arrival. Alternatively, park in West Wittering village and walk down the entrance road off Pound Road.

The sandy nature of the site makes it largely unsuitable for wheelchair and pushchair users, although observing the Brent Geese flock can be done from the car park.

FACILITIES: A café run by the West Wittering Estate is found in the car park. At certain times of the year, other food stalls are open. Toilets (including accessible facilities) are available between East Head and the car park at West Wittering. There are good, sheltered picnic spots in the dunes.

CALENDAR

October–February: Brent Geese, Shelduck, sometimes rarer geese species, seaduck, Great Northern and Red-throated Divers, Water Rail, Sanderling, Golden Plover, Greenshank, Grey Plover, Knot, Jack Snipe, Peregrine, chance of Snow Bunting.

March–September: Sandwich Tern, Little Tern, Wheatear on passage.

---

83 FISHBOURNE CREEK

OS Explorer 120
OS grid ref: SU 835035
Postcode: PO20 7EE

HABITAT

The easternmost of four thin channels that weave up from Chichester Harbour, Fishbourne Creek is a narrow area of intertidal mudflats with pockets of saltmarsh and flanked by farmland. There are small areas of reed and scrub as well. A sewage works is situated near Apuldram, where the River Lavant flows into the creek. The whole area receives surprisingly poor coverage.

SPECIES

A nice variety of birds can be enjoyed on a winter's day. A wader roost occurs on a rising tide and Fishbourne is particularly reliable for wintering Greenshank and Spotted Redshank. More regular are species such as Ringed and Grey Plovers, Dunlin, Black-tailed Godwit, Curlew and Redshank. The area of saltmarsh north of the sewage works is worthy of attention – Jack Snipe winters here and one or two may be seen among the Snipe that are flushed by the rising tide. It's also a good area for Rock Pipit and, occasionally, Water Pipit, which is a rare Sussex bird.

The area of reeds by the sewage works holds Cetti's Warbler. The works have occasionally held wintering Firecrest, a species also found in the winter near Apuldram church. South of the works, where the River Lavant flows out into the creek, is a good area for collecting waders and wildfowl, sometimes including Pintail. The deeper parts of the creek are favoured by Red-breasted Merganser and Goldeneye, with Long-tailed Duck not unheard of – the southern part of the creek near Dell Quay is best for diving duck. Kingfisher often shows well in this area in the winter, too. Decent numbers of gulls gather in the area. Mediterranean is virtually guaranteed – sometimes in impressive numbers – and there's an outside chance of Yellow-legged Gull.

Typical passage species include Whimbrel and occasional Osprey. Scarcer passage waders might feature the odd Curlew Sandpiper or Little Stint, at least in the autumn.

Summer can be quiet, though a large herd of Mute Swans is usually present, sometimes with a few summering Brent Geese. Yellowhammer used to be common but has declined; it is still possible around the horse paddocks.

Further south, towards Chichester Channel, is Chichester Marina. The creek there holds many of the species previously described, while the large reedy pool (which has a hide) has irregularly hosted Bittern and Bearded Tit. More likely are roosting Marsh Harriers in the winter, and Cetti's and Reed Warblers.

TIMING

A rising tide is best for good views of waders, though low tide is ideal for checking the saltmarsh. The area can be popular with walkers at the weekends.

ACCESS

Fishbourne station is less than half a mile away. From the station, head south along Salthill Road, then east at the T-junction along the A259 then take a right onto Mill Lane. Walk to the south of Mill Lane, to the pond, where public footpaths take you to the creek. Bus service 700 from Brighton to Southsea stops at Fishbourne.

Parking is best opposite the Crown and Anchor Pub in Dell Quay (SU 836028). You can walk north from here along the sea wall. Alternatively, park at the south end of Mill Lane (SU 838045) in Fishbourne. There is limited parking at Dell Quay and the churches at Apuldram and Fishbourne. There is also parking at Chichester Marina.

Fishbourne Creek is not suitable for visitors with limited mobility, although a view of the creek can be obtained from the Crown and Anchor Pub car park.

FACILITIES: There is a small hide overlooking the pool at Chichester Marina.

CALENDAR
All year: Little Egret, Mediterranean Gull, Cetti's Warbler, Yellowhammer.

April–September: Occasional Osprey, a chance of passage waders such as Little Stint and Curlew Sandpiper.

October–March: Brent Goose, Shelduck, Wigeon, Pintail, Goldeneye, Red-breasted Merganser, Kingfisher, Ringed and Grey Plovers, Dunlin, Jack Snipe, Black-tailed Godwit, Greenshank, occasional Whimbrel and Spotted Redshank, Rock Pipit, occasional Water Pipit.

84 CHIDHAM

OS Explorer 120
OS grid ref: SU 799044
Postcode: PO18 8TE

HABITAT

At the head of Bosham Channel, two creeks split off: Cutmill Creek on the Chidham side, and Colner Creek on the Bosham side. The habitat here is similar to Fishbourne Creek: intertidal mudflats with small patches of saltmarsh, flanked by farmland. Most of the Chidham peninsula is farmland, with small areas of woodland. It's generally poorly birded.

SPECIES

A similar array of wildfowl and waders to Fishbourne Creek can be found in the winter, usually including a few Greenshank. There is an impressive Mute Swan flock here than often reaches three-figures. Usually, a non-native Black Swan is among them. Resident species on the farmland that are easier to find in the winter include Grey Partridge (Cobnor Farm) and Yellowhammer. In strong southerlies a Great Northern Diver or scarcer grebe may be detectable off Cobnor Point.

Late summer and early autumn can see impressive build-ups of gulls in Cutmill Creek, involving hundreds of Mediterranean. It's also a hotspot for Yellow-legged Gull, with double-figure counts possibly between July and September.

More unusual waders such as Curlew Sandpiper and Spotted Redshank are possible on passage, which can also be good for passerine movement involving chats, flycatchers and Wheatear.

TIMING

A rising tide is best for good views of waders. Low tide is probably best for gulls, which loaf on the saltmarsh and mudflats at Cutmill Creek.

ACCESS

Fishbourne train station is 2.5km away. From the station, walk south on Broad Road before heading east along the A259 until you reach the head of Cutmill Creek at SU 797052.

If travelling by car, there is a free car park in a small copse at the southern end of Chidham village at SU 793035. Footpaths from here go south to Cobner Point and east to Bosham Channel. Alternatively, there is room for a couple of cars in a lay-by along the A259 at SU 796052.

FACILITIES: The only toilets are at The Old House at Home pub in Chidham village.

CALENDAR

All year: Grey Partridge, Mediterranean Gull, Yellowhammer.

April–September: Occasional Osprey, Yellow-legged Gull between July and September, passage waders and passerines.

October–March: Brent Goose, Shelduck, Wigeon, Pintail, Goldeneye, Red-breasted Merganser, Kingfisher, Ringed and Grey Plovers, Dunlin, Black-tailed Godwit, Greenshank, occasional Whimbrel, Rock Pipit.

85 THORNEY ISLAND

OS Explorer 120
OS grid ref: SU 759036
Postcode: PO10 8DB

HABITAT

In the far south-western corner of West Sussex, just a stone's throw from the Hampshire border, lies Thorney Island – perhaps the best birding locale within the Chichester Harbour area. Although not an easy site to access, the diverse array of habitats and species on offer here make it very much a worthwhile destination for any birder.

Thorney is now a peninsula, not an island, owing to it being connected to the mainland by sea walls in the 1830s and the saltmarshes in between being reclaimed. Most of the 'island' is dominated by a military base, which is off-limits to the public, so there is only a perimeter path available to walkers and birders. From this path, though, you can enjoy all that the area has to offer, both in terms of waders and other coastal specialities along the shoreline and out to sea, and scrub, marsh and grassland species on the inland side. Do bear in mind that the full loop is just over 11km in total, so this is not one for the faint-hearted or anyone short on time!

SPECIES

The estuary birding on offer at Thorney is among the best in South-East England, with some impressive wader gatherings in the winter months, particularly towards the south-eastern side in the area around Pilsey Island (an RSPB reserve with no public access). All around the perimeter path one can enjoy close views of throngs of Dunlin, Ringed Plover, Turnstone, Curlew, Redshank, Bar-tailed Godwit, Grey Plover and Oystercatcher, along with generally smaller numbers of Black-tailed Godwit, Greenshank, Sanderling, Avocet, Lapwing, Golden Plover and Knot. In passage season these may be joined by the likes of Spotted Redshank, Curlew Sandpiper, Whimbrel, Green and Common Sandpiper. Spotted Redshank, Greenshank and Common Sandpiper sometimes overwinter in small numbers.

Scanning out into the open waters of Thorney Channel to the east and Emsworth Channel to the west will often yield small rafts of Red-breasted Merganser and Shelduck in the winter, the latter gathering in their hundreds here outside of the breeding season. As with any site along this stretch of coast in the winter, hundreds or even thousands of Brent Geese can be seen commuting to and from the fields on the island or bobbing about on the water offshore. Little and Great Crested Grebes will often be seen just offshore too. Even more infrequently, species such as Long-tailed Duck and Eider may be sighted and a Great Northern Diver or two sometimes winters in the harbour. The wildfowl action is not limited to the open water, as the Deeps which separate the main 'island' from the mainland also host a variety of ducks, especially in the winter. Gadwall, Wigeon, Pintail and Teal, and smaller numbers of Mallard and Shoveler can be seen here, along with more Little Grebe often gathering around the Great Deep sewage outflow sluice just south of Thornham Point. Look out for Goldeneye as well, either offshore or on Great Deep, although in increasingly smaller numbers as the wintering population of this species continues to dwindle.

Historically, the site has been very good for egrets, famous for its three-figure Little Egret roost at the start of the twenty-first century, when the species was still highly desirable. Unsurprisingly, both Great and Cattle Egrets have now been added to that cast and are almost as likely to be encountered in the area as their precursory cousin. Thorney Island is perhaps the most reliable location along the West Sussex coast for Osprey, especially in the autumn, and a scan west from Thornham Point towards the Deeps will often reveal one perched on the old landing lights. The site attracts many other raptors, including Kestrel, Peregrine, Buzzard, Sparrowhawk and Marsh Harrier. Short-eared Owl and Merlin often patrol the grassland areas in the winter and there is always a chance of a Hen Harrier.

All manner of passerines can be encountered along the route around the island, from Skylark, Stonechat, Meadow and Rock Pipits in the open marshland to winter thrushes, tits and other woodland species in the hedgerows and areas of tree cover. The sewage works along Thornham Lane attracts Chiffchaff in the winter, and Firecrest can be found here too. As with any coastal spot of this nature, regular migrant passerines such as Wheatear, Whinchat and Yellow Wagtail will often be encountered in spring and autumn. Listen out for the machine gun song of Cetti's Warbler in the reedbeds in the Deeps and possibly the pinging call of Bearded Tit, which turns up here most years.

As you might imagine for a coastal site offering so much quality habitat, Thorney has an impressive list of rarities to its name, with more recent highlights including Pallid Harrier, Semipalmated Plover, Kentish Plover and Red-breasted Goose. Rather more frequent vagrants and passage migrants include Wryneck, Red-backed Shrike and Red-necked Phalarope.

TIMING

For the greatest numbers of waders and wildfowl, visit between November and March and, for the former, choose a time when the tide is on the way in and head down to Longmere Point looking out across Pilsey Sands. Here you will be able to enjoy the spectacle of the many thousands of shorebirds which congregate in the winter months.

For the best chance of passage migrants, visit in April or May or August to October. Autumn visits will likely produce more in the way of grounded passerines, hirundines and the best chance of an Osprey or two.

ACCESS

By car, turn south onto Thorney Road, roughly halfway between Emsworth and Southbourne on the A259. There is room for a few cars by the junction of Thorney Road and Thornham Lane, and on Prinsted Lane, just south of Prinsted village, at SU 765050. Either of these parking spots offer a good starting point for a full circuit of the island, offering footpath access in both directions.

The perimeter path is mostly flat and well surfaced, although do allow at least half day for this as, once you are halfway round, there is no shortcut back! Thorney is well placed for access by train, with Emsworth and Southbourne stations both just a couple of kilometres away; Emsworth is closer for accessing the western side of the island and Southbourne for the eastern side. Bus route 700 (Stagecoach South) stops at various points along the A259 between Emsworth and Prinsted, while route 27 (First Portsmouth, Fareham and Gosport) from Rowlands Castle terminates on North Street in Emsworth.

FACILITIES: The nearest facilities are in Emsworth and Southbourne.

CALENDAR

All year: Shelduck, Little Grebe, Little Egret, Cattle Egret, Oystercatcher, Ringed Plover, Grey Plover, Black-tailed Godwit, Curlew, Greenshank, Redshank, Turnstone, Sandwich Tern, Marsh Harrier, Cetti's Warbler, Skylark, Bearded Tit, Reed Bunting.

November–March: Brent Goose, Red-breasted Merganser, Goldeneye, Wigeon, Teal, Dunlin, Redshank, Great Northern Diver, Barn Owl, Short-eared Owl, Merlin, Avocet, Golden Plover, Lapwing, Rock Pipit.

March–May: Osprey, Whimbrel, Mediterranean Gull, Little Tern, Common Tern, Barn Owl, Whinchat, Wheatear, various warblers.

August–October: Osprey, early returning wildfowl, Hobby, Bar-tailed Godwit, Curlew Sandpiper, Greenshank, Spotted Redshank, Common Sandpiper, hirundines and migrant passerines.

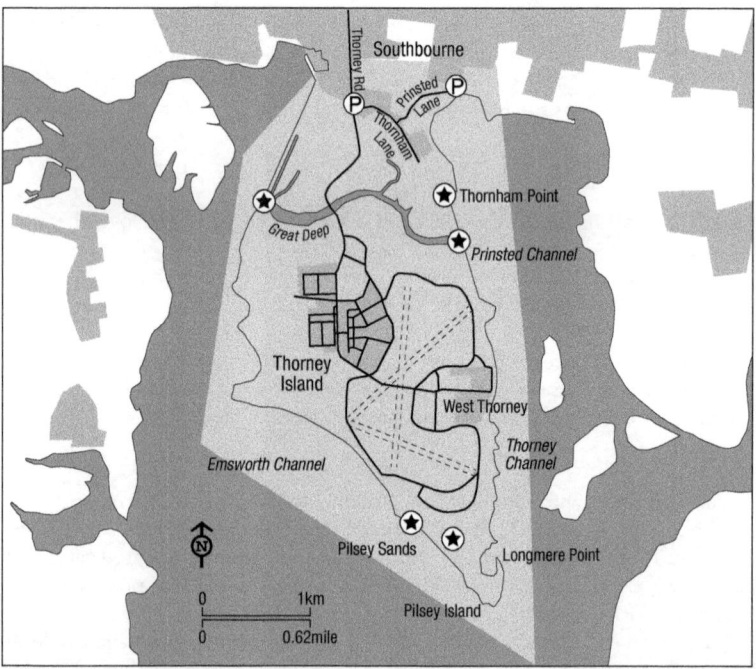

OTHER SITES IN SOUTH WEST SUSSEX

H1 KINGLEY VALE NNR

Just north-west of Chichester, this intriguing site is dominated by a steep-sided, south-facing bowl covered in Yew trees, attracting Hawfinches, which probably breed, as well as Ring Ouzel in autumn. Turtle Dove may still be encountered in the summer and the evening flypast of gulls in winter will often include good numbers of Mediterranean Gull. There is a large car park at SU 824087.

H2 LEVIN DOWN

Twenty-seven hectares of (steep!) chalk grassland and scrub can be explored on this Sussex Wildlife Trust reserve just north-east of Singleton. Breeding species include Skylark, Yellowhammer and Bullfinch, and there are occasional records of Hawfinch. An impressive butterfly list, numbering some 38 species, includes Small Blue and Brown Hairstreak. Limited parking is available just off Charlton Road at SU 887130.

H3 BIGNOR HILL

Between Sutton and Madehurst, Bignor Hill is a rare example of chalk heathland, like Lullington Heath in East Sussex (site 62), offering good downland birding, especially for passage migrants in the autumn. It scores regular records of Ring Ouzel, as well as good counts of Redstart, Wheatear, Whinchat and especially Spotted Flycatcher (including counts of up to 30). A car park is located at SU 973129.

H4 LORD'S PIECE AND SUTTON COMMON

A triangle of heathland (SU 994170) roughly halfway between Coldwaltham and Duncton, and not far from Burton Mill and Chingford Ponds (site 63). Breeding species include Nightjar, Tree Pipit, Woodlark, Yellowhammer, Willow Warbler, Firecrest, Marsh Tit and sometimes Crossbill. Brambling is possible in the winter. Ample car parking is available at the designated car parks on the western and southern sides.

H5 PARHAM PARK

A 350ha deer park (TQ 059144) just south of Pulborough, which is mostly private but with a public footpath running past a lake that can produce Mandarin Duck and sometimes Goosander. Breeding birds include Spotted Flycatcher, Raven and Little Owl in the wider estate. The nearest train station is in Amberley and limited car parking is available on Rackham Street on the west side of the estate.

H6 HOUGHTON FOREST

This 235ha block of woodland north of Arundel offers a pleasing selection of woodland passerine species including Marsh Tit, Firecrest, Siskin and Hawfinch. A good chance of Goshawk too. Brambling can be found in the winter, sometimes in large flocks. A large pay-and-display car park and café are located at Whiteways on the A29 at TQ 001108.

H7 KITHURST AND CHANTRY HILLS

Another good area of well-managed downland farmland, just south of Storrington. Corn Bunting and Yellowhammer breed and can form large flocks in winter. There are also good numbers of Red Kites and Ravens, as well as the occasional Hen Harrier in the winter. Passage migrants in spring and autumn include Wheatear, Ring Ouzel and occasionally Wryneck. There are two small car parks at TQ 087119 and TQ 069124.

H8 REWELL WOOD

Another great woodland site near Arundel holding Marsh Tit, Firecrest and Hawfinch, with sometimes good counts of the latter in the winter, especially around Sherwood Rough (SU 998092) and Fairmile Bottom LNR (SU 989093). It's also a great site for butterflies, including Pearl-bordered Fritillary – the largest remaining population of the species in West Sussex is found here. The nearest station is Arundel while the closest car parking spots are the various lay-bys along the A29 or on Baycombe Lane (SU 970085).

H9 SWANBOURNE LAKE

A long, thin lake tucked away in the trees (TQ 016079) between Arundel Park and WWT Arundel (site 69), strikingly blue in colour owing to it being fed by a chalk stream, and with a 2km circular trail right around. This popular site for dog walkers and families can be surprisingly good for birds, particularly Gadwall, Tufted Duck, Mandarin Duck and gulls. Close views can be enjoyed, particularly from near the café, offering good photographic opportunities. The flanking woodland has a healthy population of Firecrests. The lake is just a 15-minute walk from Arundel station, while parking is available on Mill Road.

H10 CHANCTONBURY RING

Another Iron Age hillfort (TQ 139120), like Cissbury Ring (site 74), rising to 242m above sea level. The likes of Yellowhammer, Skylark and Marsh Tit can be found year-round, but the site is particularly good for autumn migration, especially Pied Flycatcher and Ring Ouzel. Hen Harrier is possible in the winter. The car park is located along Chanctonbury Ring Road at the grid reference above.

H11 BEEDING BROOKS

Just north of Upper Beeding, Beeding Brooks offers similar birding to the much larger Henfield Levels (site 70) which is situated a little way to the north. Species found here include various wildfowl, Lapwing, Snipe, Barn Owl and, in good winters for the species, Short-eared Owl. Parking is best on Pound Lane (TQ 197111), from where you can join the footpath north into the Brooks.

H12 BEEDING AND TRULEIGH HILLS

Both these great downland sites offer breeding Skylark, Yellowhammer and Corn Bunting, as well as potential for falls of passage migrants, especially Wheatear in spring and Ring Ouzel in autumn. Hen Harrier is possible in the winter. Beeding Hill has hosted Red-footed Falcon in June before. A small car park can be found at TQ 207096.

H13 BROOKLANDS PLEASURE PARK

This popular spot (TQ 173034) for dog walkers and families is another that can be surprisingly good for birds, with recent records of Pallas's Warbler and overwintering Dartford Warbler. The adjoining golf course is being rewilded which should enhance the wildlife value. The main lake attracts plenty of common wildfowl and occasionally waders such as Common Sandpiper and Black-tailed Godwit. The park is around a 25-minute walk from Lancing station and there is ample parking on the south-west side.

H14 ELMER ROCKS

This site (SZ 986998) will occasionally produce a scarcer grebe offshore, such as Red-necked, or even Black Guillemot. Great Northern Divers are usually present in the winter. The rocks themselves are attractive to Black Redstart outside the breeding season. Bus number 600 (Stagecoach South) stops on Templesheen Road, otherwise the nearest parking is on Blakes Road in Felpham.

H15 BARNHAM BROOKS

This extensive area of under-watched marshland and farmland (SU 957027) south of Barnham is attractive to wildfowl and wintering waders, plus raptors (especially Marsh Harrier), owls and breeding farmland species. Both Bewick's and Whooper Swans have been recorded in the winter. It's around a 30-minute walk from Barnham railway station (where there is also a good-sized car park) to the Brooks.

NORTH WEST SUSSEX

For the purposes of this book, North West Sussex is defined as everything north of the 18 northings line from Harting Downs east to the county border with East Sussex near Burgess Hill.

The landscape here is a far cry from the coastal and downland habitats of South West Sussex. It is much drier – indeed, half of the main sites featured here are heathlands, while the others are largely woodland or open country except for Warnham LNR and Ardingly Reservoir. Rivers are generally less of a feature in this region. The Wey and Arun Canal and the upper reaches of the River Arun run through the north of the county here, meeting just east of Loxwood, but these are narrower and far less wild stretches than further south down the river valley. Similarly, the River Adur is little more than a trickle for much of its length until it meets the Lancing Brook at the Knepp Castle Estate and swells further downstream. Work at Knepp to encourage seasonal flooding of the Adur is encouraging and will hopefully increase the wetland habitat there. Away from the rivers, all the waterbodies featured here are artificial; Ardingly Reservoir is the only large reservoir, while former mill ponds form the bulk of the wetland interest at Knepp, Warnham LNR and Lurgashall Mill Pond.

What this region lacks in wetland it more than makes up for with its excellent array of woodland and heathland sites, offering some of the best opportunities in Sussex to connect with some of the specialist species associated with these habitats. Woolbeding Common, for example, is arguably the most reliable and

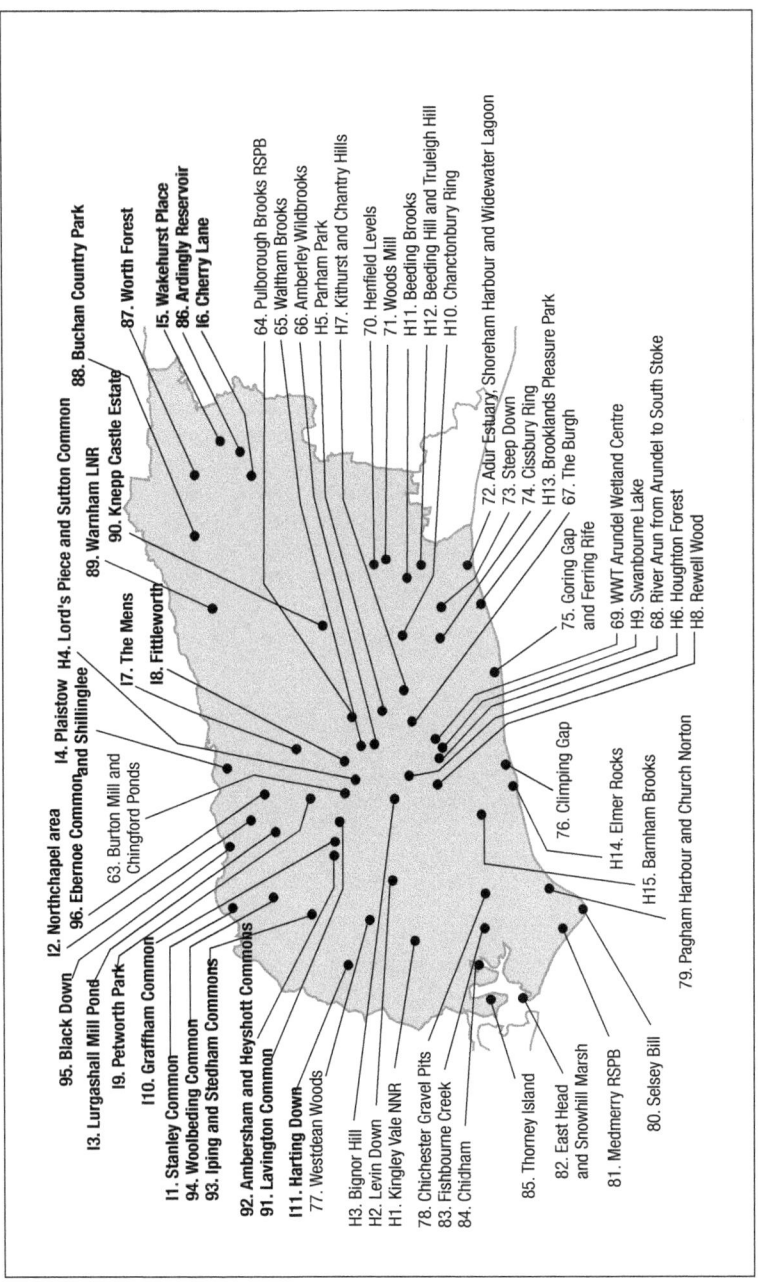

95. Black Down
I2. Northchapel area
I4. Plaistow H4. Lord's Piece and Sutton Common
96. Ebernoe Common and Shillinglee
89. Warnham LNR
88. Buchan Country Park
90. Knepp Castle Estate
I5. Wakehurst Place
86. Ardingly Reservoir
87. Worth Forest
I6. Cherry Lane
I3. Lurgashall Mill Pond
I9. Petworth Park
I10. Graffham Common
I7. The Mens
I8. Fittleworth
63. Burton Mill and Chingford Ponds
I1. Stanley Common
94. Woolbeding Common
93. Iping and Stedham Commons
92. Ambersham and Heyshott Commons
91. Lavington Common
I11. Harting Down
77. Westdean Woods
H3. Bignor Hill
H2. Levin Down
H1. Kingley Vale NNR
78. Chichester Gravel Pits
83. Fishbourne Creek
84. Chidham
85. Thorney Island
82. East Head and Snowhill Marsh
81. Medmerry RSPB
80. Selsey Bill
79. Pagham Harbour and Church Norton
H15. Barnham Brooks
H14. Elmer Rocks
76. Climping Gap
H8. Rewell Wood
H6. Houghton Forest
68. River Arun from Arundel to South Stoke
H9. Swanbourne Lake
69. WWT Arundel Wetland Centre
75. Goring Gap and Ferring Rife
67. The Burgh
H13. Brooklands Pleasure Park
74. Cissbury Ring
73. Steep Down
72. Adur Estuary, Shoreham Harbour and Widewater Lagoon
H10. Chanctonbury Ring
H12. Beeding Hill and Truleigh Hill
H11. Beeding Brooks
71. Woods Mill
70. Henfield Levels
H7. Kithurst and Chantry Hills
H5. Parham Park
66. Amberley Wildbrooks
65. Waltham Brooks
64. Pulborough Brooks RSPB

269

accessible Honey Buzzard viewpoint in Sussex, while Goshawk can be almost guaranteed at Black Down, The Mens and Worth Forest in early spring. Around a third of all heathland coverage in Sussex is found here, mostly to the west and north-west of Petworth, with Ambersham and Heyshott and Iping and Stedham Commons forming the largest contiguous blocks after Woolbeding and Pound Commons. In turn, these support crucial populations of the likes of Nightjar, Woodlark and Dartford Warbler.

North West Sussex is a region of contrast in terms of the distribution and concentration of its human population, which of course has a knock-on effect in terms of habitats and birding. The towns of Horsham and Crawley towards the far north-east have both grown rapidly in recent decades – to the extent that they are now almost a single conurbation – and are cumulatively home to around 200,000 people. The areas to the north and west, meanwhile, are some of the most rural and sparsely populated in this part of the country, as well as offering some of the most extensive and unbroken areas of woodland and heathland. The patchwork of tiny rural villages in this area offers its own novel opportunities for birding, with the stretch of the River Rother through Fittleworth and Lurgashall Mill Pond offering the chance to connect with the likes of Lesser Spotted Woodpecker, Kingfisher and Spotted Flycatcher in a landscape more reminiscent of the countryside of old England – a sharp contrast from the urban sprawl of the more built-up areas.

Despite the growing population, it remains one of the more under-watched areas of Sussex, suffering as its North-East counterpart does somewhat in terms of being an area that many birders travel through to reach the coast. This is unfortunate and in no way a reflection of the quality of birding on offer here, and there is much to be discovered for anyone willing to put in the hours in the field.

86 ARDINGLY RESERVOIR

OS Explorer: 135
OS grid ref: TQ 334287
Postcode: RH17 6SQ

HABITAT

Just south-east of the village of Balcombe and around 4km north of Haywards Heath, nestled in among 80ha of High Weald ancient woodland, lies Ardingly Reservoir. In addition to being a crucial source of drinking water for Mid-Sussex, the site is also a popular destination for day trippers; be they watersports enthusiasts, dog walkers or simply sun worshippers, who congregate around the banks and shoreline in summer. The southern section has been designated as an LNR.

Although somewhat eclipsed by the larger and more well-watched Weir Wood Reservoir a few kilometres to the north-east, Ardingly is still attractive to birds, although disturbance does often mean they don't hang around quite as long as they might otherwise. The north-east arm of the reservoir, above the Mill Lane causeway, is owned and managed by Kew/Wakehurst as part of the Loder Valley Nature Reserve (site 15). Access to this area is for permit-holders only (see Access section).

SPECIES

Common wildfowl can be enjoyed here year-round, sometimes joined by a handful of Pochard in the winter and possibly a Goldeneye. While the depth of the
reservoir makes it rather unappealing to dabbling ducks, a few Mandarin will
usually be found, although never in any great number. Egyptian Geese, on the
other hand, can be numerous, with winter counts here of 70 or more not uncommon. Great White Egret is gradually increasing in its frequency as a visitor here
too. Great Crested Grebes are present year-round.

The winter gull roost at Ardingly can hold thousands of individuals (sometimes
3,000+), mostly Black-headed and Herring Gulls, although usually a few Lesser
Black-backed and Common Gulls, and sometimes small numbers of Great
Black-backed and Mediterranean Gulls.

As with any large inland waterbody, the site tends to come alive in the passage
months, when a headwind and drizzly weather may push down large numbers of
hirundines and perhaps a Little Gull. Common Terns nest on a purpose-built raft.
Garganey has occurred on occasion. Common Sandpiper is the most likely wader
to see here on passage; other species are possible but not to be expected.
Lapwings can be expected in small numbers in later summer to autumn.

Ardingly is a reasonably reliable place to find an Osprey, especially in late
summer and autumn when southbound birds may linger in the area for a day or
more. Hobby may be spotted in the summer too. Other raptors can be seen
year-round, including Peregrine and Goshawk, which is becoming an increasingly
common sight in this region.

The site is not particularly known for being a passerine hotspot, though all the
expected species of woodland and scrub breed in the surrounding landscape,
including Marsh Tit and Garden Warbler. Wood Warbler historically bred in this area
and there have been occasional passage records of the species in recent years, so
it is one to keep in mind. Reed Warblers breed in the reedbed, which also holds a
winter Reed Bunting roost. Kingfishers can be seen at any time of year.

TIMING

Winter is best for flocks of wildfowl and the gull roost. Summer for hirundines and
breeding waterbirds. The passage season is best for finding waders, terns or a
passing Osprey. Bear in mind this site can become very busy on sunny weekends
and summer holidays, so an early visit is recommended if possible.

ACCESS

The nearest railway station is Balcombe, while bus route 272 stops at Hapstead
Hall in Ardingly village, about a 25-minute walk from the reservoir entrance/car
park. If travelling by car, take the turning off College Road about 1km south of
Ardingly village. This will take you onto Shell Brook which eventually winds its way
to the reservoir car park.

The 4km Kingfisher Trail from the car park winds its way up the east side of the
reservoir from the car park – note there is no circular trail right the way around. The
north-east arm is accessible to permit-holders only, which can be obtained from
Wakehurst (site 15), although part of it can be viewed from Mill Lane.

The dam end of the reservoir is well surfaced and fairly flat, other parts of the trail
are rather steep and can be muddy in winter.

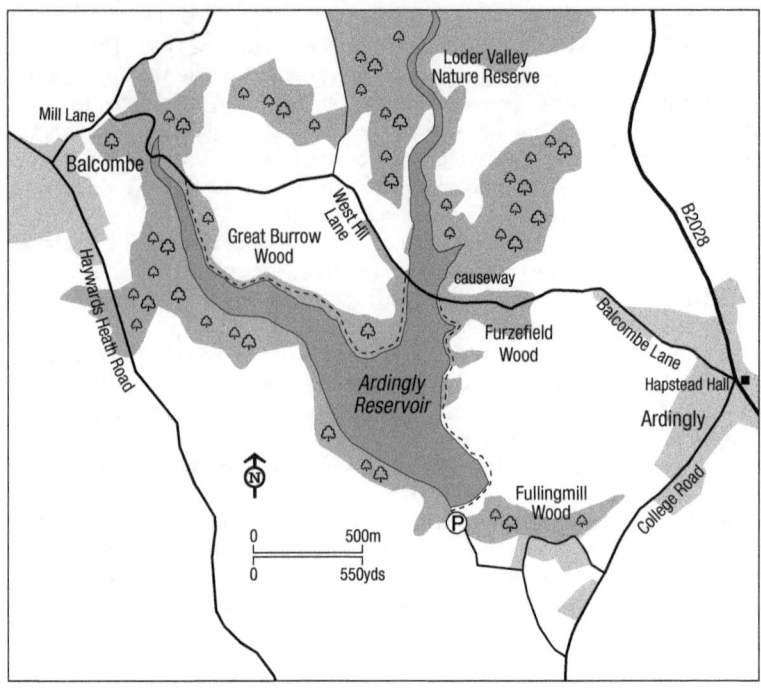

FACILITIES: The car park is open from 9:00 am to 5:30 pm from 31st March to 30th September and 9:00 am to 4:00 pm in winter. Parking is also possible on Mill Lane/West Hill, heading west from Ardingly towards Balcombe, which is useful if the main car park is full and also allows quicker access to the north section of the reservoir (generally the best bit for birds). An accessible car park for blue badge holders (accessed with a RADAR key) is available nearer the reservoir. A café and toilets are located in the Ardingly Activity Centre in the summer months. Otherwise the nearest facilities are in Ardingly village.

CALENDAR

All year: Egyptian Goose, Mandarin Duck, Little Grebe, Great Crested Grebe, Little Egret, Goshawk, Kingfisher, Raven, Marsh Tit, Grey Wagtail, Yellowhammer, Reed Bunting (roost in winter).

April–September: Lapwing, Common Sandpiper, chance of more unusual waders dropping in, Common Tern, chance of passage tern species passing through, Hobby, warblers including Sedge, Reed and Garden, common passage migrants – Yellow Wagtail, Wheatear, etc.

October–March: chance of White-fronted Goose, large numbers of gulls in roost – chance of Mediterranean, Great Black-backed or possibly something rarer.

87 WORTH FOREST

OS Explorer OL34
OS grid ref: TQ 310350
Postcode: RH10 4SD

HABITAT

Just to the south of the M23, and a stone's throw from the hustle and bustle of Crawley town centre, lies an enticing patchwork of woodland, with Worth Forest forming the largest contiguous block. Different parts of this forest have their own names such as Worthlodge Forest, Cowdray Forest and Oldhouse Warren. An area of ghyll woodland of just under 44ha in the south-western part is a designated SSSI. Although classified as ancient woodland, much of the woodland is conifer plantation, though there is a rich mix of broadleaved species and numerous veteran trees, mostly oaks and Beech. There is an extensive network of rides and areas of the plantation are periodically clearfelled, creating additional habitat.

Once a hunting forest for Saxon kings, the area now supports a wealth of birds and other wildlife. Some interesting insects can be found here in the summer, including Glow-worms, White Admiral, Silver-washed Fritillary and Purple Hairstreak butterflies, as well as Keeled Skimmer and Golden-ringed Dragonflies.

SPECIES

The area offers a pleasing selection of woodland species, especially in summer and during migration periods. Worth Forest was one of the last reliable spots in West Sussex for Willow Tit, but these sadly no longer occur here. Marsh Tit is common though, and its sneezing calls regularly soundtrack a walk in the area. Firecrest, Willow Warbler, Chiffchaff, Garden Warbler and Blackcap breed and a few Firecrests can be found in the winter too. Several pairs of Redstart breed in the south-western section. Tree Pipit once bred, so may still occur from time to time on passage. Spotted Flycatchers are present in late summer and may breed. Grey Wagtail will often be encountered along the various streams that criss-cross the forest. Crossbills sometimes nest and may be sighted at any time during influx years. In winter, look out for Lesser Redpoll, Brambling and Hawfinch, all of which may occur when there is good food availability. The south-western part of the forest occasionally produces records of Lesser Spotted Woodpecker (Cowdray Forest is best), so there is always a chance of meeting this delightful but declining species.

Higher, open areas are good for raptor watching, with a fine day in late winter or early spring very likely to produce Goshawk, along with Buzzard, Sparrowhawk, Red Kite and sometimes Peregrine. A good location to scan across the forest is from the Worth Way track to the north of Worthlodge Forest. Woodcock is present year-round, with the breeding resident population supplemented by migrants in the winter. A dusk watch from any viewpoint overlooking the forest may yield flight views of these enigmatic and elusive waders. Along the eastern boundary of Worthlodge Forest, south of The Burches, there are some areas of arable land that can be worth scanning in the winter as they can attract flocks of finches, winter thrushes, Skylarks, Meadow Pipits and, sometimes, Woodlark. Woodlark may also be encountered in the Whitely Hill area (TQ 306340) with up to 10 recorded here on occasion in the winter, and pairs have bred in some of the clearfell areas.

TIMING

Although potentially good at any time of year, visit in summer for the best selection of breeding species.

ACCESS

Worthlodge Forest is rather tricky to access, with the nearest available parking on Church Road near St Nicholas' Church at roughly TQ 301362. From there, follow the Worth Way footpath to the north-east, across the footbridge over the M23, past Worth Lodge Farm, then follow Standinghall Lane south until you reach the junction with another footpath that takes you back north-west. This 5.5km trail around and through Worthlodge Forest eventually comes out on the B2036, near where it goes over the M23, and from there you can walk the short distance back to Church Road. Another parking spot is at Worth Abbey School from where the footpath goes north into the forest at TQ 314343.

For access to Cowdray Forest, on the west side of the B2036, there is a moderately sized car park at TQ 306330, with various permissive footpath routes radiating out from there. This area is often very busy with people and dogs and best visited very early in the morning. Bus routes 100 and 643 (Metrobus) both stop on Allyington Way in Maidenblower, just a short walk from Church Road. Route 62 (Compass Travel) from Crawley to Haywards Heath makes a couple of stops along the Balcombe Road (B2036), not far from Cowdray Forest. The nearest railway station is in Crawley, around a 2km walk from the Church Road entrance to Worth Forest.

Unfortunately, the muddy and uneven nature of the paths does not make this a particularly favourable site for wheelchair or pushchair users.

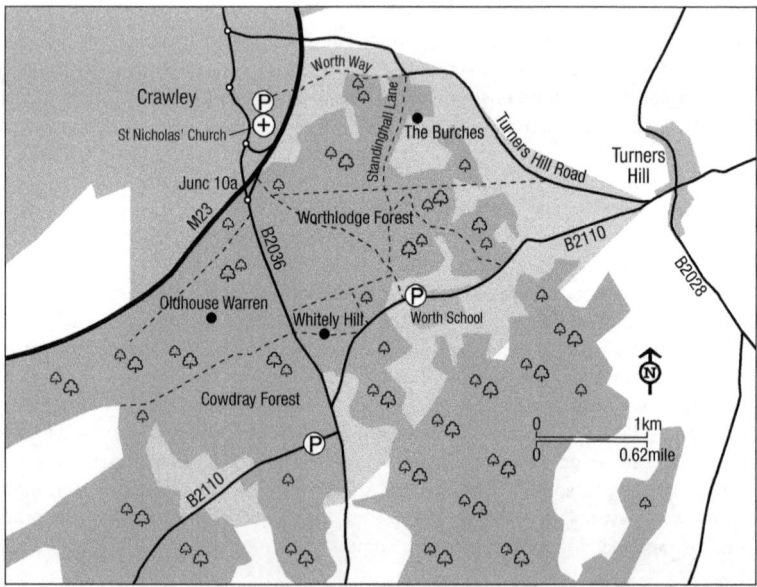

FACILITIES: The nearest facilities are in Crawley.

CALENDAR

All year: Woodcock, Goshawk, Red Kite, Lesser Spotted Woodpecker, Raven, Marsh Tit, Firecrest, Crossbill, Siskin.

April–July: Hobby, Spotted Flycatcher, Redstart.

October–March: Woodlark, Brambling, Hawfinch, Lesser Redpoll.

88 BUCHAN COUNTRY PARK

OS Explorer OL34
OS grid ref: TQ 240342
Postcode: RH11 9HQ

HABITAT

With its diverse mix of woodland, heathland and ponds, the 69ha Buchan Country Park offers a surprisingly rewarding nature-watching experience, especially given its location immediately on the outskirts of Crawley. The two main ponds, Douster Pond and Island Pond, attract a reasonable selection of waterbirds which can be enjoyed at close quarters. The ponds here are rich in invertebrate life, including 23 dragonfly species at the time of writing. Purple Emperor butterflies can also be seen in June and July. Target Hill Heath to the eastern edge of the park is an interesting pocket for heathland; it has benefited from many years of restoration work and is now home to breeding Nightjars, as well as Adders and other reptiles.

SPECIES

The first port of call once you arrive will likely be Douster Pond, as it greets you immediately as you cross the bridge from the main car park and entrance. This and the slightly smaller Island Pond to the south host a selection of waterbirds including Mandarin Duck, Kingfisher and Great Crested Grebe. Grey Wagtails are a common sight around the outflow.

The woodland areas around the main trails are popular with all manner of common resident woodland passerines. In the summer, these will be joined by Chiffchaff, Willow Warbler, Blackcap and sometimes Spotted Flycatcher, and Siskin and Lesser Redpoll in the winter. Lesser Spotted Woodpecker is still recorded here on occasion, but is characteristically elusive and best looked or listened for in early spring before the leaves emerge.

Nightjars breed on the heathland at Target Hill and Tree Pipits still turn up occasionally on passage. Buchan also held breeding Wood Warbler until the early 2010s, so there is a small chance of a passage bird passing through. Marsh Tits are reasonably common and can often be seen visiting the bird feeders behind the visitor centre.

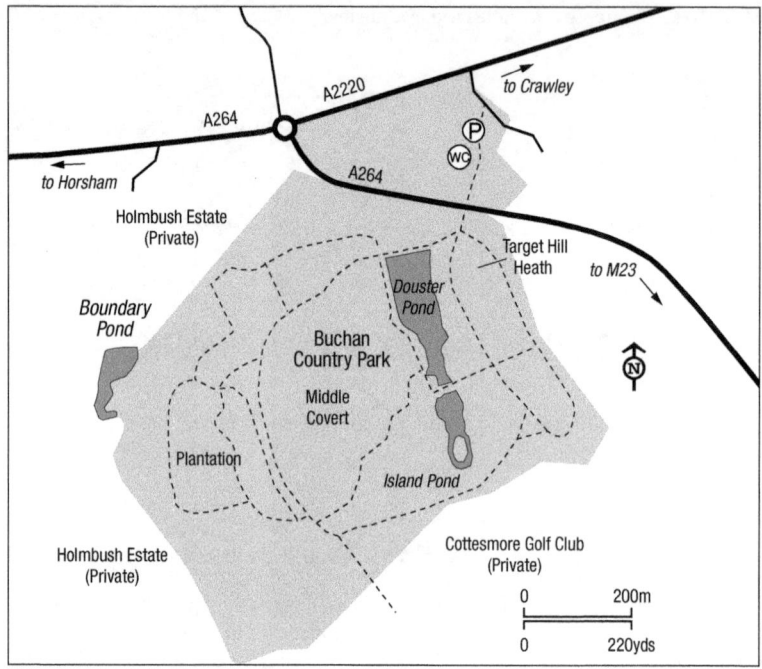

TIMING

The car park and toilets are open every day from 8:00 am to 9:00 pm, including Christmas Day and bank holidays. Note the car park can get very busy on weekends and bank holidays.

ACCESS

The nearest railway station is in Ifield, approximately a 2km or 40-minute walk away to the north. The Crawley circular bus route 1 (Metrobus) stops at various locations in the Broadfield area, while the Fastway bus 10 stops in Bewbush, from where it's just a 10-minute walk on well-signposted tarmac footpaths to the country park. If travelling by car, the car park and entrance are accessed from a narrow minor road on the south side of the A2220. From the car park, follow the main path south, over the bridge across the A264, then you will be in the CP proper and will see Douster Pond just ahead of you.

The main trails are all hard surfaced, suitable for wheelchairs and pushchairs, and well signposted.

FACILITIES: There is a car park (open 8:00 am–6:00 pm or 9:00 pm in the summer, free/voluntary donation), a visitor centre with coffee kiosk and toilets (open Wednesday–Sunday, 9:00 am–4:00 pm), an accessible toilet and baby-changing facilities, picnic benches and barbecue stands.

CALENDAR

All year: Mandarin Duck, Great Crested Grebe, Kingfisher, Marsh Tit, Firecrest, Grey Wagtail.

April–September: Nightjar, Willow Warbler, Tree Pipit.

October–March: Lesser Redpoll, Siskin.

89 WARNHAM LOCAL NATURE RESERVE

OS Explorer OL34
OS grid ref: TQ 167323
Postcode: RH12 2RA

HABITAT

Situated just 3km north-west of Horsham town centre and owned and managed by Horsham District Council, Warnham was designated as a Local Nature Reserve in 1988. Though relatively small at just 37ha, the site offers a diverse selection of habitats. The main feature is the 7ha mill pond, which is surrounded by woodland, reedbeds and marginal vegetation. A well-surfaced, level Discovery Trail meanders around the reserve and there are five hides and two viewing screens, including one positioned in front of a feeding station, allowing for accessible and close viewing of a range of species. Indeed, Warnham is a popular destination for photographers hoping to capture close shots of otherwise elusive species such as Water Rail, Marsh Tit and Kingfisher.

SPECIES

For its size, Warnham offers a pleasant selection of bird species that are attainable on a fairly short walk. The mill pond attracts a multitude of wildfowl including Shoveler, Gadwall, Pochard and Wigeon in the winter, plus Little Grebe and Great Crested Grebes. Common Terns breed on the mill pond in the summer, when Reed Warblers and Reed Buntings can be found nesting in the reeds. Grey Heron also breeds and Kingfisher is a familiar sight. Another resident is Water Rail – this usually shy bird can be uncharacteristically showy from the hides and viewing screens.

The muddy margins of the mill pond sometimes attract waders, most frequently Little Ringed Plover and Common Sandpiper in the spring and Snipe in the winter, although Jack Snipe, Dunlin, Avocet, Greenshank, Curlew and Whimbrel have been recorded on occasion. Bittern has occasionally wintered but is not to be expected, in keeping with its rare status in West Sussex.

The woodland areas offer a range of passerine species, from migrant warblers such as Blackcap, Garden Warbler and Whitethroat to common residents such as Marsh Tit and Bullfinch. Siskins and Lesser Redpolls frequent the areas of Alders in the winter, particularly to the north and east of the mill pond. The woodland feeding stations are a good place to get close views of various species, including Reed Bunting, Marsh Tit, Chaffinch and sometimes Brambling in the winter.

Recent scarce species to occur here include Scaup, Tundra Bean Goose, Iceland Gull and a long-staying Little Bunting.

TIMING

The reserve and car park are open 10:00 am–5:00 pm, April to September, and 10:00 am–4:00 pm, October–March (closed Christmas Day and Boxing Day). At the time of writing, entry to the reserve is £3 per person, or an annual membership allowing unlimited entry is £18.

ACCESS

There is a good-sized car park just off the B2237. The reserve is just a half-hour walk or 15-minute cycle ride from Horsham Town Centre. The nearest railway station is Horsham and bus route 51 from Horsham stops at Redford Avenue/ Saxon Crescent, which is just a 10-minute walk from the reserve. All the main trails are wide, level and well surfaced which, combined with the viewing screens and hides offering close views of the birds, make this an ideal reserve for wheelchair and pushchair users. Dogs are not allowed on the reserve.

FACILITIES: Facilities include a free car park, visitor centre, café, toilets, accessible toilets, baby changing, picnic area and outdoor seating.

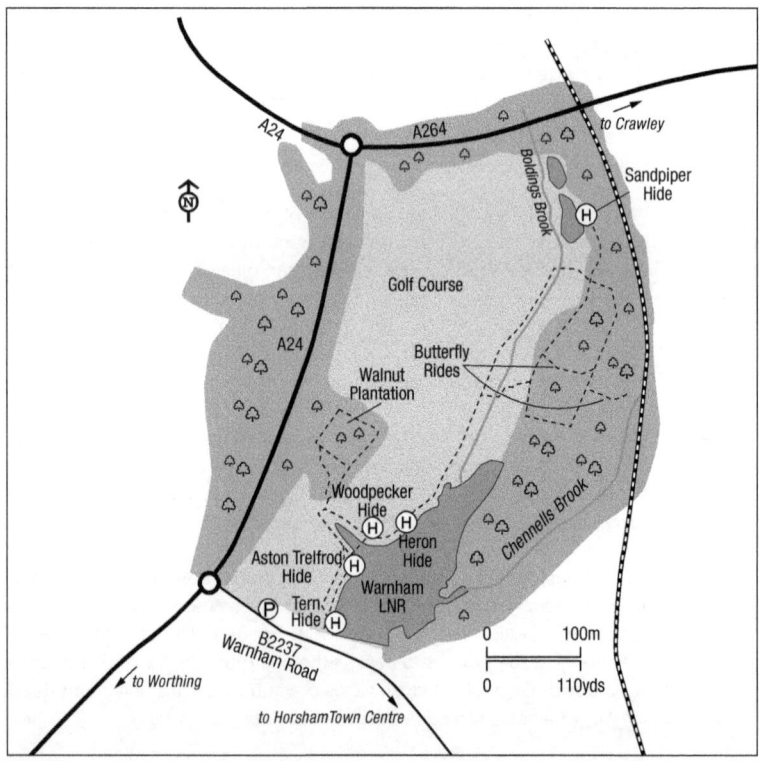

90 KNEPP CASTLE ESTATE

OS Explorer OL34
OS grid ref: TQ 150204
Postcode: RH13 8NN

HABITAT

The Knepp rewilding project a few kilometres south of Horsham is arguably one of the most exciting and ambitious nature recovery experiments in Europe. Much of the 1,400ha estate has transformed from intensive arable farmland to a mix of scrub and savannah-like wood-pasture, driven by the removal of fences and the introduction of free-roaming large herbivores such as longhorn cattle, Tamworth pigs, Exmoor ponies and Red and Fallow Deer.

The estate has been owned by the Burrell family for generations and the current owners, Charles Burrell and his wife Isabella Tree, made the decision in the early 2000s to transform their land and turn previous farming efforts on their head. Fields that were formerly monocrops of wheat and barley are now bursting with wildflowers and scrub, while the previously flailed farm hedgerows have been left to grow out, providing an incredible display of blossom in spring and summer followed by a rich berry banquet in the autumn.

The estate is split into three blocks, all offering a degree of access via public and permissive footpaths (25km in total). It is the Southern Block, however, which offers the most open access and is generally the destination of choice for walkers and wildlife watchers, as it provides the best all-round Knepp experience.

A large Hammer Pond in the Southern Block is viewable from the public footpath that runs north from New Barn Farm to Hammer Farm, at approximately TQ 147208. It can be viewed from the footpath which runs east–west from Shipley to the A24, and there is also a hide on the south-west side at TQ 156212.

The rewilding project has been running for two decades now and in that time has seen some spectacular booms in wildlife, including several red-listed bird species and three-quarters of the UK bat species, but also a significant population of Purple Emperor butterflies, Vagrant Emperor dragonflies and a host of other rare and scarce invertebrates.

SPECIES

Knepp is perhaps most famous for three bird species: Turtle Dove, Nightingale and White Stork. The latter were introduced to the site in 2016, when rehabilitated individuals were shipped in from Warsaw Zoo. These soon attracted the attention of passing wild birds, at least one of which paired with a released bird and, in 2020, the species bred on-site for the first time. Turtle Dove can be elusive but is generally relatively easy to find in the spring and early summer, usually located by its song. Both the Hammer Pond and the tree platform at Tory Copse (TQ 148197) are worth spending some time at to give the best chance of hearing or possibly seeing one. Dozens of pairs of Nightingale breed all over the Southern Block, again most easily found by their song in spring. *Sylvia* warblers proliferate in the dense scrub, with the songs of Blackcap, Whitethroats, Lesser Whitethroats and Garden Warblers all much in evidence in the spring. The abundance of breeding passerines, especially Dunnock and Reed Warbler, attract good numbers of Cuckoos which all add to the rich soundscape from April to June. Indeed, the volume of the dawn chorus at Knepp has been described as rattling in the listener's ribcage!

Both the Hammer and Mill Ponds attract a variety of wildfowl such as Gadwall, Tufted Duck, Shoveler, Wigeon and Teal, as well as Great Crested Grebes and Kingfishers. A hide is situated towards the south-west corner of the Mill Pond. The scrape in the Southern Block can be viewed from the tree platform at TQ 138205 and sometimes attracts waders such as Green Sandpiper on passage or in the winter, when Snipe and Woodcock may also be flushed from wet areas. The Mill Pond is the site of the third-largest heronry in Sussex, so Grey Heron is an ever-present feature, while Great White Egret is becoming an increasingly regular visitor and there is always the chance of a passing Osprey in spring and autumn. Many birds of prey including Buzzard, Red Kite, Sparrowhawk, Kestrel, Hobby and Peregrine all breed on the estate too, and there's always the possibility of a Goshawk as the species continues to increase in West Sussex.

Knepp has attracted quite a few scarcities and rarities in recent years including Whooper Swan, Black Stork, Golden Oriole, Hoopoe and lingering Red-backed Shrike. The latter has been mooted as another reintroduction species here.

TIMING

For the most impressive dawn chorus experience, an early morning visit in early May is recommended. All the public and permissive footpaths are open for walkers all year-round, although be warned that the whole site gets very wet and muddy in the winter so wellies are essential!

ACCESS

Knepp is not currently well served by public transport routes. The nearest railway station is at Horsham, some 13km away. Bus route 23 from Worthing to Horsham regularly stops along the A24 at Buck Barn, but this is around a 45-minute walk from the site. The safari base and campsite are at New Barn Farm off Swallows Lane near Dial Post (RH13 8NN). Parking is available here for campers and safari-goers, while for walkers and day visitors there is a car park at Swallows Farm (TQ 156200). Maps of the footpaths can be obtained from the campsite reception at New Barn Farm.

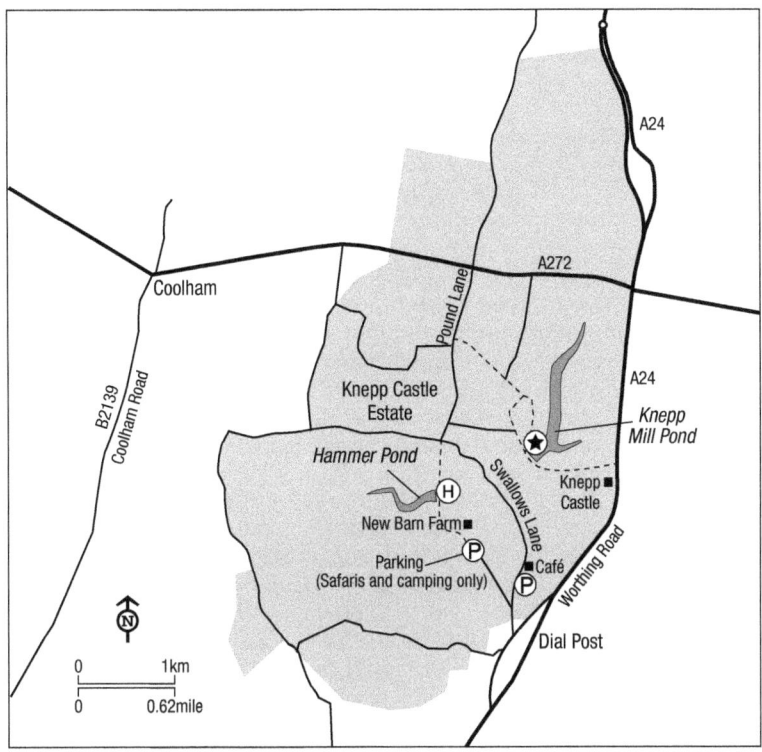

Many of the main trails around the Southern Block are hard-surfaced and suitable for wheelchairs and pushchairs, although please note cycling is only permitted on the bridleways.

FACILITIES: There is a farm shop, café and toilets at Swallows Farm just north of Dial Post, along with ample car parking (£5 donation for a day at the time of writing). There are also charging points for electric vehicles here.

CALENDAR

All year: Gadwall, Great Crested Grebe, White Stork, Little Owl, Barn Owl, Raven, Yellowhammer.

April–August: Turtle Dove, Cuckoo, Hobby, Reed Warbler, Garden Warbler, Lesser Whitethroat, Spotted Flycatcher, Nightingale.

October–March: Shoveler, Wigeon, Teal, Pochard, Snipe, Woodcock.

91 LAVINGTON COMMON AND PLANTATION

OS Explorer OL10
OS grid ref: SU 949188
Postcode: GU29 0QB

HABITAT

Situated some 4km south-west of Petworth, Lavington Common is a precious island of lowland heath surrounded by conifer plantation. The common and plantation combined cover 65ha and the northern section has been designated as an SSSI due to the quality of the heathland and its rare lichen communities. The plantation south of the road was planted with Scots Pine in the 1950s and is now gradually being cleared and restored to heathland by the National Trust, which took ownership of the site in 2000.

SPECIES

Most of the heathland specialist birds can be encountered here, depending on the timing of your visit. Dartford Warbler and Stonechat may be found all year, especially in the newly cleared areas of the plantation, which are also attractive to Woodlark. Tree Pipit, Nightjar and Woodcock all breed, with the latter two species usually encountered on summer evening visits, which may also produce the begging calls of juvenile Tawny Owls, as this species breeds in the wooded edges. The woodland areas also support Marsh Tit, Firecrest and Goldcrest, while both Snipe and Jack Snipe may be flushed from boggier areas out in the open. Finch flocks will often be found in the wooded edges in the winter, especially Lesser Redpoll and Siskin, but also sometimes Brambling and Crossbill in influx years. More unusual species such as Ring Ouzel and Pied Flycatcher have occurred on migration, on occasion, and Great Grey Shrike has wintered.

TIMING

As with most heathlands, spring and summer tend to be the most productive, as it can get very quiet in the winter months. A visit in April will likely produce singing Dartford Warbler, Woodlark, Tree Pipit and Stonechat, while a slightly later evening visit, in May or June, should yield churring Nightjar and roding Woodcock. This site is fairly popular with dog walkers so early morning visits are recommended (unless you're hoping for Nightjar).

ACCESS

There is a free car park on the minor road that runs between the common and the plantation. From the A285, turn west onto the road signposted towards Graffham. Follow this road for 1.5km, then look out for the car park on the right-hand side. Bus route 99 stops nearby, although must be arranged by pre-booking. Phone 01903 264776 at least 15 minutes before departure and ask the driver to let you alight at the route 412 stop for Lavington Common. Some paths can get very boggy in the winter, so wellies are advisable! Due to the narrow paths and lack of any hard surfacing here, it is not a suitable site for wheelchair or pushchair users.

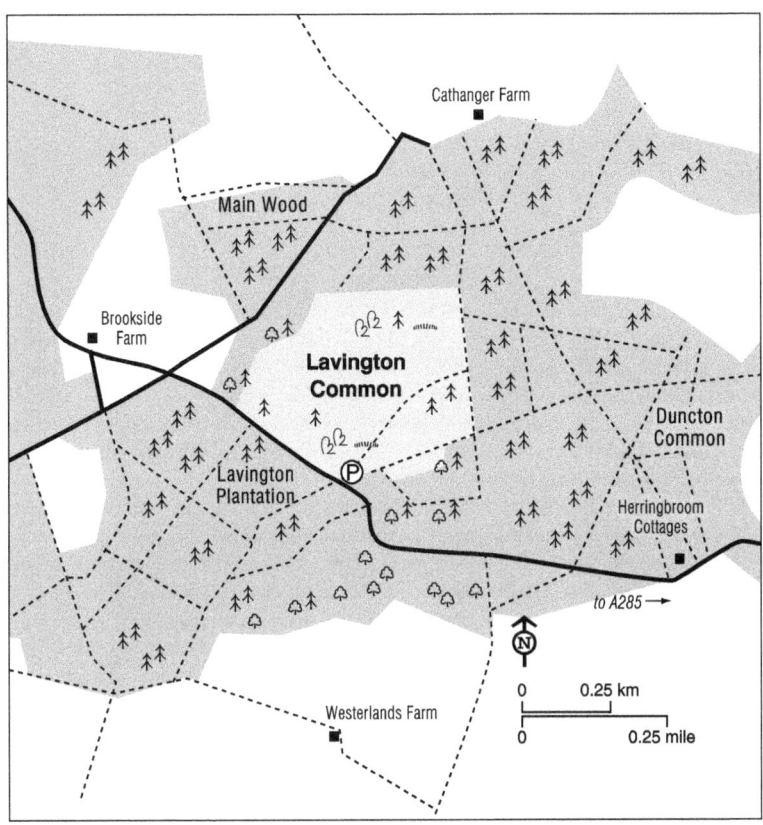

FACILITIES: The nearest facilities are in Petworth.

CALENDAR

All year: Woodcock, Tawny Owl, Raven, Marsh Tit, Woodlark, Firecrest, Stonechat, Siskin, Yellowhammer.

April–September: Cuckoo, Nightjar, Willow Warbler, Garden Warbler, Tree Pipit.

October–March: Snipe, Jack Snipe, Brambling, Lesser Redpoll, Crossbill.

92 AMBERSHAM AND HEYSHOTT COMMONS

OS Explorer 121
OS grid ref: SU 912194
Postcode: GU29 0DA

HABITAT

Owned by the Cowdray Estate, the 140ha of Ambersham and Heyshott Commons represent part of the mere 1 per cent of remaining heathland cover in the South Downs National Park and lie just east of halfway along a disjointed line of remnant heaths stretching from Pulborough to Bordon in Hampshire. Although a little tricky to find, being situated down windy country lanes a couple of kilometres south-east of Midhurst, this area is very much worth the effort for some of the finest heathland birding on offer in West Sussex.

The pockets of largely dry heathland here are home to the full suite of specialist bird species in the breeding season, but the sites also support populations of Sand Lizards and Glow-worms and an impressive community of plants, including Lesser Skullcap, Cotton-grass, Bog Pimpernel and Oblong-leaved Sundew.

The commons are surrounded on all sides by mixed tree cover, with a block of largely coniferous woodland separating the two, just to the east of New Road. The whole landscape is dotted with interesting historical features, from a dismantled railway to the north, a Second World War pillbox, a disused sandpit to the south side and five bowl barrows on Heyshott Common which are scheduled ancient monuments, thought to date back as far as 2400 BC.

SPECIES

By mid-April, Ambersham and Heyshott Commons are filled with the sound of a host of heathland species including Woodlark, Tree Pipit, Stonechat, Redstart and Dartford Warbler. Cuckoos are often heard and sometimes seen too. Dusk visits from May to August will almost always produce the distinctive churring of Nightjar. Similarly, keep an eye and ear out for the roding flight of a Woodcock patrolling the boundaries of its territory, though this species is so sadly in decline on the West Sussex heaths. Tawny Owls can be heard duetting in the wooded areas and a Hobby can sometimes be seen hawking for insects or even bats, right up until last light. Ravens are an increasingly common sight and sound here now, as the species continues to increase across the south-east.

Marsh Tit can be encountered in the more wooded areas, most easily located by their sneezing call. Listen out too for the high-pitched song of Firecrest. Spotted Flycatchers can be found in the summer, and it's always worth scouring through mixed flocks of passerines in late summer for a Pied Flycatcher. Other songbirds one can expect to encounter here include Yellowhammer, Whitethroat, Blackcap, Willow Warbler and Chiffchaff.

In the winter months the landscape falls largely silent, although little parties of Stonechats may still be found, perhaps with a Dartford Warbler in tow. Redwings and Fieldfares may be seen feasting on berries and flocks of finches assemble in the wooded areas including Siskin, Lesser Redpoll, Chaffinch and sometimes

Brambling and Crossbill. Hen Harrier is occasionally sighted in the winter and, rarely, a Great Grey Shrike.

TIMING

Spring and summer for the best selection of heathland species. Early morning is ideal to beat the dog walkers.

ACCESS

Access to the area is challenging by public transport, with no train stations nearby. Midhurst Community Bus route Y1 stops at Down Farm in Hoyle, just 500m south of Heyshott Common, but note this service only runs on Mondays. By car, follow the A272 and take the turning just east of Benbow Pond (SU 913222), south onto Ambersham Hollow Road, which becomes New Road just south of the hamlet of South Ambersham. There are four car parks dotted along New Road, with the largest being the most northerly one, nearest the disused railway. Footpaths from here are clearly marked, taking you either west onto Heyshott Common or east onto Ambersham Common.

The paths are all unsurfaced, so it's worth noting that the sandy soil is unsuitable for wheelchairs and pushchairs.

FACILITIES: Aside from the car parks, all other facilities can be found in nearby Midhurst.

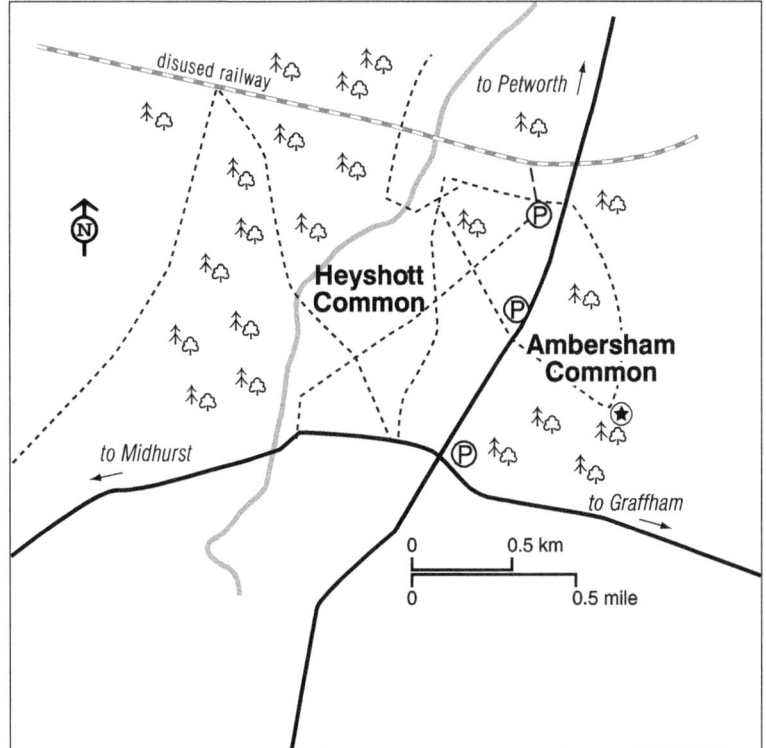

93 IPING AND STEDHAM COMMONS

OS Explorer OL33
OS grid ref: SU 852220
Postcode: GU29 0JR

HABITAT

Iping and Stedham Commons, along with Ambersham, Heyshott and Woolbeding, comprise the most extensive area of heathland in West Sussex, and a rewarding experience awaits any visiting birder exploring the mosaic of wet and dry and wooded and open heath, *Sphagnum* bog, grassland and scrub. Stedham Common is rather more wooded than Iping, but a programme of tree and scrub clearance by Sussex Wildlife Trust since the 1990s has seen a substantial amount of heather regeneration take place.

First designated as an SSSI in 1954 and updated in 1980, the sites are not just of note for their birdlife, although they do offer the full suite of heathland birds to be found in this part of the country. Species such as Field Cricket and Heath Tiger Beetle were reintroduced in the early 2000s and are now well established. Silver-studded Blue butterflies can also be found here from late May to early July.

SPECIES

Dartford Warbler, Stonechat, Siskin and Woodlark can be encountered at all times of the year, with the latter singing as early as January during clement weather. Iping and Stedham Commons really come alive, though, in spring and summer with the return of the various migratory breeding species. In March and April, the resident songbird chorus is gradually supplemented by the arrival of Chiffchaff, Blackcap, Willow Warbler, Whitethroat and Garden Warbler. Nightjars return in May and put on excellent displays churring and wing clapping on fine evenings throughout the summer. Tree Pipits are a regular sight and sound, giving their parachuting display flight from April into the summer months. Redstart has sadly fallen from the list of breeding species here in recent years, but you may still be lucky enough to find one.

In the autumn and into winter, good finch years can produce arrivals of Brambling, Lesser Redpoll and even Hawfinch to the area. Crossbill can sometimes be numerous here in influx years too, when some may stay to breed. Where there are Rowan trees, such as at SU 840219, there is always a chance of Ring Ouzel stopping off on migration to feed in October. Great Grey Shrike is a possibility in the winter months, though is rare.

There is no great list of rarities found here, owing more to the area being under-watched than anything else, but Little Bunting has wintered before, a record that stands as testament to the quality of the site.

TIMING

Spring and summer are best for the full set of heathland species. As with all commons in this region, Iping and Stedham can be busy with dog walkers on weekends and bank holidays.

ACCESS

By car, the reserve car park and main entrance lie just a couple of kilometres west of Midhurst, on the Elsted Road just off the A272. The nearest railway station is at Petersfield, around 10km to the west. Stagecoach bus routes 91 and 92 both travel from Petersfield to Midhurst and stop at Iping Lane on the A272 just opposite the junction of Elsted Road, a short walk away to the north of the car park.

As with other heathland sites in the region, the unsurfaced paths here can be muddy in winter and are not generally suitable for wheelchair or pushchair users.

FACILITIES: The nearest facilities are in Midhurst.

CALENDAR

All year: Raven, Marsh Tit, Woodlark, Dartford Warbler, Stonechat, Crossbill, Siskin, Yellowhammer.

April–September: Cuckoo, Nightjar, Woodcock, Hobby, Willow Warbler, Garden Warbler, Spotted Flycatcher, Redstart, Whinchat, Wheatear, Tree Pipit.

October–March: Great Grey Shrike, Brambling, Hawfinch, Lesser Redpoll.

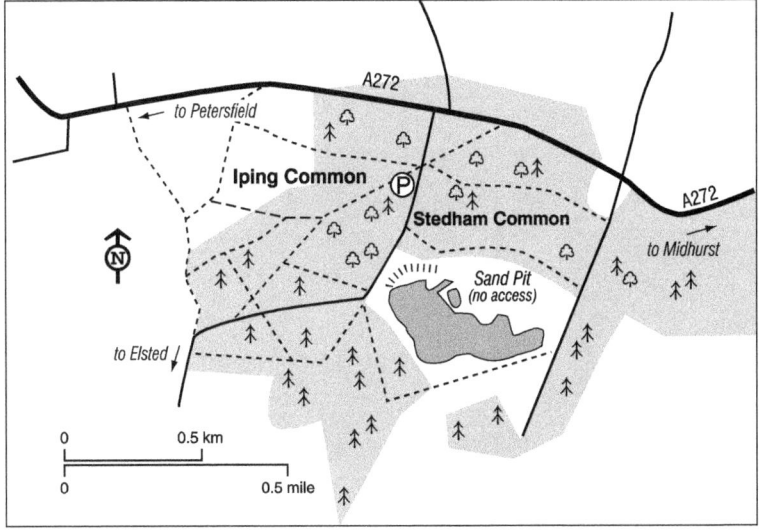

94 WOOLBEDING COMMON

OS Explorer 133
OS grid ref: SU 869259
Postcode: GU29 9RR

HABITAT

Woolbeding and Pound Commons is a 171ha lowland heath SSSI situated just to the north of Midhurst on the western Wealden Greensand, managed in part by the National Trust. Heather and gorse take up much of the heath, which is grazed by cattle and is home to a variety of interesting invertebrates. Part of the Serpent Trail – a 103km long-distance footpath from Haslemere to Petersfield – runs through the site. The commons are situated in a broader heavily wooded landscape. There are several scenic vistas, including one immediately west of the car park.

SPECIES

At the time of writing, this is one of the most reliable sites in the South-East for Honey Buzzard. The west-facing ridge west of the car park (see Access) offers far-reaching views and this is the place to watch from, preferably with a telescope. Late May to mid-June is the best time to visit, with a chance of seeing the species' iconic butterfly display. Normally only one or two birds will show, though occasionally there are more. Activity can drop throughout midsummer – although

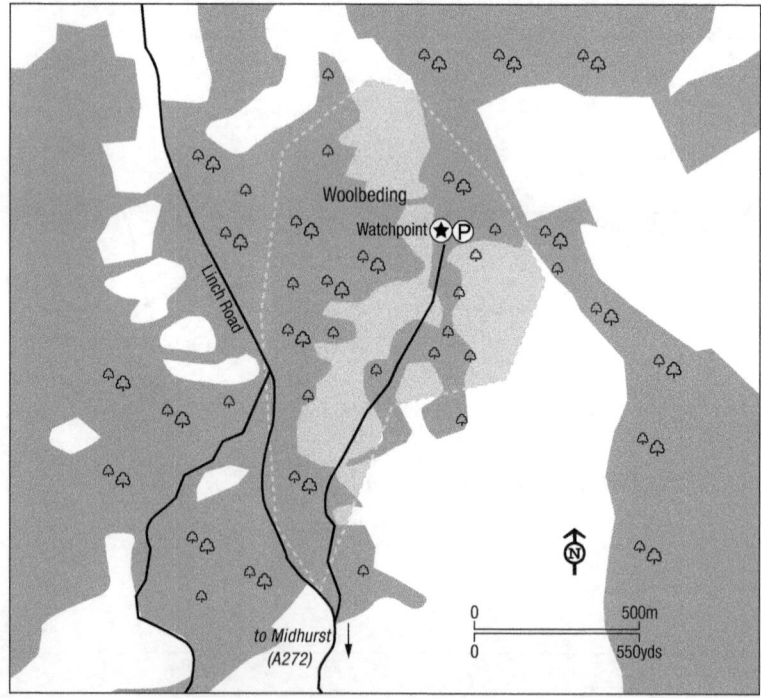

there's still a reasonable chance of seeing a bird in late June and July – and picks up again from mid-August through early September, when juveniles can be on the wing.

Other possibilities while skywatching include Goshawk, Hobby and Raven. In spring and summer, the wider heathland is home to Cuckoo, Nightjar, Woodlark, Dartford and Willow Warblers, Stonechat, Tree Pipit and Yellowhammer – all of which make a fine soundscape while waiting for raptors. Marsh Tit and Spotted Flycatcher may be found in the more wooded areas. Crossbill occasionally visits too, but is less likely in the summer.

TIMING

Honey Buzzards can sometimes be active early in the morning, but a good bet is to be in position from 9:30/10:00 am when birds become airborne. Display is most likely to occur late morning but this of course is variable. There is less chance of views from 3:00 pm or 4:00 pm onwards. It's best if it's dry and still, and ideally bright and warm.

ACCESS

From Linch Road, there is a track running north-east at SU 865250. Follow it until you reach the car park at SU 869260. From here the vantage to the west is obvious – sit anywhere on the open ridge (immediately opposite the car park is good), taking care not to disturb any habitat. A telescope is recommended. The proximity of the viewpoint to the car park means it's possible for those with limited mobility to find a suitable place to scan from.

This is quite a remote site and travelling by public transport is unfortunately not practical.

FACILITIES: The nearest facilities are in Midhurst.

CALENDAR

All year: Goshawk, Marsh Tit, chance of Crossbill.

April–August: Cuckoo, Woodcock, Honey Buzzard, Hobby, Woodlark, Dartford Warbler, Spotted Flycatcher, Tree Pipit, Yellowhammer.

95 BLACK DOWN

OS Explorer OL33
OS grid ref: SU 918301
Postcode: GU27 3BJ

HABITAT

One of the highest points in southern England – and the highest in Sussex at some 280m – Black Down is a large National Trust site with a range of mixed woodland, heath and scrub. Geologically part of the Greensand Ridge and historically a grazed common before Scots Pine took over, the National Trust now carries out a programme of tree-felling and controlled burning to maintain and regenerate areas of open heath. The reserve has been fenced so that conservation grazing with cattle can be used as a management tool.

The pine- and heather-covered slopes have numerous paths and one can spend many hours walking them. Given its elevation, there are multiple vantages from which to scan the area and wider countryside, with seats and benches dotted along the paths. The county border with Surrey bisects the area but the majority of the site is in Sussex.

SPECIES

With its elevation and multiple vistas, Black Down is a good place for skywatching. This is especially the case for raptors, of which a variety can be seen at different times of the year and with some patience. One of the best spots to scan from is the seat near the Serpent Trail at SU 917293, where a near 360-degree view of the wider area can be obtained. A telescope is recommended. Other seats at SU 913302 and SU 916296 are good options too, along with the 'reservoir track' at SU 915305. The Temple of the Winds offers a stunning view but is less ideally situated. From February through to April, fine spring-like conditions – ideally with a bit of a breeze and a little cloud – are useful for getting *Accipiter*s on the wing. Black Down is a good site for Goshawk and you have a fair chance of encountering this species here. From May through to August raptors are less fussy, but hot days with little wind can make for difficult viewing. Honey Buzzard is often present at this time of year, but birds can range widely at this site, so patience is required. A headwind can be useful for spotting passage species such as Osprey and Marsh Harrier, which are annual visitors.

From April, the classic heathland species of the region can be found in the open areas: Dartford Warbler, Nightjar, Redstart, Stonechat and Tree Pipit all breed and can be readily encountered. Marsh Tit and Spotted Flycatcher breed in the areas of deciduous woodland. Black Down is a reliable site for Crossbill, year-round, though they are best looked for in winter or during bumper periods for the species. Parrot Crossbill has occurred here in the past in a large flock of Crossbills. Black Down is an exceptional site for Firecrest – between April and June double-figure counts of singing males can easily be attained, especially in dense stands of holly on the slopes. A species that's gone in the opposite direction on the Greensand Ridge is Wood Warbler, and this delightful *Phylloscopus* species is now a rarity in Sussex. In some years a male can be seen here but it's not one to be expected.

Passerine passage can be decent in August and September, with Pied Flycatcher almost annual among the more regular Wheatear and Whinchat. From late September, Ring Ouzel moves through – Black Down is a great site for this thrush between then and the end of November. The aforementioned 'reservoir track' is one of the best places to seek it out, though any area with berry-laden vegetation is worth a check. Approach any feeding birds with caution as they are usually very skittish. Vis-mig can be productive, too, with movements of finches, thrushes and Woodpigeon detectable. Winter is quiet, though Great Grey Shrike will occasionally drop in.

TIMING

Spring or summer is the best time to visit. Suitable conditions from February through to August can be productive for raptors; a stakeout any time between 10:00 am and 3:00 pm is worthwhile. April through to June is best for the breeding heathland specialists; October is the best month for Ring Ouzel. Due to its elevation, windy days are best avoided. Black Down is a popular locale with the general public and so early mornings are ideal; it can be very busy on weekends, especially later in the day if the weather is fine.

ACCESS

By car, bike or on foot, head out of Haslemere on B2131, turn right up Haste Hill, follow onto Tennyson's Lane and head south-west until you come to the main free NT car park. Black Down is ideal for cycling with numerous easy-going bridleways, as well as more challenging terrain. Haslemere train station is only 3km away and has excellent connectivity with London and Portsmouth.

A circular trail from the car park is accessible for wheelchair users and pushchairs.

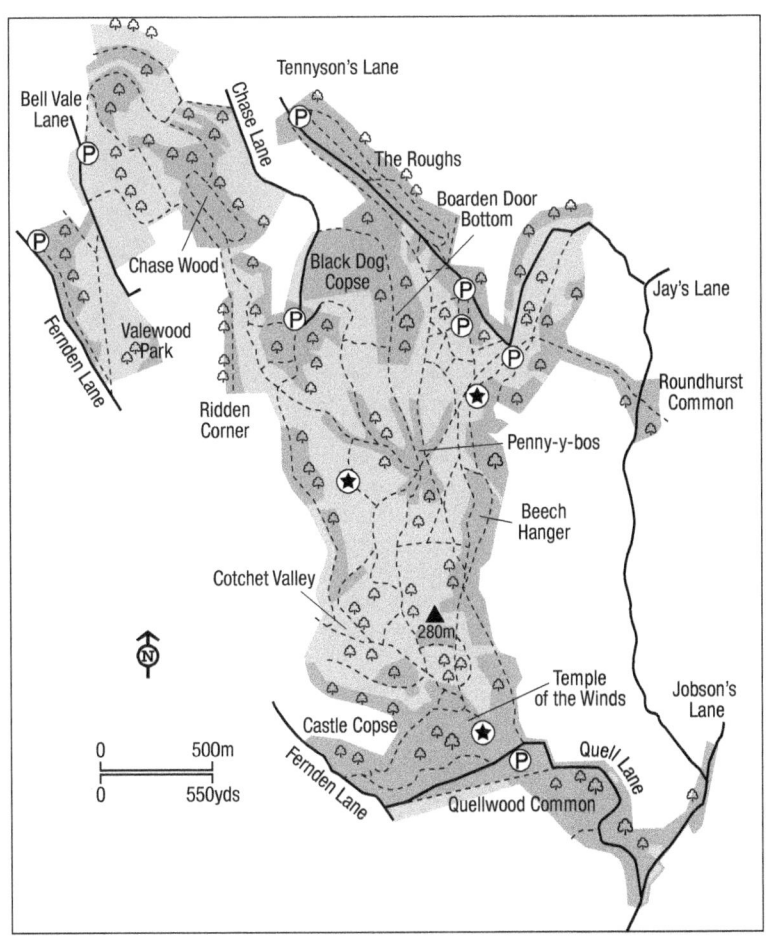

FACILITIES: Two free car parks are to be found on Tennyson's Lane (SU 920308 and SU 923306). Other free non-National Trust car parks can be found on Quell Lane and Bell Vale Lane. The nearest toilets and other facilities are in Haslemere.

CALENDAR

All year: Goshawk, Marsh Tit, Dartford Warbler, Stonechat, chance of Crossbill.

April–July: Cuckoo, Nightjar, Hobby, Redstart, Spotted Flycatcher, Firecrest, Tree Pipit.

August–November: passage raptors, migrant chats in August and September, Ring Ouzel from late September.

November–March: Brambling, Lesser Redpoll, chance of Great Grey Shrike.

96 EBERNOE COMMON

OS Explorer 133
OS grid ref: SU 976270
Postcode: GU28 9LD

HABITAT

A tranquil site nestled between the villages of Northchapel (site 12) and Balls Cross, Ebernoe Common is an SSSI that has been managed by the Sussex Wildlife Trust since 1980. Extensive ancient woodland pasture is dotted with ponds, streams, meadows and reclaimed arable land, which is being reconverted into woodland. It is nationally important for lichens, with over 100 species, and for fungi, with seven Red Data Book species. Butterfly species include Purple Emperor, Purple Hairstreak, White Admiral and Silver-washed Fritillary. The glades and rides support flora such as Devil's-bit Scabious, Adders'-tongue Fern, Sneezewort and various orchids, while the areas of meadows are brightened up by Quaking-grass, Pepper-saxifrage and Betony during the summer months.

Across the site there is evidence of historical industrial use including a brick-works, ponds, an iron furnace site, clay pits and a small quarry. There are also the remains of several cottages. To the south and the east are the Butcherlands fields and woods – these were added to the reserve in 2000.

SPECIES

Spring and summer are the best times to visit Ebernoe, with the number of species recorded breeding here sometimes approaching 50. It's an excellent site for Nightingale, with several males holding territory. Dawn chorus can be an uplifting experience, with singing Cuckoo joining the excellent populations of many common woodland birds along with good numbers of *Sylvia* warblers such as Garden Warbler, and Firecrest. A few Skylarks can be found in the wildflower meadows. There is also a chance of Turtle Dove, too, although this species is rare; Lesser Spotted Woodpecker is an outside possibility as well. There is a record of a Melodious Warbler singing on one early June morning at Butcherlands – so keep an open mind. Ebernoe is not well known for raptors but both Goshawk and Honey Buzzard have been recorded before.

Winter is much quieter. Marsh Tit is resident and vocal from late January. A Woodcock may be flushed from cover and the Furnace Pond may have a few Mandarin Duck. Finch flocks may include a few Bramblings and, as with any wooded site in the Weald, there is a chance of bumping into a Hawfinch.

TIMING

Spring and summer are the best times to visit. Early mornings are best for bird-song, with dusk also good for Nightingales. It rarely gets too busy here.

ACCESS

There is a car park at Ebernoe Church (SU 975278), off Streel's Lane, from which you can access the reserve. The ground is flat but can be muddy, so may not be suitable for wheelchair or pushchair users.

This is quite a remote site and travelling by public transport is not practical.

FACILITIES: The nearest facilities are in Northchapel.

> **CALENDAR**
> **All year:** Chance of Lesser Spotted Woodpecker, Skylark, Marsh Tit, Firecrest.
> **April–July:** Cuckoo, chance of Turtle Dove, Hobby, Nightingale, warblers.

OTHER SITES IN NORTH WEST SUSSEX

I1 STANLEY COMMON
Tucked right in the north-western corner of West Sussex, just a stone's throw from the Surrey and Hampshire borders and part of the wider 122ha Linchmere Commons LNR, the wooded heath of Stanley Common holds breeding Cuckoo, Woodlark, Tree Pipit, Redstart and Firecrest, as well as sometimes good numbers of Brambling and Crossbill in the winter. Goshawk is another species to keep an eye out for. There is a small car park at SU 848301.

I2 NORTHCHAPEL AREA
The farmland and woodland surrounding this village, which is situated along the A283 north of Petworth (SU 952292), hold some interesting species, including a small and elusive population of Turtle Dove. Skylark, Linnet and Yellowhammer are far more likely to be encountered. The odd Nightingale may be found in some of the scrubbier woodland, while Goshawk is possible at Frith Wood.

I3 LURGASHALL MILL POND
Although the old mill building is gone, the pond – created by damming of the River Lod centuries ago – remains. Kingfisher and Grey Wagtail may often be seen here, while Pochard sometimes occurs in the winter. The woodland edges may hold Lesser Spotted Woodpecker and Spotted Flycatcher in the summer. There is space to park a car or two on the minor road on the eastern side of the pond (SU 940259).

I4 PLAISTOW AND SHILLINGLEE
Close to the Surrey border, a little west of Plaistow, is Birchfield Copse and Kingspark Wood (SU 988307), which are part of the Chiddingfold Forest complex. This is a relatively reliable spot for Turtle Dove, though Nightingale, Firecrest and Marsh Tit are more likely. Three fishponds at Shillinglee Park include The Lake (SU 968310), which occasionally records Osprey on passage and has breeding Great Crested Grebes. There is limited parking on Shillinglee Park Road (SU 965321) and the various minor roads of Shillinglee Road and Dunsfold Road.

I5 WAKEHURST PLACE
This sister site of Kew Botanic Garden is also home to 60ha Loder Valley Nature Reserve. Ponds hold a selection of wildfowl year-round, sometimes including Great Crested Grebe and Kingfisher, and have even hosted Garganey and Shag on occasion. The woodland areas produce good selection of passerines including Marsh Tit and Firecrest, and can be good for Hawfinch in autumn and winter. There is a large car park (TQ 341316) off Selsfield Road and the nearest train station is in Haywards Heath.

I6 CHERRY LANE

This area of arable farmland near Balcombe can prove very attractive to finches in the winter, with triple-figure counts of Chaffinch and Linnet possible, and sometimes a few Bramblings too, plus thrushes, Woodlark and Yellowhammer. Red Kites roost in the winter and Hawfinches occur to the north-west of the parking area at TQ 304280. Great Bentley Farm flood meadows a short way east of here (TQ 315279) can be productive for waders, ducks and geese.

I7 THE MENS

This is a 205ha SSSI of ancient woodland 5km east of Petworth, largely managed by the Sussex Wildlife Trust. It has a real air of wildness and enchantment, rather likes parts of the New Forest or the Forest of Dean, and supports many woodland species including Goshawk, Marsh Tit, Tawny Owl, Firecrest and Spotted Flycatcher, as well as spectacular displays of fungi and butterflies such as Purple Emperor and White Admiral. Car parking is available at TQ 023237.

I8 FITTLEWORTH

The stretch of the River Rother through Lower Fittleworth can be productive for Kingfisher and Grey Wagtail, as well as the scarce Common Clubtail dragonfly. The wet meadows nearby can attract Snipe and Little Egret in the winter. Fittleworth Common (TQ 015188) to the north holds Marsh Tit and Firecrest. There is space for a few cars by the bridge at TQ 010183 and the nearest station is in Pulborough.

I9 PETWORTH PARK

This 283ha parkland (SU 969224) designed by 'Capability' Brown in the 1700s is now managed by the National Trust. Known for its ancient trees and herds of free-roaming Fallow Deer, the park also supports some interesting birdlife. The larger and more southerly of the two ponds, confusingly known as Upper Pond, regularly hosts wintering Goosander as well as Egyptian Goose, Great Crested Grebe and Tufted Duck. There are two car parks and some street parking available along the A272 to the south.

I10 GRAFFHAM COMMON

This relatively small site between Lavington Common (site 91) and Ambersham and Heyshott Commons (site 92) is the Sussex Wildlife Trust's newest reserve and is being restored from pine plantation to heath. A good selection of heathland species occur, including Nightjar, Woodlark, Woodcock, Stonechat, Cuckoo, Yellowhammer and Hobby. A visit can be combined with a look at nearby Graffham Down (SU 924163), which can be good for Hawfinch and migrant passerines (although it is steep!). Limited parking is available along the road between Selham and Graffham (SU 931192).

I11 HARTING DOWN

This National Trust site is one of the largest areas of ancient chalk grassland in Sussex and is designated as an SSSI. Good for open-country and downland species including Skylark, Yellowhammer, Cuckoo, Meadow and Tree Pipits and migrant Wheatears. It has also produced Quail in the past. The north-facing slopes also hold Duke of Burgundy butterflies. There is a car park on the B2141 at SU 790180 which costs £3 a day (free for NT members) at the time of writing.

TOP DISABLED ACCESS AND PUBLIC TRANSPORT SITES

Where possible in the main text, we have tried to outline how practical access is at each main site for those who are disabled or have limited mobility, and those who travel via public transport. The reality is many of the better, more productive birding sites are in rural, remote locations – often with limited facilities or far from train stations, railways or bus stops. On this page we have put together two lists: one of what we deem to be 12 of the best reserves or sites for disabled access, and another comprised of 12 that offer relatively easy access by public transport. The sites deemed accessible via public transport are all within a reasonable walking or cycling distance from a bus stop or train station, with further details laid out in the site text. Most of the disabled access sites either have ramp access, maintained/flat pathways, accessible hides or are viewable from a car. Of course, there are other sites to be considered, but we hope these two shortlists of optimum places within these categories can help you plan where to go, and quickly take you to the relevant page numbers.

12 TOP RESERVES FOR DISABLED ACCESS

Richmond Park, Surrey, site 5, pages 36–38

WWT London Wetland Centre, Surrey, site 6, pages 38–41

Holmethorpe area, Surrey, site 10, pages 48–50

Tice's Meadow, Surrey, site 29, pages 100–103

Weir Wood Reservoir, East Sussex, site 36, pages 120–122

Rye Harbour, East Sussex, site 45, pages 147–152

Arlington Reservoir, East Sussex, site 61, pages 190–193

Pulborough Brooks RSPB, West Sussex, site 64, pages 204–208

WWT Arundel Wetland Centre, West Sussex, site 69, pages 219–221

Woods Mill, West Sussex, site 71, pages 224–226

Selsey Bill, West Sussex, site 80, pages 250–253

Warnham Local Nature Reserve, West Sussex, site 89, pages 277–279

12 TOP RESERVES FOR ACCESS BY PUBLIC TRANSPORT

Staines Reservoir, Spelthorne, site 1, pages 22–25

WWT London Wetland Centre, Surrey, site 6, pages 38–41

Beddington Farmlands, Surrey, site 8, pages 44–46

Bookham and Leatherhead, Surrey, site 11, pages 50–52

Rye Harbour, East Sussex, site 45, pages 147–152

Pett Level, East Sussex, site 46, pages 152–155

Cuckmere Haven, East Sussex, site 53, pages 171–174

Seaford Head, East Sussex, site 54, pages 174–176

Pulborough Brooks RSPB, West Sussex, site 64, pages 204–208

WWT Arundel Wetland Centre, West Sussex, site 69, pages 219–221

Adur Estuary, West Sussex, site 72, pages 226–229

Pagham Harbour (including Church Norton), West Sussex, site 79, pages 244–249

THIRTY SPECIES TO SEE IN SURREY AND SUSSEX

In this section, we outline 30 species considered especially desirable in Surrey and Sussex. The list includes various scarce and sought-after birds, as well as some commoner species that are either popular or tricky to locate. Inevitably there are a few difficult omissions – fine-tuning this list took quite a while and various birders from across the region were consulted! We tried to focus on species that either Surrey or Sussex is known for being 'good' for, as well as those that are notable within the context of South East England or even Britain as a whole. To some extent a list like this is subjective, though, and we know people may disagree with some inclusions or exclusions.

Various common, elusive or shy species are also not included – Cuckoo, Water Rail, Woodcock and Barn Owl being examples. Many scarce migrants are omitted, such as Wryneck, Yellow-browed Warbler and Red-backed Shrike. Although they occur annually and may be more reliable at some sites than others, they are not predictable in where they turn up. This list is very much a snapshot in time, too. If it was drawn up 20 years ago, we would doubtless include some species not featured here – take Mediterranean Gull, Cattle Egret and Firecrest as examples. And in 20 years' time, would we feature birds like Black-winged Stilt and Serin? On the other hand, will some species – perhaps Bewick's Swan or Great Grey Shrike – even occur enough to be mentioned in the future? Similarly, once-common and formerly widespread species like Turtle Dove are now rare enough to be considered here, while some birds previously associated with the region, such as Smew, are already no longer reliable enough to be included.

BRENT GOOSE

Dark-bellied Brent Goose is an emblematic winter species of the southern English coast, returning in October/November and departing between January and March. In Sussex, it is the western harbours that support virtually the entire county population of around 12,000 birds, though birds can be encountered anywhere along the coast in small numbers. Pagham Harbour (site 79) is perhaps the best place, with Medmerry RSPB (site 81) and Thorney Island (site 85) also recommended. The car park at West Wittering (site 82) can often be a good place for close-range views. It is a scarce bird inland.

BEWICK'S SWAN

The Arun Valley is one of the last places in southern England for this beautiful winter visitor. Traditionally a sizeable herd would be present from mid-November until late February, but these days birds often don't arrive until after Christmas and there are normally only a handful present. The best places to look are various water meadows along the River Arun: Arundel to South Stoke (site 68), namely Burpham, Offham and Warningcamp. If the water levels are high enough, the birds may frequent Amberley Wildbrooks (site 66). Occasionally the swans will roost at WWT Arundel (site 69) or Pulborough Brooks RSPB (site 64), but they are best searched for at daytime feeding sites.

GOOSANDER

This attractive sawbill is rather particular about which waterbodies it frequents. A few Surrey sites are reliable, with Cutt Mill Ponds (site 23) perhaps the best in the county. Small numbers are also reliably found on the River Mole at Leatherhead (site 11), Pennymead Lake (site C5) and Walton-on-Thames area reservoirs (site 4). In Sussex, it is slightly less predictable and can be encountered at a range of sites (including coastal ones), though the most reliable are Petworth Park (site 19) and various spots in the Arun Valley.

GREY PARTRIDGE

This declining farmland species is synonymous with much of the South Downs in West Sussex, where The Burgh (site 67) and surrounding downland can be considered the best place to connect, although Grey Partridge can be encountered on any stretch of the Downs in this part of the county. Steep Down (site 73) is another reliable locale. There are some small coastal populations at places like Chidham (site 84), Climping (site 76) and Rye Harbour (site 45).

NIGHTJAR

The heaths of both counties are excellent places to encounter Nightjar. In Surrey, the western sites are best, with Ash Ranges (site 31), Chobham Common (site 32), Crooksbury Common (site 22), Farnham Heath RSPB (site D3), Hankley Common (site D4) and Thursley Common (site 25) recommended. In the east of the county Leith Hill (site 17) is the best place. In recent years Esher Common (site B2), close to London, has had churring males. In West Sussex, Black Down (site 95), Ambersham and Heyshott Commons (site 92), Iping and Stedham Commons (site 93) and Lavington Common (site 91) are reliable, along with Wiggonholt Common (site 64). In East Sussex, Ashdown Forest (site 37) supports a healthy population, with Broadwater Warren RSPB (site 38) and Chailey Common (site 43) good locations as well.

TURTLE DOVE

Sadly, Turtle Dove is now restricted to a handful of sites in both counties. Knepp (site 90) is by far the best place, with as many as 20 purring males each spring. Woods Mill (site 71) is another fairly reliable locale, birds are possible at Ebernoe Common (96), Northchapel (12), and, to a lesser extent, at Henfield Levels (site 70) and Rye Harbour (site 45). Both sides of the county border at Chiddingfold Forest (sites 28 and 14) may support a few individuals in good years.

BLACK-NECKED GREBE

Staines Reservoir (site 1) is one of the easiest places in southern England to see this handsome grebe, which may be present virtually year-round, though the winter months are best with several birds present. Elsewhere in Surrey it is rare, though birds sometimes pitch up at Tice's Meadow (site 29) or the Walton-on-Thames area reservoirs (site 4). In Sussex, Black-necked Grebe can be tricky and is unpredictable. Rye Harbour (site 45) can sometimes host birds, while small numbers appear on the sea and on inland waterbodies in spring.

LITTLE RINGED PLOVER

Little Ringed Plover is rather a scarce breeder in the two counties. In Surrey, Tice's Meadow (site 29) is a particularly regular site, as is WWT London Wetland Centre (site 6). It is regular, though harder to find, at Beddington Farmlands (site 8). In Sussex, regular locations include Medmerry RSPB (site 81), Pett Level (site 46), Pulborough Brooks RSPB (site 64) and Rye Harbour (site 45). On passage it can occur at a variety of freshwater sites.

PURPLE SANDPIPER
This highly localised winter visitor is only reliable at three Sussex sites: Shoreham Fort (Adur Estuary, site 72) in West Sussex and Brighton Marina (site 58) and Newhaven Tide Mills (site 56) in East Sussex. Occasionally birds will pop up elsewhere, usually among Turnstones, at places like Toe End at Medmerry RSPB (site 81).

SPOTTED REDSHANK
Pagham Harbour (site 79) is an excellent site for Spotted Redshank, with a few birds usually wintering. From July to September, it can be readily encountered, especially in and around Ferry Channel. Fishbourne Creek (site 83) and, to a lesser extent, Thorney Island (site 85) are also reliable. In East Sussex, Rye Harbour (site 45) is a good site for passage birds.

CASPIAN GULL
Beddington Farmlands (site 8) has long been a hotspot for this large gull, with small numbers visiting between midsummer and early spring. Another reliable Surrey site is Rotherhithe (site 7). They are more sporadic at other locales close to the Thames, including WWT London Wetland Centre (site 6) and Walton-on-Thames area reservoirs (site 4). In Sussex, Caspian Gull remains rather scarce in the west, with Goring Gap (site 75) producing the majority of recent records. It is more regular in East Sussex but rather random in its appearances. Scanning the loafing gull flocks at Cuckmere Haven (site 53) is easily the best way to find the species in the whole of Sussex.

LITTLE TERN
This charming tern has enjoyed a bit of a recovery in Sussex, where breeding colonies can be observed at two sites: Church Norton (site 79) and Rye Harbour (site 45). Little Terns are also encountered in small numbers at various seawatching sites in spring.

POMARINE SKUA
This iconic species of South Coast spring seawatching is always a popular draw in late April and early May. With good fortune and south-easterly winds, it can be encountered at any watchpoint along the coast, but three main sites are highly recommended: Selsey Bill (site 80), Beachy Head (site 52) and, best of all, Splash Point (site 55).

FULMAR
Several East Sussex cliffs support breeding colonies of this elegant tubenose, which can be back at nest sites as early as December. Excellent views can often be had at Peacehaven, Newhaven and Telscombe Cliffs (Newhaven Harbour area, site 56) and Seaford Head (site 54). Other sites include Beachy Head (site 52), Rottingdean (site G10) and Ovingdean (site G11). Fulmar is often seen on seawatches along the coast on windier days.

HONEY BUZZARD

This desirable, rare summer visitor can be found at a handful of woodland sites in Sussex. The best places to look are Ashdown Forest (site 37) and Woolbeding Common (site 94). Black Down (site 95) is another decent location. Other possible sites include Chailey Common (site 43), Ebernoe Common (site 96), Pulborough Brooks RSPB (site 64), Penhurst Lane (site 41), Sheffield Park and Garden (site 42), Weir Wood Reservoir (site 36) and Westdean Woods (site 77). In Surrey, there are no reliable sites, but Leith Hill (site 17) offers some chance of an encounter. Beachy Head (site 52) is a regular exit point for passage birds on fine days in early autumn.

GOSHAWK

This elusive *Accipiter* has increased in Surrey and Sussex. In the former county, Effingham Forest (site 14) is a reliable place to see birds. Leith Hill (site 17) is another recommended spot, with Box Hill (site 18) and Winterfold (site 13) other possible sites. There are many sites in Sussex, some of the best being Ashdown Forest (site 37), Black Down (95), Broadwater Warren (site 38), Penhurst Lane (site 41), Stanley Common (site 11), Westdean Woods (site 77), Woolbeding Common (site 94) and Worth Forest (site 87).

HEN HARRIER

Hen Harrier is a winter and passage visitor to the region. There are normally a few on the South Downs in West Sussex, with The Burgh (site 67) the most reliable site. These birds often visit the Arun Valley, too, including Amberley Wildbrooks (site 66) and Pulborough Brooks (site 64). They are less regular on the coast. Hen Harrier occurs with less regularity in Surrey, though occasionally a bird will winter at Thursley Common (site 25) or perhaps Ash Ranges (site 31).

WHITE-TAILED EAGLE

Released birds from the Isle of Wight release programme have spread into West Sussex and some have become resident in the Arun Valley. Amberley Wildbrooks (site 66), Pulborough Brooks (site 64) and The Burgh (site 67) are the best places to look.

SHORT-EARED OWL

This rather erratic winter visitor is a popular one. In Surrey, the Papercourt area (site 34) can be excellent – or devoid of birds – depending on the year. Staines Moor (site 2) is another semi-reliable location. In West Sussex, the Arun Valley and The Burgh (site 67) are good bets, as is Thorney Island (site 85). Cuckmere Haven (site 53), Rodmell Brooks (site 60) and Pevensey Levels (site 49) occasionally support birds in East Sussex.

LESSER SPOTTED WOODPECKER

Lesser Spotted Woodpecker has become a rare and localised resident in Surrey and Sussex. In Surrey, birds may still be found at Richmond Park (site 5), Thundry Meadows (site D5), Ashtead Common (site 12) and Puttenham Common (site 23). In Sussex, some of the more reliable sites include Ashdown Forest (site 37), Burton and Chingford Ponds (site 63), Pulborough Brooks (site 64), Lurgashall Mill Pond (site 13) and Worth Forest (site 87; especially Cowdray Forest).

HOBBY

Hobby is quite widespread inland, but heaths are often the most reliable places. In Surrey, these include Ash Ranges (site 31), Crooksbury Common (site 22), Frensham Common and Ponds (site 21), Ockham and Wisley Commons (site 33) and Thursley Common (site 25). In Sussex, some of the better areas are Ashdown Forest (site 37), Black Down (site 95), Ambersham and Heyshott Commons (site 92), Iping and Stedham Commons (site 93), Lavington Common (site 91), Broadwater Warren RSPB (site 38) and Chailey Common (site 43). Passage birds can be found at various waterbodies, as well as coastal sites.

GREAT GREY SHRIKE

Although historically cyclical in its winter status, there seems to have been a decline in Great Grey Shrike in both counties. In Surrey, the South-West commons – chiefly Thursley (site 25), Frensham (site 21) and Hankley (site D4) – are traditional haunts. Ash Ranges (site 31) is also a historically reliable site. In Sussex, the best locale is Ashdown Forest (site 37). Black Down (95), Iping and Stedham Commons (site 93) and Waltham Brooks (site 65) have supported wintering birds before as well.

WOODLARK

Woodlarks are readily found on heathlands across both counties. Some of the sites with the biggest populations include Ambersham and Heyshott Commons (site 92), Ashdown Forest (site 37), Ash Ranges (site 31), Black Down (site 95), Crooksbury Common (site 22), Farnham Heath RSPB (site D3), Leith Hill (site 17) and Thursley Common (site 25).

DARTFORD WARBLER

Like Woodlark, Dartford Warbler is found at a range of heathland sites, including Ambersham and Heyshott Commons (site 92), Ashdown Forest (site 37), Ash Ranges (site 31), Black Down (site 95), Broadwater Warren RSPB (site 38), Chobham Common (site 32), Crooksbury Common (site 22), Frensham Common and Ponds (site 21), Hankley Common (site D4), Iping and Stedham Commons (93), Leith Hill (site 17) and Thursley Common (site 25). They can also be found at some coastal sites, including Beachy Head (site 52), Hastings Country Park (site 47) and Medmerry RSPB (site 81), mainly in winter.

RING OUZEL

This popular passage visitor can occur in a variety of places, but there are several sites that can be considered hotspots, even if its appearance at some of these is unpredictable. In Surrey, sites include Leith Hill (site 17), Little Woodcote (site B13), Epsom and Walton Downs (site B16) and Winterfold (site 13). Almost any coastal site can attract a bird or two, but more reliable places (where birds tend to linger) are often inland in Sussex, and include Black Down (site 95) and Cissbury Ring (site 74). Beachy Head (site 52), Hastings Country Park (site 47) and Sheepcote Valley (site 57) are further recommended sites.

NIGHTINGALE

There are still a few strongholds for this mesmerising songster in both counties. In Surrey, Chiddingfold Forest (site 28) and Milford Common (site 26) are the best locations. Other spots include Bookham Common (site 11), Capel and Newdigate (site C9), Cranleigh area (site C7) and Painshill Farm (site D11). Knepp (site 90) is an excellent place to experience this species in Sussex. It is also pleasingly easy to encounter at Abbott's Wood (site G13), Brede High Woods (site 40), Ebernoe Common (site 96) and Pulborough Brooks RSPB (site 64 is particularly productive for this species), among others.

REDSTART

Like Woodlark and Dartford Warbler, this attractive chat is most typically found on heathland sites. These include Ambersham and Heyshott Commons (site 92), Ashdown Forest (site 37), Black Down (site 95), Crooksbury Common (site 22), Frensham Common and Ponds (site 21), Hankley Common (site D4), Leith Hill (site 17), Puttenham Common (site 23) and Thursley Common (site 25).

WATER PIPIT

Water Pipit is a localised bird in both counties. In Surrey, the best location for good views is Staines Moor (site 2) and WWT London Wetland Centre (site 6). Good numbers occur at Beddington Farmlands (site 8) but viewing the species is currently tricky there. In Sussex, Combe Valley Countryside Park (site 48), Pett Level (site 46) and West Rise Marsh (site 51) are among the few places that can be considered regular, though a few elusive birds also winter in the Arun Valley and upper Adur Valley.

HAWFINCH

This elusive forest inhabitant can be found with relative ease at sites in both counties. Effingham Forest (site 14) is the most reliable site in Surrey, where Box Hill and Headley Heath (site 18), Chiddingfold Forest (site 28) and Leith Hill (site 17) may also produce sightings. In Sussex, Westdean Woods (site 77) is the best-known site. Other reliable locations include Dallington Forest and Penhurst Lane (site 41), Harting Down (site I11), Kingley Vale (site H1) and Rewell Wood (site H8).

CROSSBILL

Crossbill numbers fluctuate with each year, but both counties can be considered good ones for the species. In Surrey, Effingham Forest (site 14) and Leith Hill (site 17) are the most reliable, along with Ash Ranges (site 31), Crooksbury Common (site 22), Hindhead Common (site 20), Thursley Common (site 25) and Winterfold (site 13). In Sussex, Ashdown Forest (site 37) is a favoured haunt, along with Black Down (site 95), Broadwater Warren (site 38), Stanley Common (site I1) and Westdean Woods (site 77).

LIST OF ORGANISATIONS AND OTHER USEFUL LINKS

Abbreviations used in the text are indicated in brackets.

NATIONAL

British Trust for Ornithology (BTO) – bto.org

Royal Society for the Protection of Birds (RSPB) – rspb.org.uk

Wildfowl and Wetlands Trust (WWT) – wwt.org.uk

National Trust (NT) – nationaltrust.org.uk

Natural England (NE) – natural-england.org.uk

Forestry England (FE) – forestryengland.uk

BirdGuides – birdguides.com

eBird – ebird.org

SURREY

Surrey Bird Club (SBC) – surreybirdclub.org.uk

Surrey Wildlife Trust (SWT) – surreywildlifetrust.org

Surbiton & District Bird Watching Society – surbitonbirds.org

Tice's Meadow Bird Group – ticesmeadow.org

Beddington Farm Bird Group (BFBG) – bfnr.org.uk / beddingtonfarmlands.sightings@gmail.com

London Bird Club – londonbirders.fandom.com/wiki/LatestNews

London Natural History Society (LNHS) – lnhs.org.uk

Haslemere Natural History Society – haslemerenaturalhistorysociety.org.uk

BTO Surrey – btosurrey.co.uk

RSPB Guildford & District Local Group – rspbguildford.org.uk

RSPB North West Surrey Local Group – rspb.org.uk/groups/nwsurrey

RSPB Croydon Local Group – rspb.org.uk/groups/croydon

RSPB East Surrey Local Group – group.rspb.org.uk/eastsurrey

RSPB Dorking & District Local Group – rspb.org.uk/groups/dorkinganddistrict

RSPB Richmond & Twickenham Local Group – rspb.org.uk/groups/richmond

SUSSEX

Sussex Ornithological Society (SOS) – sos.org.uk

Sussex Wildlife Trust (SWT) – sussexwildlifetrust.org.uk

Shoreham and District Ornithological Society (SDOS) – sdos.org

Henfield Birdwatch – henfieldbirdwatch.co.uk

Rye Harbour – rye.sussexwildlifetrust.org.uk

RSPB Brighton & District Local Group – group.rspb.org.uk/brighton

RSPB Chichester Local Group – group.rspb.org.uk/chichester

RSPB Crawley & Horsham Local Group – group.rspb.org.uk/crawleyandhorsham

RSPB East Grinstead Local Group – group.rspb.org.uk/egrinstead

RSPB Eastbourne & District Local Group – group.rspb.org.uk/eastbourne

RSPB Hastings & St Leonards Local Group – group.rspb.org.uk/hastings

OTHER USEFUL LINKS

Bus timetables – bustimes.org

Birdability – birdability.org

Birding For All – birdingforall.com

GLOSSARY

ORNITHOLOGICAL TERMS

Aythya – see also Diving duck: Genus of diving ducks including Tufted Duck and Pochard.

Common birds of prey – see also Raptor: Unless otherwise stated, Buzzard, Sparrowhawk, Kestrel and Red Kite.

Common passerine – see also Passerine: Unless otherwise stated, and depending on the context, generally the likes of Blackbird (and other common thrushes), Wren, Starling, House Sparrow, common finches (Chaffinch, Goldfinch, Greenfinch) and Pied Wagtail.

Common woodland species: Unless otherwise stated, Blue Tit, Great Tit, Coal Tit, Long-tailed Tit, Goldcrest, Nuthatch, Treecreeper and Great Spotted Woodpecker.

Corvid: Birds of the crow family.

Dabbling duck: Duck species that feed by dabbling just below the water's surface, or upending, and which generally favour freshwater, such as Mallard, Wigeon, Teal and Shoveler.

Diving duck: Duck species that dive below the water's surface to find food, such as Tufted Duck, Pochard, Scaup and Common Eider.

Fall: A mass arrival of migrant landbirds grounded by adverse weather conditions.

High summer: Colloquially speaking, the hottest part of the summer but, in birding parlance, it is generally used to refer to the slight lull between the end of spring migration and the start of return passage.

Hirundine: A member of the swallow or martin family – in a British context, Sand Martin, Swallow and House Martin.

In-off: Used particularly by seawatchers to refer to birds visibly arriving on the coast from out at sea.

Insectivore: In the context of this book, any bird species that has a diet consisting largely or entirely of insects, e.g. warblers and hirundines.

Landbird: Bird species that spend all or most of their lives on or over land (though they can and will migrate over the sea).

Migrant: Any bird that readily migrates to and from – or through – the region.

Migrant chat: Any routinely migratory member of the chat family, e.g. Wheatear, Whinchat, Redstart, Black Redstart.

Passage – see also Passage migrant: The movement of migratory bird species, particularly in spring and autumn.

Passage migrant: Birds that routinely migrate to through a given area but don't stay to breed in the context of this region, such as Ring Ouzel, Whinchat, Pied Flycatcher and various wader species.

Passerine: All songbirds or, indeed, any bird that has feet adapted for perching, including the corvids.

Raptor – see also Common birds of prey: Generally refers to diurnal (active in the daytime) birds of prey such as Sparrowhawk, Common Buzzard, Kestrel, etc.

Rarity: Generally denotes a rare bird for which sightings of some species are referred to the British Birds Rarities Committee for vetting.

Roding: Territorial flight of Woodcock, usually at dusk and dawn.

Scarce grebe: The less common members of the grebe family to occur in Britain, i.e. Slavonian, Black-necked and Red-necked.

Scarcity: Species which, while perhaps not qualifying for vetting by regional and national rarities committees, are nonetheless noteworthy in the region or at a particular site.

Seabird: Refers to birds of the open sea which under normal circumstances only come close inshore to breed, i.e. shearwaters, storm-petrels, Gannet, skuas, Kittiwake and auks.

Seaduck: Duck mostly confined to the sea, i.e. Common Eider, Long-tailed Duck, scoter species and Red-breasted Merganser.

Seawatching: Persistent scanning of the sea from land to observe birds moving past the coast.

Tubenose: Any seabird of the order Procellariformes, from storm-petrels to albatrosses.

Vagrant: a species outside of its typical range, normally a rarity.

Vis-mig/Visible Migration: The observation of the diurnal (daytime) movement of birds.

Warbler, *Acrocephalus:* A species of warbler normally associated with reedbeds and marsh vegetation, e.g. Sedge and Reed Warblers.

Warbler, *Phylloscopus:* A 'leaf warbler' normally associated with woodlands, e.g. Chiffchaff and Willow Warbler.

Waterbirds – see also Waterfowl and Wildfowl: All birds associated with wetland habitats.

Waterfowl – see also Waterbirds and Wildfowl: Refers to wildfowl and some other birds associated with water, e.g. divers, grebes, Cormorant, herons, rails, etc.

White-winger: Colloquial term for scarce white-winged gulls, i.e. Glaucous and Iceland Gulls (generally not used in reference to Mediterranean Gull, however).

Wildfowl – see also Waterbirds and Waterfowl: Birds belonging to the family Anatidae, i.e. swans, geese and ducks.

ACRONYMS

AONB   Area of Outstanding Natural Beauty

BTO   British Trust for Ornithology

DEFRA   Department for Environment, Food and Rural Affairs

GPs   Gravel pits

LNR   Local Nature Reserve

MOD   Ministry of Defence

NNR   National Nature Reserve

NT   National Trust

RSPB   Royal Society for the Protection of Birds

SAC   Special Area of Conservation

SF   Sewage farm

SNCI   Site of Nature Conservation Interest

SPA   Special Protection Area

SSSI   Site of Special Scientific Interest

SWT   Surrey Wildlife Trust or Sussex Wildlife Trust

WeBS   BTO Wetland Bird Survey

HABITAT TERMS

Chalk grassland/downland: A rare and declining habitat, only found in North-West Europe, where shallow soils on chalk bedrock are kept open through grazing and human management. These habitats support some very specialist plant and invertebrate species.

Floodplain – see also Marsh: Low-lying land in a river valley, prone to flooding – and often managed intentionally for this purpose.

Ghyll woodland: A unique habitat to this region, where narrow streams cut through wooded sandstone ravines.

Marsh – see also Floodplain: An area of low-lying land that floods in wet weather or at high tide and remains waterlogged for long periods of time.

Mire: In the context of this book, any area of wet, boggy ground.

Ramsar site: A wetland site designated as being of international importance under the Ramsar Convention of 1971.

Saline: Contains salt (usually in water).

Saltmarsh: A coastal habitat of grassland regularly flooded by seawater.

Water meadows: Riverside meadows which are regularly flooded.

Weald: Generally a heavily wooded area, from the Old English for 'forest'.

SURREY AND SUSSEX BIRD LIST

Checklist of birds recorded in Surrey and Sussex

1 Brent Goose
2 Red-breasted Goose
3 Canada Goose
4 Barnacle Goose
5 Greylag Goose
6 Taiga Bean Goose
7 Pink-footed Goose
8 Tundra Bean Goose
9 White-fronted Goose
10 Mute Swan
11 Bewick's Swan
12 Whooper Swan
13 Egyptian Goose
14 Common Shelduck
15 Ruddy Shelduck
16 Mandarin Duck
17 Garganey
18 Blue-winged Teal
19 Shoveler
20 Gadwall
21 Eurasian Wigeon
22 American Wigeon
23 Mallard
24 Pintail
25 Eurasian Teal
26 Green-winged Teal
27 Red-crested Pochard
28 Common Pochard
29 Ferruginous Duck
30 Ring-necked Duck
31 Tufted Duck
32 Greater Scaup
33 Lesser Scaup
34 King Eider
35 Common Eider
36 Surf Scoter
37 Velvet Scoter
38 Common Scoter
39 Long-tailed Duck
40 Goldeneye
41 Smew
42 Hooded Merganser
43 Goosander
44 Red-breasted Merganser
45 Ruddy Duck
46 Black Grouse
47 Grey Partridge
48 Golden Pheasant
49 Common Pheasant
50 Quail
51 Red-legged Partridge
52 Common Nighthawk

53 Nightjar
54 Alpine Swift
55 Common Swift
56 Pallid Swift
57 Great Bustard
58 Little Bustard
59 Great Spotted Cuckoo
60 Yellow-billed Cuckoo
61 Common Cuckoo
62 Pallas's Sandgrouse
63 Feral Pigeon
64 Stock Dove
65 Woodpigeon
66 Turtle Dove
67 Collared Dove
68 Water Rail
69 Corncrake
70 Sora
71 Spotted Crake
72 Moorhen
73 Coot
74 Baillon's Crake
75 Little Crake
76 Crane
77 Little Grebe
78 Pied-billed Grebe
79 Red-necked Grebe
80 Great Crested Grebe
81 Slavonian Grebe
82 Black-necked Grebe
83 Stone-curlew
84 Oystercatcher
85 Black-winged Stilt
86 Avocet
87 Lapwing
88 Sociable Plover
89 Golden Plover
90 Pacific Golden Plover
91 American Golden Plover
92 Grey Plover
93 Ringed Plover
94 Semipalmated Plover
95 Little Ringed Plover
96 Killdeer
97 Kentish Plover
98 Lesser Sand Plover
99 Greater Sand Plover
100 Dotterel
101 Upland Sandpiper
102 Eurasian Whimbrel
103 Hudsonian Whimbrel
104 Curlew

105 Bar-tailed Godwit
106 Black-tailed Godwit
107 Turnstone
108 Knot
109 Ruff
110 Broad-billed Sandpiper
111 Stilt Sandpiper
112 Curlew Sandpiper
113 Temminck's Stint
114 Long-toed Stint
115 Sanderling
116 Dunlin
117 Purple Sandpiper
118 Baird's Sandpiper
119 Little Stint
120 Least Sandpiper
121 White-rumped Sandpiper
122 Buff-breasted Sandpiper
123 Pectoral Sandpiper
124 Semipalmated Sandpiper
125 Long-billed Dowitcher
126 Woodcock
127 Jack Snipe
128 Great Snipe
129 Common Snipe
130 Terek Sandpiper
131 Wilson's Phalarope
132 Red-necked Phalarope
133 Grey Phalarope
134 Common Sandpiper
135 Spotted Sandpiper
136 Green Sandpiper
137 Lesser Yellowlegs
138 Common Redshank
139 Marsh Sandpiper
140 Wood Sandpiper
141 Spotted Redshank
142 Greenshank
143 Collared Pratincole
144 Oriental Pratincole
145 Black-winged Pratincole
146 Kittiwake
147 Sabine's Gull
148 Slender-billed Gull
149 Bonaparte's Gull
150 Black-headed Gull
151 Little Gull
152 Laughing Gull
153 Franklin's Gull
154 Mediterranean Gull
155 Common Gull
156 Ring-billed Gull
157 Great Black-backed Gull
158 Glaucous Gull
159 Iceland Gull
160 Glaucous-winged Gull
161 Herring Gull

162 Caspian Gull
163 Yellow-legged Gull
164 Lesser Black-backed Gull
165 Gull-billed Tern
166 Caspian Tern
167 Royal Tern
168 Lesser Crested Tern
169 Sandwich Tern
170 Elegant Tern
171 Little Tern
172 Least Tern
173 Bridled Tern
174 Sooty Tern
175 Roseate Tern
176 Common Tern
177 Arctic Tern
178 Whiskered Tern
179 White-winged Black Tern
180 Black Tern
181 Great Skua
182 Pomarine Skua
183 Arctic Skua
184 Long-tailed Skua
185 Little Auk
186 Common Guillemot
187 Razorbill
188 Black Guillemot
189 Puffin
190 Red-throated Diver
191 Black-throated Diver
192 Great Northern Diver
193 White-billed Diver
194 Storm Petrel
195 Leach's Petrel
196 Fulmar
197 Cory's Shearwater
198 Sooty Shearwater
199 Manx Shearwater
200 Balearic Shearwater
201 Black Stork
202 White Stork
203 Gannet
204 Red-footed Booby
205 Brown Booby
206 Cormorant
207 Shag
208 Glossy Ibis
209 Spoonbill
210 Eurasian Bittern
211 American Bittern
212 Little Bittern
213 Night-heron
214 Squacco Heron
215 Cattle Egret
216 Grey Heron
217 Purple Heron
218 Great White Egret

219 Little Egret
220 Osprey
221 Honey-buzzard
222 Short-toed Eagle
223 Sparrowhawk
224 Goshawk
225 Marsh Harrier
226 Hen Harrier
227 Pallid Harrier
228 Montagu's Harrier
229 Red Kite
230 Black Kite
231 White-tailed Eagle
232 Rough-legged Buzzard
233 Common Buzzard
234 Barn Owl
235 Little Owl
236 Long-eared Owl
237 Short-eared Owl
238 Snowy Owl
239 Tawny Owl
240 Hoopoe
241 Roller
242 Kingfisher
243 Bee-eater
244 Wryneck
245 Lesser Spotted Woodpecker
246 Great Spotted Woodpecker
247 Green Woodpecker
248 Kestrel
249 Red-footed Falcon
250 Merlin
251 Hobby
252 Gyr Falcon
253 Peregrine
254 Ring-necked Parakeet
255 Red-backed Shrike
256 Lesser Grey Shrike
257 Great Grey Shrike
258 Woodchat Shrike
259 Golden Oriole
260 Jay
261 Magpie
262 Nutcracker
263 Chough
264 Jackdaw
265 Rook
266 Carrion Crow
267 Hooded Crow
268 Raven
269 Waxwing
270 Crested Tit
271 Coal Tit
272 Marsh Tit
273 Willow Tit
274 Blue Tit
275 Great Tit

276 Penduline Tit
277 Bearded Tit
278 Woodlark
279 Skylark
280 Crested Lark
281 Shore Lark
282 Short-toed Lark
283 Sand Martin
284 Crag Martin
285 Barn Swallow
286 House Martin
287 Red-rumped Swallow
288 Cliff Swallow
289 Cetti's Warbler
290 Long-tailed Tit
291 Wood Warbler
292 Western Bonelli's Warbler
293 Hume's Warbler
294 Yellow-browed Warbler
295 Pallas's Leaf Warbler
296 Radde's Warbler
297 Dusky Warbler
298 Willow Warbler
299 Chiffchaff
300 Greenish Warbler
301 Great Reed Warbler
302 Aquatic Warbler
303 Sedge Warbler
304 Paddyfield Warbler
305 Blyth's Reed Warbler
306 Reed Warbler
307 Marsh Warbler
308 Booted Warbler
309 Sykes's Warbler
310 Melodious Warbler
311 Icterine Warbler
312 River Warbler
313 Savi's Warbler
314 Grasshopper Warbler
315 Blackcap
316 Garden Warbler
317 Barred Warbler
318 Lesser Whitethroat
319 Sardinian Warbler
320 Eastern Subalpine Warbler
321 Common Whitethroat
322 Dartford Warbler
323 Firecrest
324 Goldcrest
325 Wren
326 Nuthatch
327 Wallcreeper
328 Eurasian Treecreeper
329 Short-toed Treecreeper
330 Northern Mockingbird
331 Rose-coloured Starling
332 Common Starling

333 White's Thrush
334 Siberian Thrush
335 Song Thrush
336 Mistle Thrush
337 Redwing
338 Blackbird
339 Fieldfare
340 Ring Ouzel
341 American Robin
342 Rufous-tailed Scrub Robin
343 Spotted Flycatcher
344 Robin
345 Bluethroat
346 Thrush Nightingale
347 Common Nightingale
348 Red-flanked Bluetail
349 Red-breasted Flycatcher
350 Pied Flycatcher
351 Collared Flycatcher
352 Black Redstart
353 Common Redstart
354 Rock Thrush
355 Blue Rock Thrush
356 Whinchat
357 Eurasian Stonechat
358 Northern Wheatear
359 Isabelline Wheatear
360 Desert Wheatear
361 Western Black-eared Wheatear
362 Pied Wheatear
363 Dipper
364 Tree Sparrow
365 House Sparrow
366 Alpine Accentor
367 Dunnock
368 Yellow Wagtail
369 Citrine Wagtail
370 Grey Wagtail
371 Pied Wagtail
372 Richard's Pipit
373 Tawny Pipit
374 Meadow Pipit

375 Tree Pipit
376 Olive-backed Pipit
377 Red-throated Pipit
378 Buff-bellied Pipit
379 Water Pipit
380 Rock Pipit
381 Chaffinch
382 Brambling
383 Hawfinch
384 Bullfinch
385 Trumpeter Finch
386 Common Rosefinch
387 Greenfinch
388 Twite
389 Linnet
390 Common Redpoll
391 Lesser Redpoll
392 Arctic Redpoll
393 Parrot Crossbill
394 Two-barred Crossbill
395 Common Crossbill
396 Goldfinch
397 Serin
398 Siskin
399 Lapland Bunting
400 Snow Bunting
401 Corn Bunting
402 Yellowhammer
403 Pine Bunting
404 Rock Bunting
405 Ortolan Bunting
406 Cirl Bunting
407 Little Bunting
408 Rustic Bunting
409 Black-headed Bunting
410 Pallas's Reed Bunting
411 Reed Bunting
412 Dark-eyed Junco
413 White-throated Sparrow
414 Baltimore Oriole
415 Black-and-white Warbler
416 Blackpoll Warbler

SPECIES RECORDED ONLY IN SPELTHORNE

Note: Spelthorne is part of present-day Surrey, but does not form part of the Watsonian vice-county, so lies outside the Surrey recording area.

1 Sharp-tailed Sandpiper
2 Solitary Sandpiper
3 Brown Shrike

INDEX TO SPECIES

Species index listed by site number